MACROEVOLUTION
Diversity, Disparity, Contingency

Essays in Honor of Stephen Jay Gould

Edited by
Elisabeth Vrba and Niles Eldredge

Published by
The Paleontological Society

Elisabeth Vrba is a Professor in the Department of Geology and Geophysics and a Curator of the Peabody Museum of Natural History at Yale University. Niles Eldredge is a Curator in the Department of Invertebrates at the American Museum of Natural History. Both editors worked closely with Stephen Jay Gould on joint research projects and the formulation of key ideas that they incorporated together into current evolutionary theory.

Copyeditor: Natasha Atkins
Cover Design: Mary Parrish
Frontispiece: Photograph by Jerry Harasewych

Printed in the United States of America by Allen Press, Lawrence, Kansas
02 01 00 1 2 3

ISBN: 1-891276-49-2 (paperback)
ISSN: 0094-8373

Library of Congress Cataloging-in-Publication Data
Macroevolution: Diversity, Disparity, Contingency. Essays in Honor of Stephen Jay Gould / edited by Elisabeth S. Vrba and Niles Eldredge.

p. cm. –(Supplement to Volume 31, number 2 of the journal *Paleobiology*)

Includes bibliographic references.
1. Paleobiology. 2. Paleontology. 3. History of Life.
I. Vrba, Elisabeth S. II. Eldredge, Niles

Paleobiology (ISSN: 0094-8373) is published quarterly by The Paleontological Society, 810 East 10th St., Lawrence, KS 66044, USA. This book is a supplement to Volume 31, number 2.

Contents

Steve Gould and Glenn Goodfriend sifting for Holocene fossil *Cerion* snails on a bluff 2 km E of Grays Settlement on Long Island, Bahamas, December 5, 2001.

Stephen Jay Gould (1941–2002)

Preface

Steve Gould died in May, 2002. Since his death the world of paleobiology, indeed, the world of evolutionary biology in general, has just not been the same as it had been for the previous three and a half decades of his active professional life.

Steve was as known for his takes on the most general of evolutionary concepts as he was for his myriad more detailed and content-driven analyses of fossil snails, historical texts, and specific theoretical work. He argued famously, for example, for the notion of contingency in evolution—where perturbations in the system play as strong a role as the deterministic elements (like natural selection—itself only statistically deterministic) in shaping the history of life. But at the same time he was thoroughly immersed in the details of many distinct areas of evolutionary biology and paleobiology. Those of us, including the present editors (along with Steve the original—and wholly self-proclaimed—"Three Musketeers"), who identified most closely with Steve's approach each had our own specific areas of research focus. What brought us together, though, was a common passion for seeing the recurrent patterns in the history of life as themselves lawlike, and thus crying out for inclusion in causal evolutionary theory.

The 14 papers assembled here to honor the life and work of Stephen Jay Gould reflect very well the depth and breadth of Steve's interests. He wouldn't have agreed with absolutely every word (no one could—especially not the argumentative Dr. Gould). But he would have been pleased to see in this selection an embodiment of his catholic tastes—from genetics and developmental biology to turnovers and mass extinctions, from the Burgess Shale to an interdisciplinary examination of the causes of stasis, and so on.

He himself of course would have wished to have been a contributor to a similar volume, conceived for some other reason. This collection instead, stands as an homage to, and memento of, a bold and far-ranging thinker who did much to bring paleobiology closer to the rest of evolutionary biology. He continues to be missed.

Niles Eldredge
Elisabeth Vrba
Co-editors

Paleobiology, 31(2), 2005, pp. 1–16

Disparity, adaptation, exaptation, bookkeeping, and contingency at the genome level

Jürgen Brosius

Abstract.—The application of molecular genetics, in particular comparative genomics, to the field of evolutionary biology is paving the way to an enhanced "New Synthesis." Apart from their power to establish and refine phylogenies, understanding such genomic processes as the dynamics of change in genomes, even in hypothetical RNA-based genomes and the in vitro evolution of RNA molecules, helps to clarify evolutionary principles that are otherwise hidden among the nested hierarchies of evolutionary units. To this end, I outline the course of hereditary material and examine several issues including disparity, causation, or bookkeeping of genes, adaptation, and exaptation, as well as evolutionary contingency at the genomic level—issues at the heart of some of Stephen Jay Gould's intellectual battlegrounds. Interestingly, where relevant, the genomic perspective is consistent with Gould's agenda. Extensive documentation makes it particularly clear that exaptation plays a role in evolutionary processes that is at least as significant as—and perhaps more significant than—that played by adaptation.

Jürgen Brosius. Institute of Experimental Pathology, ZMBE, University of Münster, Von-Esmarch-Strasse 56, Münster, Germany. E-mail: RNA.world@uni-muenster.de

Accepted: 1 September 2004

The Promise and Perils of Molecular Genetics and Genomics to Evolutionary Biology

Although it did not appear so at the outset, rediscovery of Mendel's laws (Mendel 1866, 1870) in the early twentieth century was decisive in strengthening the field of evolutionary biology. A few decades later, the fusion of genetics with evolutionary biology led to a "Modern Synthesis" (Dobzhansky 1937; Huxley 1942; Mayr 1942; Simpson 1944; Rensch 1947; Jepsen et al. 1949; Schmalhausen 1949; Stebbins 1950). Discovery of the double-helical structure of DNA (Watson and Crick 1953) and the molecular basis of heredity promised another windfall. However, the nucleotide sequences of genes or even short segments thereof remained out of reach for some time. Nonetheless, significant advances were made indirectly, by examining the amino acid sequences of proteins (McLaughlin and Dayhoff 1973; King and Wilson 1975); and, in particular, by establishing catalogs of ribosomal RNA-derived sequence snippets. The latter effort led to the discovery of a third domain of life, the Archaea (Woese and Fox 1977). In the 1970s, Fred Sanger, Wally Gilbert, and coworkers made rapid DNA sequencing possible (Sanger and Coulson 1975; Maxam and Gilbert 1977; Sanger et al. 1977), eventually leading to the era of whole genome sequencing (Wolfe and Li 2003). Evolutionary Biology immediately profited from direct phylogenetic comparison of genomic sequences, despite some apparent problems of misinterpretation, such as placing the guinea pig and its relatives outside the order of Rodentia (Graur et al. 1991; D'Erchia et al. 1996) and *Dermoptera* within the order Primates (Arnason et al. 2002; Murphy et al. 2001). One of the difficulties in comparing sequences, especially those from mitochondria, arises from ignoring the differential base-composition between species (Schmitz et al. 2002). A purely mathematical approach without considering interfering biological factors is, thus, often treacherous. Although I agree with Maynard Smith, who claimed that "an ounce of algebra is better than a ton of words" (Maynard Smith 1989: p. 97),[1] I should add that an ounce of common sense is better than a pound of algebra. This said, a more sensible approach to molecular

[1] I saw a variant of this quote akin to "An ounce of mathematics is better than a pound of ideology" from the same author, or perhaps Haldane, but I cannot retrace its exact origin.

 0094-8373/05/3102-0001/$1.00

phylogenetic analysis, first described by Ryan and Dugaiczyk (1989) and fully developed by Okada, has led to tremendous advances in molecular phylogeny (Okada 1991; Rokas and Holland 2000; Shedlock et al. 2004). The process relies on random but virtually irreversible[2] insertion events in genomes, which usually occur by retroposition (enzymatic conversion of RNA into DNA and integration into the genome; see below for details), and has proved very useful for finally settling longstanding questions concerning primate evolution or that of the land-dwelling relatives of whales (Shimamura et al. 1997; Schmitz et al. 2001, 2002). Although even this approach calls for caution, owing to the complications of lineage sorting, the problem can be virtually excluded by analyzing several independent retropositional events. Without disregarding the value of small changes (e.g., for studies in population genetics) it is clear that for phylogenetic analysis one should heed the evolutionary bangs ("saltatory" events at the genome level) rather than the white noise of nucleotide substitutions.

New approaches to phylogeny are not the only benefit of the "second new synthesis" of molecular genetics with evolutionary biology. Consideration of evolutionary questions in a genomic context (DNA or RNA) may be useful for their clarification. Because of the power of direct analysis, obviating the prediction of genetic change by the mere observation of phenotypic consequences, this reductionist approach, pioneered by George C. Williams and Richard Dawkins, has opened new avenues to examine evolutionary principles and concepts (Dawkins 1976, 1982; Williams 1966). Nevertheless, it is apparent that the role played by extant genomes to influence survival is intertwined with that of other biological and evolutionary units and that selection acts at multiple levels: on nuons,[3] genes, genomes, individuals, kin, groups, societies, species, and superorganisms (Ghiselin 1974; Gould 1980, 2002; Wilson 1980; Sober 1984; Wilson and Sober 1989; Keller 1999).

As a means of viewing evolution through the lens of genomics, I will outline the course of hereditary material, from the necessarily highly speculative RNA world through several evolutionary transitions (Szathmáry and Smith 1995) to modern genomes. One of the most significant of these, the transition from RNA to DNA as a hereditary nucleic acid, has been remarkably drawn out over time, as this process of retroposition still plays a pervasive role in the evolution of many modern vertebrate genomes (Brosius and Gould 1992; Brosius and Tiedge 1995; Brosius 1999d). The account of genomic materials will end by recognizing, at least in the primate lineage, the dawn of yet another transition toward a form of Lamarckian evolution as well as the end of nucleic acids as hereditary material. Along this trajectory, I will address several ongoing controversies from a genomic perspective with emphasis on those where Stephen Jay Gould held a prominent position.

Time Machine Back to the RNA World

We begin our chronology of genetic material in the RNA world. When we look back to even this most primitive of cellular stages currently imaginable, it is expected that selection acted

[2] Although DNA between two similar retroposons can be deleted by unequal crossing over (Batzer and Deininger 2002) or parts of them can be deleted by other mechanisms, a clean deletion resulting in a return to the previous sequence at the integration site is virtually impossible. Application of genome rearrangements to phylogenetic analysis dates back to studies on *Drosophila pseudoobscura* (Sturtevant and Dobzhansky 1936).

[3] A nuon is any distinct nucleic acid, a defined sequence module (Brosius and Gould 1992). The term can be used with a prefix (e.g., retronuon) to designate any DNA module that was generated by retroposition. I prefer retronuon over retroposon and especially over transposable element (TE) or mobile element (ME). In fact, any RNA is a potential mobile element: if a segment of the genome is transcribed in the germline it has the potential to serve as template for retroposition (hence, *RNA* might be considered the ultimate selfish unit). However, upon integration into the genome, there is no guarantee for autonomous transcription in the germline, which results in a loss of mobility. The original transcript, however, can serve as a template for retroposition multiple times. In contrast to TE or ME, the term "retronuon" solely indicates the mode of origin, but not the potential for successive amplification. Only a minority of retronuons are true TEs or MEs, such as endogenous retroviruses or intact LINE elements [see Brosius 2003a]).

robustly on groups of RNA nuons (Brosius 2003c). Although extrapolating back in time to an RNA world and reenacting evolution in the test tube both appear to be ultra-reductionist approaches, extant DNA genomes are already so convoluted and interdependent on other evolutionary units as to necessitate such tactics. A look back at the workings in a possible RNA world with rudimentary cells emphasizes the validity of the basic principles of Darwinian evolution, but surprisingly, also illuminates even more complex issues (Brosius 2003c).

Even RNA evolution in the test tube, a procedure now quite common in laboratories, can reveal much about evolutionary principles. A randomized population of RNAs is selected for a certain property (from binding to a particular protein to catalyzing a particular reaction), re-amplified with a certain amount of variation, and once more selected. After several rounds, one often obtains a so-called RNA aptamer with the desired property (Ellington and Szostak 1990; Szathmáry 1990; Tuerk and Gold 1990). This amplification/selection, or SELEX, procedure epitomizes the three most visible pillars of Darwinian evolution: amplification, modification and selection of hereditary units—with the first two often combined in a single step during the replication of genetic material. Using this minimalistic, in vitro amplification/selection procedure, researchers documented a case of molecular cooperation. One of the evolved RNA aptamers, itself being catalytically inactive, participated in a productive intermolecular interaction with a parallelly evolving active ribozyme, thus ensuring the survival of both nuons in the nucleic acid population (Hanczyc and Dorit 1998). Furthermore, such selection/amplification schemes (Wilson and Szostak 1999; Famulok et al. 2000) may eventually lead to the establishment of man-made life forms capable of reenacting life in a primitive cell reminiscent of the RNA world (Bartel and Unrau 1999; Szostak et al. 2001). It is obvious that, although such a reconstruction cannot accurately retrace the path of evolution, the resulting RNA-based cells will certainly serve as models for life in the RNA world, and on the basis of such a primitive cell (man-made or primordial[4]) one may test other evolutionary principles than the apparent cycles of amplification, modification, and selection. For example, this approach has already underscored the significance of group selection[5] in primitive, RNA-based proto-cells, and the model is consistent with the early presence of more subtle and controversial phenomena including sexual reproduction, evolutionary conflict and cooperation, as well as the virtual absence of species barriers (Brosius 2003c).

Replacing Genomic RNA with DNA

In the course of the RNA world, a major takeover (Cairns-Smith 1982) or transition (Maynard Smith and Szathmáry 1995) took place, when templated protein biosynthesis significantly enriched the enzymatic repertoire of a cell (Brosius 2001, 2003c, 2005a, Woese 2001). A further major transition occurred when, with the aid of one of the acquired enzymes, RNA was gradually converted into DNA, the genetic material of extant cells (Darnell and Doolittle 1986; Brosius 1999d). Interestingly, this conversion process has not ended, even over the long intervening period of perhaps 2–3 billion years (Brosius and Tiedge 1995; Jurka 1998; Brosius 1999c). Retroposition uses any type of cellular RNA as a template for reverse transcription concomitant with the more or less random integration of the resulting cDNA into genomes (Weiner et al. 1986; Brosius 1999c). In modern

[4] Whether an RNA world, in which RNA contained the genetic information *and* carried out structural and catalytic tasks in the cell (Cold Spring Harbor Laboratory Press 1987; Gesteland et al. 1999), existed at all is—although widely accepted after the groundbreaking findings that RNA has catalytic activity (Kruger et al. 1982; Guerrier-Takada et al. 1983)—by no means certain (Dworkin et al. 2003). Furthermore, the possibility of an RNA world does not mean that other informational macromolecules, such as PNA—peptide nucleic acid (Nelson et al. 2000)—or inorganic templates did not precede RNA (Cairns-Smith 1982; Wächtershauser 1992).

[5] Perhaps the concept of group selection is easier to grasp for evolutionists interested in team sports; Gould loved baseball, whereas my personal preference is football (soccer). In either case, selection occurs at least at two levels. Players compete to be members of teams and teams compete with each other in championships. For every game the most suitable players of a team are selected; depending on the opposing team and other factors, the selection might differ from game to game.

genomes we recognize autonomously retroposing modules such as retroviral-like elements, which contain flanking LTRs (long terminal repeats)[6], or non-LTR LINE elements (long interspersed elements). Although often truncated, full-length LINEs are more than 6 kb (kilobases) long and, in the human genome, contribute more than 20% of the total 3 $\times 10^9$ bp (base pairs). Some of the full-length LINEs likely provide the enzymatic activity (reverse transcriptase) for retroposition *in trans* (Kajikawa and Okada 2002; Boeke 2003; Deininger et al. 2003; Dewannieux et al. 2003; Hagan et al. 2003). Because the retroposed RNAs are usually devoid of a gene's regulatory elements such as transcriptional promoters, the retropositions often lead to inactive retropseudogenes. Small non-messenger RNAs (snmRNAs) such as tRNAs, tRNA-related RNAs, SRP RNA, snRNAs, and rRNAs frequently serve as templates for reverse transcription, some with extremely high efficiencies, leading to accumulation of so-called SINEs (short interspersed elements) in genomes (Singer 1982). SINE copy numbers can range from hundreds to a million, as exemplified by the primate-specific Alu elements that, despite their moderate lengths of about 300 bp, occupy more than 10% of the human genome. Although some promoter elements are internal to RNA polymerase III-transcribed snmRNAs, most of these modules are transcriptionally dead on arrival, unless they fortuitously integrate near appropriate upstream promoter elements, as do mRNA-derived retrogenes. Overall, a stunning 42% of the human genome is composed of recognizable retronuons (Lander et al. 2001). Owing to near randomization by base substitutions and short indels (depending on the organism and genomic locus, sequences may randomize toward certain characteristics, such as base composition [Bernardi 2004]), retronuons are only discernible for around 100 million years. However, as retroposition is an ancient process, one can extrapolate that the bulk of mammalian genomes (except for the portions generated by segmental duplication, slippage of DNA polymerase during replication, or DNA transpositions[7]) has been contributed by retroposition, suggesting that the process has continued nonstop since its inception (Brosius 1999c,d, 2003a, 2005a).

Causation or Bookkeeping of Genes

The likelihood that modern cells evolved from more primitive cells wherein RNA macromolecules had both genotypic and phenotypic impact invites a look at the controversy of whether genes are pulling the strings of evolutionary processes and thus are causal, or are merely keeping track of them and thus have only a bookkeeping role (Dawkins 1976, 2003; Gould and Lloyd 1999; Gould 2002). Obviously, in the RNA world, the causation component was decidedly prominent, but the "division of labor" precipitated by major evolutionary transitions (Szathmáry and Smith 1995) leading to recruitment and templating of additional macromolecules (proteins and DNA) suggests a specialized role of genes in bookkeeping. As human beings evolve the near-Lamarckian capabilities of gene-therapeutics and other directed changes affecting the germ-line (Brosius 2003b,c), we are approaching the other end of the continuum, with genes having little to do with direct causation and almost exclusively with bookkeeping.

Retronuons: Major Agents of Genomic Change

Over relatively short evolutionary time spans, the activities of retronuons can lead to remarkable differences in the appearance of genome sequences even among species within the same mammalian order (Fig. 1) (see also Kurychev et al. 2001: Fig. 1) depicting a comparison of the same locus among four primate species). Since the early days of molecular biology, it was known that even closely related species can have drastically different genome sizes, a phenomenon known as the C-value paradox—a paradox because no correlation

[6] Often solo-LTRs are the only genomic remnants of such elements, owing to unequal homologous recombination between two LTRs.

[7] Less than 3% of the human genome consists of recognizable DNA transposons.

between organismal complexity and genome size could be identified (Gregory and Hebert 1999; Gregory 2001). Molecular genetics confirmed that the differences in C-values are largely due, perhaps with the exception of whole genome duplications, to differential amounts of apparently nonfunctional sequences.

The odd architecture of many genomes from multicellular organisms (pleiocytes) with relatively few functional modules interspersed with an excess of seemingly nonfunctional modules or randomized sequences is permissive for the events leading to the C-value paradox or, if expansion and contraction keep a balance, to notable sequence turnover. Among two or more mammalian genomes, some 5–7% of the total genomes are conserved as sequence blocks (summarized in Brosius 2005a). The degree of sequence similarity is such that one can infer a conserved function. These islands of similarity are designated by a variety of terms including conserved sequence blocks (CSBs) (O'Brien and Murphy 2003), multispecies conserved sequences (MCSs) (Margulies et al. 2003; Thomas et al. 2003), or clusters of orthologous bases (COBs) (Kirkness et al. 2003). They include nuons that are building blocks of genes, such as open reading frames (ORFs) and regulatory regions, as well as conserved non-genic sequences (CNGs) (Dermitzakis et al. 2003). In mammals, a mere 1.5%[8] of the total genome represents ORFs of the estimated 25,000 protein-encoding genes. An equal percentage of the genome could be occupied by non-protein-coding genes, large or small RNAs whose significance has only recently become more widely appreciated (Hüttenhofer et al. 2001; Pasquinelli and Ruvkun 2002; Mattick 2003; Nelson et al. 2003; Numata et al. 2003; Brosius and Tiedge 2004; Herbert 2004; Novina and Sharp 2004; Brosius 2005b). At least 1% of mammalian genomes consists of regulatory elements, such as promoters, enhancers, and other binding sites for regulatory proteins. The remainder, C"N"Gs, may comprise thus far unidentified ORFs, untranslated RNAs, regulatory elements, or additional elements under evolutionary constraint; this last group may occupy understudied or unrecognized roles perhaps in the modulation of chromatin structures that, in turn, alters the expression states of genes. On the other hand, we cannot yet exclude the possibility, that a sizeable fraction of CNGs are serendipitously conserved. Conserved functional sequence blocks, as a whole, appear to be islands in a sea of seemingly "useless" (but see below) DNA, introns and intergenic sequences for the most part (Fig. 1).[9] In summary, it should now be clear that genomic disparity involving superfluous DNA and largely mediated by retroposition contributes to C-value fluctuations in different species and, once two lineages separate, can bring forth change much more rapidly than the slower action of nucleotide exchanges and small indels.

Retronuons: Potential Exaptations

Opinions concerning the functional significance of retronuons have experienced wild

[8] This does not count the 5′ and 3′ untranslated regions of messenger RNAs that generally are much less conserved and add perhaps another 2% (W. Makalowski personal communication 2004).

[9] An extension of Wally Gilbert's metaphor "exons in a sea of introns" (Gilbert 1978). Functional nuons are islands in a sea of nonfunctional (nonaptive) sequences. Nevertheless, any of those sequences has the potential to be exapted into novel functions (Brosius and Gould 1992; Balakirev and Ayala 2003). While "plate tectonics," or exon shuffling, occasionally leads to rearrangements of existing functional nuons (Gilbert 1978), retroposition, the major force in the plasticity of genomes, which in our analogy is more akin to volcanic eruptions, frequently creates new nuons. Initially, most nuons (islands) are barren (nonfunctional, nonaptive) but have the potential to be fertilized by some microevolutionary base changes or short indels and exapted as *functional* nuons. Nonfunctional nuons erode over time and the islands disappear in the sea of anonymous sequences. An interesting example is the recruitment of part of an Alu retronuon as an alternative exon in an isoform of the cytokine tumor necrosis factor receptor. Insertion of the Alu element occurred after *Anthropoidea* split from prosimians and a subsequent point mutation generated an ATG start codon. This base substitution alone, however, was not sufficient for exaptation of the Alu element as a protein-coding exon, as this sequence is nonaptive (not used as part of an alternative mRNA) in Platyrrhini. Only two additional small changes in the lineage leading to *Catarrhini* including apes, a C→T transition to generate a GT 5′ splice site and a 7-bp deletion to provide translation into the next exon in the correct reading frame, led to generation and exaptation of this alternative exon (Singer et al. 2004).

swings. Initially, every repeat was ascribed a function in gene regulation (Britten and Davidson 1969, 1971), whereas a decade later they were considered mere functionless, "selfish DNA" (Doolittle and Sapienza 1980; Orgel and Crick 1980). Gould (1981) subscribed neither to the former: "these adaptive explanations cannot account for all repetitive DNA. There is simply too much of it, too randomly dispersed. . . . " nor to the latter: "I do not believe that all repetitive DNA is selfish DNA" (Gould 1983: p. 169).[10] In a landmark paper, Gould and Vrba (1982: p. 11) formulated the following: we might suggest that repeated copies are nonapted features, available for cooptation later, but not serving any direct function at the moment. When coopted, they will be exaptations in their new role (with secondary adaptive modifications if altered).

As stated above, any cellular RNA can serve as a template for reverse transcription. The first isolated reports of novel functional protein genes (McCarrey and Thomas 1987; Dahl et al. 1990) and putative functional small nonmessenger RNA genes (DeChiara and Brosius 1987; Martignetti and Brosius 1993; Tiedge et al. 1993) that arose by retroposition instead of segmental gene duplication were considered mere oddities. Only later did some authors recognize the advantages of retroposed genes, namely that, although the odds of generating a functional gene are low, genes duplicated via retroposition can recombine with alternative regulatory elements (Brosius 1991). Furthermore, the magnitude of this process was finally appreciated when it was evident that perhaps all of the many intronless genes were derived by retroposition[11] and that re-

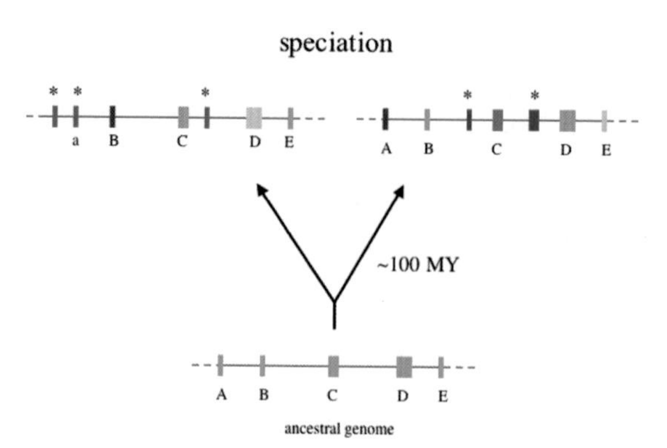

FIGURE 1. Evolutionary changes in eukaryotic genomes over time. Functional nuons A–E (open reading frames of exons, regulatory elements, RNA genes) are shown as gray boxes, the remaining, seemingly nonfunctional 95% or so as lines (olive in the ancestral genome). Speciation, itself occasionally triggered by the spread of novel retronuons, goes hand in hand with small, mainly neutral changes in conserved sequences (slight variations to different shades of gray = microevolution). Drastic changes are apparent in the remaining, the majority of sequences of the two diverging lineages (blue or magenta lines). These radical sequence changes over a period that can be less than about 100 million years are a combination of retronuon insertions, segmental gene duplications, deletions of large blocks, as well as nucleotide substitutions/small indels. Moreover, from this sequence disparity, novel functional nuons, including alternative exons and novel regulatory regions for spatiotemporal expression of genes, can be exapted (blue or magenta boxes with asterisks), sometimes replacing ancestral sequences (nuon A with nuon a, in the lineage leading to the upper left lineage). At the level of the genome, these could be considered macroevolutionary changes.

[10] Although it is not widely appreciated, Gould seemed quite comfortable taking the middle of the road. In his first book for a general audience, Gould stated: "Nature is so wondrously complex and varied that anything possible does happen" and "A person who wants clean, definitive, global answers to the problem of life must search elsewhere, not in nature" and "Really big questions succumb to the richness of nature—change can be directed or aimless, gradual or cataclysmic, selective or neutral. I will rejoice in the multifariousness of nature and leave the chimera of certainty to politicians and preachers" (Gould 1977a: p. 271). A quarter-century later his position had not changed: "our deepest puzzles and most fascinating inquiries often fall into a no-man's land not clearly commanded by either party. . . ." (Gould 2002: p. 1212). Clearly, Gould's initial intention was not to polarize—perhaps he was co-opted to do so (Sterelny 2001)—as if Gould, although mostly occupying the middle ground, was pushed by his opponents into an extreme corner. When placed in that corner, instead of stressing his moderate positions, he opted to fight. This role was, perhaps, not always pleasant for him; he had to withstand a lot of abuse from colleagues, and the creationists then misappropriated the controversies by claiming that even evolutionary biologists disagreed on the *basics* of evolution. On the other hand, the polarization crystallizing around his person catalyzed a tremendously active and enlightening debate, which is summarized in his opus magnus, *The Structure of Evolutionary Theory* (Gould 2002), and which will continue to fertilize the field for decades and centuries to come. There could hardly be a better statement than the one from Richard Dawkins: "What a brilliant way for a scholar to go. I shall miss him" (Dawkins 2003: p. 222).

[11] In the human genome about 15% of protein coding genes are devoid of introns (Sakharkar et al. 2002): http://sege.ntu.edu.sg/wester/intronless/data.html).

tronuons, including LTRs, LINEs, and SINEs, were potential mobile regulatory elements capable of altering the expression of resident genes or contributing novel domains to protein-coding genes (Brosius and Gould 1992). In the meantime, a plethora of such retronuon contributions have been reported and amply collected and reviewed (Makalowski et al. 1994; Brosius and Tiedge 1995; Britten 1996, 1997; Kidwell and Lisch 1997; Jurka 1998; Brosius 1999b,c, 2003a,d, 2005a; Makalowski 2000, 2003).

Although it is clear that theoretically, like point mutations, retronuons are probably more often a disadvantage than an advantage to the affected individual and in most cases they have no effect at all, it is remarkable that 25% of analyzed promoter regions in the human genome contain retronuon-derived sequences (Jordan et al. 2003) and that the 5′ ends of a large proportion of mRNAs contain parts of retronuons, thus indicating a role of the respective retronuons in gene regulation (van de Lagemaat et al. 2003; Oei et al. 2004; see also Franchini et al. 2004). A further striking discovery is that up to 5% of human genes harbor sequences from Alu retronuons in their protein-coding regions that arose mainly via alternative splicing (Nekrutenko and Li 2001; Sorek et al. 2002; Lev-Maor et al. 2003; Kreahling and Graveley 2004; Singer et al. 2004), although it needs to be established what percentage of the alternatively spliced mRNAs encode functional protein variants. Despite this unexpected functional potential underlying retronuon insertions, the figures still likely represent an underestimate of the significance of retronuons for the exaptation of gene regulatory elements or novel protein sequence domains over longer evolutionary time frames.

In summary, events that shape eukaryotic genomes overwhelmingly support the proposal by Gould and Vrba (1982) to raise exaptation to the same level of evolutionary importance as adaptation and to use aptation as a general term to refer to both. Actually, in comparison to adaptation, exaptation appears to play *the more prominent* role in both genome and species evolution and highlights the modular architecture and evolution of genomes.[12]

Why Retroposition Continues to Survive

In the transition of RNA → DNA as genetic material, we have a good explanation for the origin of retroposition (Darnell and Doolittle 1986; Brosius 1999). The *persistence* of retroposition is more difficult to rationalize, however. It is possible that reverse transcription has an unidentified function in the cell,[13] and retroposition, with its propensity to inflate genomes, persists as an epiphenomenon. The contribution of retroposition to evolutionary plasticity may be compelling, yet evolution does not have this foresight and many authors rightfully urge caution:

> Transposons provide genetic material on which natural selection can act, and so it is not surprising that new beneficial functions for these sequences evolve. This, however, should not be confused with the primary selective forces that resulted in their origin and best explain the maintenance of most transposons, and their ability to self replicate despite their harmful effects on the host. (Hurst and Werren 2001: p. 604)

Nevertheless, persistence of reverse transcriptase activity in eukaryotic lineages could be equated with replication errors, simple point mutations. Clearly, their "primary selec-

---

[12] A concept of the modular architecture and evolution of genomes was first published in the early 1990s with Gould (Brosius and Gould 1992). Although the accompanying terminology has not caught on, the concepts have received overwhelming support through the various genome projects. For example, Graur (1993) assaulted our terminology but ignored our concept. With the discoveries of the exonization of Alu repetitive elements, which were predicted by us (Brosius and Gould 1992), he, with Gil Ast and colleagues, is now contributing exciting examples for potential exaptation at the genomic level (Dagan et al. 2004; Sorek et al. 2002).

[13] After initiating the transition from RNA to DNA genomes, reverse transcriptase may have been under selection for the maintenance of chromosome ends (telomeres). Later, perhaps, the enzyme specialized as telomerase and concurrently lost the ability to reverse transcribe other RNAs (Shippen-Lentz and Blackburn 1990; Blackburn 1991; Ware et al. 2000). By then a paralogue of the primordial promiscuous reverse transcriptase might have begun its independent "life" by being part of retroviruses, retrotransposons, retroelements, or similar retronuons without possessing a function in cellular metabolism. For further discussion see Brosius 2003a.

tive forces" were not "new beneficial functions" either. Presumably, they still occur because a totally error-free mode of replication would be too costly. In any event, a lineage evolving perfect replication would be doomed. Prokaryotes are likely not dependent on reverse transcriptase mediated processes anymore to enhance genomic plasticity. This may be different, however, in multicellular organisms, especially those with relatively long generation times. I argued before that not being overly tidy with one's "junk"[14] may translate into a significant benefit over time (Brosius and Tiedge 1995; Brosius 1999a, 2003a, 2005a). In contrast, *Homo sapiens*, as individuals, customarily favor immediate gain over long-term benefit, a cardinal problem that will likely prevent us from surviving the average 4-million-year life span of species.

Diversity and Disparity at the Genomic Level: Retroposition as "Saltatory" Events?

At the level of species and above, diversity refers to the number of species and disparity to the more profound differences in body plan (Jaanusson 1981; Runnegar 1987; Gould 1989). Can we make such a distinction at the genomic level and can we correlate morphology with genomic events? Above, two major forces of variation among genomes were described: the well-known and omnipresent base substitutions and small indels on the one hand and the yin and yang of retroposition[15] or segmental duplications and large deletions[16] on the other. At the genomic level, we might distinguish between diversity and disparity (Fig. 1) and the relative contributions of these two forces. However, given enough time, even the slow but constant "microevolutionary" forces of nucleotide exchanges and small indels would have comparable effects (randomization of nonaptive sequences) on the vast majority of genomic sequences that are not under purifying selection. Conversely, punctuated retropositions can take several if not tens of millions of years to become exaptations, "awaiting" additional small changes that, for example, create a functional splice site or an open reading frame (Singer et al. 2004). Importantly, despite the large degree of genome disparity, (e.g., the majority of nucleotide sequences have been exchanged between mammals; Fig. 1), significant differences in body plans did not emerge.

Nevertheless, expression of the same gene at different times in development or in different cell types has long been suggested to be a key event in speciation (Wilson 1975; Zuckerkandl 1975; Gould 1977b),[17] and newly inserted retronuons are well capable of inducing such alterations (Brosius and Gould 1992; Brosius and Tiedge 1995; Britten 1996; Brosius 1999a,d; Makalowski 2000). Without ignoring the potential of chromosomal rearrangements (see once more footnote 16) or even point mutations in a single gene, single retroposition events and, more likely, the combined impact of a newly arisen retronuon family (see also below) are reasonable scenarios that set the

[14] Stephen J. Gould concurred (Gould 2002: p. 1269) and I almost endorsed (Brosius 2003c) Sidney Brenner's suggestion that the use of the term "junk" DNA is acceptable if one divides rubbish into two subcategories. Garbage is rubbish that we throw away, and junk is something we end up keeping (Brenner 1998). However, with the advent of junk e-mail, something you would like to get rid of as soon as possible, I am returning to our earlier position (Brosius and Gould 1992) of banning the misleading term from discussions of the genome, or at least using it in quotation marks.

[15] Although jumping retroposons can be viewed as small "saltations" at the genomic level, these events cannot be equated with the evolutionary term "saltation," which applies to saltations at higher taxonomic levels.

[16] Insertions of retronuons often lead to subsequent deletions between retronuons of similar sequences by unequal homologous recombination (Batzer and Deininger 2002; Bailey et al. 2003) Genome size reduction via illegitimate recombination also plays a significant role in opposing genome size inflation (Devos et al. 2002).

[17] "The classical data of heterochrony have been widely ignored and regarded as old-fashioned. But I believe that they have a central role to play in the growing discussion on the evolutionary significance of changes in gene regulation . . . I also believe that an understanding of regulation must lie at the center of any rapprochement between molecular and evolutionary biology; for a synthesis of the two biologies will surely take place, if it occurs at all, on the common field of development . . . The data of heterochrony are data about regulation; they define changes in timing of development for features held in common by ancestors and descendants . . . Throughout this book, I have tried to demonstrate that heterochrony is extremely important in evolution—both in frequency of occurrence and as the basis of significant evolutionary change. I hope that I have added thereby some support for the belief that alterations in regulation form the major stuff of evolutionary change" (Gould 1977b: pp. 408–409).

course for speciation (Bingham et al. 1982; Rose and Doolittle 1983; Ginzburg et al. 1984; McDonald 1990; Brosius 1991; Brosius and Gould 1992; Brosius and Tiedge 1995). Apart from the significance of Hox genes and other developmental switches, I see the likely role of retroposition in speciation as a partial vindication of Richard B. Goldschmidt's proposals concerning species-level saltations (Goldschmidt 1940; Gould 2002; Ronshaugen et al. 2002; Dietrich 2003; Wagner et al. 2003).

A major problem with some of the issues discussed is only of a semantic nature. The genomic view shows that even *apparent* saltations, such as the accumulation of retroposons in genomes, are prevalently of a gradual nature, although on an evolutionary timescale this might happen quite rapidly. One should keep in mind that environmental stresses trigger increased transcription of SINE RNAs that, in turn, could lead to bursts of retroposition (Li et al. 1999; Nevo 2001). Likewise, when considering major evolutionary transitions (Szathmáry and Smith 1995) only the outcome is viewed, such as the appearance of novel macromolecule classes. The roads toward transitions are usually long, winding, and gradual. The evolution of templated translation had primitive beginnings and gradually led to what is considered the most sophisticated of all cellular machines (Brosius 2001). Significantly, the transition from an RNA genome to a DNA genome is still not completed (see above). The most recent evolutionary transition, the evolution of a central nervous system with the ability to transfer information vertically and horizontally, has been a gradual process, perhaps beginning with the continuous growth of the primate cortex, yet the effect is conceived as punctuated (see below).

Contingency

In his book *Wonderful Life: The Burgess Shale and the Nature of History,* Stephen Jay Gould (1989) argued convincingly that were we to go back in evolutionary time and rerun the tape, we would end up with different species, even if environmental conditions were identical. He cited cases of geographical separation, where it is apparent that evolution moved along different trajectories. His view matches the consensus of others; that evolution is a chaotic, nondirectional phenomenon. In contrast, Simon Conway Morris (1998, 2003) argued for the *inevitability* of arriving at species akin to *Homo sapiens* when the tape is rerun and even if life evolved in distant parts of the universe (although he favored the idea that we are alone.) One of his main arguments for evolutionary patterns and the inevitability of intelligent life is molecular and organismal convergence but he also invoked channeling.

At first sight it is remarkable that sophisticated structures such as the eye evolved independently dozens of times (Mayr 2001: pp. 204–206), a notion that has recently been seriously challenged (Gehring and Ikeo 1999; Arendt and Wittbrodt 2001). Even if eye evolution is polyphyletic, devices that allow perception of and fast reaction to the environment are of tremendous selective value, and the limitations presented by the physics involved do not permit much leeway with respect to the structure of an organ with these particular requirements. As amplification/selection schemes with RNA sequences have shown (see above), if the molecular building blocks or their precursors are attainable, selection can reveal them in independent experiments, but also two or more molecular structures can be selected simultaneously. Therefore, we shall find similar sequence space in independent cellular and organismal developments, in part because these and not other sequences are available.[18] But, we also observe that, in different phyla, unrelated proteins (different sequence space but similar properties) were exapted, for example, to impart the optical properties of the eye lens (Piatigorsky 1998).

When comparing amplification/selection experiments to the roughly 4 billion years of evolution of life on this planet, one must keep in mind that the former are specifically geared to achieving a defined outcome to which, giv-

[18] One of the numerous statements of Gould (1983: p. 157) that also are relevant at the genomic level: "Organic material is not putty and natural selection is not omnipotent. Each organic design is pregnant with evolutionary possibilities, but restricted in its paths of potential change."

en the experimental limitations, there is sometimes only one apparent solution. Natural evolution of life, in contrast, is aimless, nondirected; there is no final destination at all. Evolution itself does not care if life on this planet or the whole universe ceases. Up to this point, life just continued, sometimes toward higher but also sometimes toward lower complexities. As mentioned in the beginning of this paper, over this long time period major evolutionary transitions occurred, initially triggered by seemingly insignificant, chanced-upon exaptations, some of which affected the replication systems themselves. Like eyes, efficient and chemically stable replication and information storage systems that led to the primate central nervous system are of enormous aptive value. However, one would be hard pressed to prove that the *only* possible trajectory leading to memes (see below) would be the one via RNA and later DNA.

One only has to consider the permanent gauntlet of extinction that life had to run over the course of 3–4 billion years, to appreciate the significance of the chance factor (Raup 1991). It is unclear whether, after a major catastrophic event, the survivors would still have the potential to evolve like their ancestors. If, for example, Bacteria and Archaea remained, might they not be too *far* evolved and specialized to once more undergo the evolutionary transitions leading to multicellular organisms?[19] Even if one hypothetically turned back the clock to the same starting ancestors (e.g., to the common ancestors of mammals prior to the mammalian radiation or to the common ancestor of chimpanzee and human) and the identical environmental conditions, it is far from clear whether the same mammals or the ape, with its sophisticated central nervous system that stores, processes, and replicates information, now pondering these possibilities would evolve. It is perhaps the seemingly insignificant flutter of a butterfly's wing that potentially changes the course of evolution in a drastic way. If the evolution of life and forms like *Homo sapiens* were not contingent, why did life start only once in four billion years?[20]

We have seen that, at the genomic level, one point mutation or one retronuon insertion could correspond to that flutter of wing. Think once more of the biogenesis of a founder RNA for a SINE (in theory by a single point mutation that turns an RNA from a poor template for retroposition into a highly efficient one) that shapes genomes and alters the evolutionary course of entire lineages (Fig. 1). What phenotype might primates have without one million Alu retronuons, that all began with a single founder element, in their genomes? Would our central nervous system have evolved to such a lofty level? What subtle yet distinguishing capability would the brains of the respective primate species be lacking if, in the lineage leading to *Anthropoidea*, a single monomeric Alu retronuon had not been exapted as a neuron-specific, dendritic, small non-messenger RNA (Tiedge et al. 1993; Kurychev et al. 2001)?

Now that humans, the "pride of creation" have arrived at their "destination," does evolution stop? Although it is rather fruitless and perhaps even contemptible trying to predict the direction of evolution in the future or from a certain time point in the past by rerunning the tape, let us consider this point for a moment merely for heuristic purposes. After all, it may be not for a lack of fantasy that science fiction writers and movie makers continue to portray intelligent aliens as resembling the image of man. They have merely anticipated the "scientific" assumption that all evolutionary paths must lead to organisms akin to

[19] In his *Origin of Species* Charles Darwin already realized that "We can clearly understand why a species when once lost should never reappear; even if the very same conditions of life, organic and inorganic, should recur. For though the offspring of one species might be adapted (and no doubt this has occurred in innumerable instances) to fill the exact place of another species in the economy of nature, and thus supplant it; yet the two forms—the old and the new—would not be identically the same; for both would almost certainly inherit different characters from their distinct progenitors (Darwin 1859: p. 315)."

[20] One explanation could be that once life highly dependent on metabolites evolved and spread to all possible niches, a fledging event would have little chance of surviving, as it would immediately be devoured. (See also Sober and Steel 2002.) Another explanation could be that possibly our preconceptions regarding life has kept us from finding less obvious forms of life.

Homo sapiens (Conway Morris 1998, 2003)[21]. So, it is assumed, then—in a typical anthropocentric overestimation of position—that *H. sapiens* is indeed the end point of evolution. However, in a very short time, on an evolutionary scale, our lineage has gone through one major evolutionary transition and we are at the brink of at least one additional one. The first is, through evolution of the central nervous system in primates, in addition and potentially in succession to RNA and DNA, the advent of an additional system to store, replicate, and transmit information in a vertical *and* a horizontal mode. In recognition of the evolutionary significance of this new medium for information storage and transfer, Dawkins coined (in analogy to genes) the term memes (Dawkins 1976; Blackmore 1999). Soon thereafter, Gould (1980: pp. 83–84) acknowledged that Cultural evolution has progressed at rates that Darwinian processes cannot begin to approach. Darwinian evolution continues in *Homo sapiens*, but at rates so slow that it no longer has much impact on our history. This crux in history has been reached because Lamarckian processes have finally been unleashed upon it. Human cultural evolution, in strong opposition to our biological history, is Lamarckian in character. What we learn in one generation, we transmit directly by teaching and writing. Acquired characters are inherited in technology and culture. Lamarckian evolution is rapid and accumulative. It explains the cardinal difference between our past, purely biological mode of change, and our current, maddening acceleration toward something new and liberating—or toward the abyss.

One can only concur[22] and add the following: "Only at the stage of memes are we able to observe Neo-Lamarckism: a learned behavior can be transmitted from individual to individual not only in a vertical but also in a horizontal transfer. . . . Of course, this knowledge or meme could not direct the alteration of genetically based behavior—up to now" (Brosius 2003c: p. 11). Presently, we are about to witness yet another major evolutionary transition. Through our advances in biology we are now able to transmit knowledge and experimental experience into the germ line of virtually all living species including our own. We will be able to correct the genetic causes of hereditary diseases and implant desired traits into future generations. In 3.5 billion years of evolution, life was perhaps never so close to some form of Lamarckian mechanism as now (Brosius 2003c); whether this is a desirable development is, of course, yet another question (Brosius and Kreitman 2000; Brosius 2003b).

On the horizon, one can already sense the harbingers of additional transitions. Several authors, among them my valued Münster colleague Kazem Sadegh-Zadeh, have noted that man's existence is already heavily intertwined with and dependent on machines and artificial intelligence (Mazlish 1993; Stock 1993; Sadegh-Zadeh 2000). One possible scenario is that human societies will develop into superorganisms not unlike those of insects (Wilson 1971). This would be a transition as dramatic as the one from single- to multicellular organisms. The professional specializations of humans and attempts to centrally control individuals and societies are some of the writings on the wall.

Life depends on stable, efficient, high-capacity systems of information storage and on information that replicates at high fidelity. We

[21] Although it is highly beneficial to occasionally challenge entrenched concepts, I wonder whether this a poorly disguised attempt to let religion participate in evolutionary thought: if you can't fight evolution—join it? Instead of catering to the ultra-naive creationists, Conway Morris (2003) appeared to target a more intelligent segment of our non-rationalist population, perhaps those who should know better but cannot liberate themselves from infantile imprinting and religious indoctrination, those who seem to possess a full deck of cards but are unable or unwilling to use them all—at least in certain games.

[22] "The legacy of brilliant men includes undeveloped foresight. English biologist J. B. S. Haldane probably anticipated every good idea that evolutionary theorists will invent during this century" (Gould 1977a: p. 262). This insight may well be extended to our new century and to several other evolutionists, including Haldane's student John Maynard Smith, Carl R. Woese, W. Ford Doolittle, Eörs Szathmary, and Michael Ghiselin—just to name a few—and, of course, Stephen Jay Gould himself.

need not venture beyond our own solar system to be keenly aware that our carrier of genetic information along with its "survival machines" function well in the environments on Earth with which they have evolved, but are useless under different physicochemical conditions.

The success of unmanned space exploration machines such as the Hubble Space Telescope and the Mars Exploration Rovers, Spirit and Opportunity, permits a glimpse into possible directions that the evolution of life may take. Perhaps originating from quite robust inorganic modes of replication (Cairns-Smith 1982; Wächtershauser 1992) life may return to a variation of its inorganic roots. Aided by man's "Lamarckian" capabilities, silicone-based information systems, including the instructions to replicate machines from highly inert inorganic and organic materials, may evolve. So long as the replication of these robust organisms is not error free and permits free variations of the "design," similar to current life forms, the *genetic* information as well as its "survival machines" may be able to live and evolve far beyond the limited range that can sustain current forms of life. With the help of a transient carrier of information (i.e., the human central nervous system and its extensions), an evolutionary transition may occur that utilizes *genetic* material much more stable than RNA or DNA and more suitable to survive billions of years in the expanse of the universe. To these new life forms, we humans may be just what *A. afarensis*, *H. heidelbergenis*, and *H. neanthalensis* are to us—dead branches of the evolutionary tree. On the other hand, humans may store their digitized genetic information (and that of other sequenced organisms) in the early versions of these future life forms. From time to time, when conditions are right, these life forms might rebuild the organic organisms—not so much unlike what a virus based on RNA as genetic material does with DNA-based organisms. This is no guarantee for the survival of *Homo sapiens*: the future life forms might not want to waste energy and resources on replicating "parasites" and find ways to eliminate our digitalized genetic information.[23] Perhaps the new life forms will keep the information on how to rebuild RNA/DNA/protein-based organisms and from time to time generate a few individuals for an "Alluvial Park." Is this our inevitable future—or can we flutter our wings to take a different direction?

Acknowledgments

I would like to dedicate this paper also to Harry F. Noller on the occasion of his 65th birthday. Harry, almost three decades ago, took me on a trip to the RNA/RNP world from which I have never returned. Thanks to M. Bundman for editorial help. The author is supported by the German Human Genome Project through the Bundesministerium für Bildung und Forschung (01KW9966), a specific targeted research project (STREP) of the European Commission (RIBOREG, program 503022), and the National Genome Research Network (NGFN, EP-S32T01).

Literature Cited

Arendt, D., and J. Wittbrodt. 2001. Reconstructing the eyes of Urbilateria. Philosophical Transactions of the Royal Society of London B 356:1545–1563.

Arnason, U., J. A. Adegoke, K. Bodin, E. W. Born, Y. B. Esa, A. Gullberg, M. Nilsson, R. V. Short, X. Xu, and A. Janke. 2002. Mammalian mitogenomic relationships and the root of the eutherian tree. Proceedings of the National Academy of Sciences USA 99:8151–8156.

Bailey, J. A., G. Liu, and E. E. Eichler. 2003. An Alu transposition model for the origin and expansion of human segmental duplications. American Journal of Human Genetics 73:823–834.

Balakirev, E. S., and F. J. Ayala. 2003. Pseudogenes: are they "junk" or functional DNA? Annual Review of Genetics 37: 123–151.

Bartel, D. P., and P. J. Unrau. 1999. Constructing an RNA world. Trends in Cell Biology 9:M9–M13.

Batzer, M. A., and P. L. Deininger. 2002. Alu repeats and human genomic diversity. Nature Reviews Genetics 3:370–379.

Bernardi, G. 2004. Structural and evolutionary genomics: natural selection in genome evolution. Elsevier, Amsterdam.

[23] If life forms from other parts of the universe visited Earth, it would not be easy to tell whether their genetic material went through one or several major transitions. Conversely, should, after our genetic information was purged and forgotten, individuals of these new life forms revisit planet Earth and still encounter Bacteria, Archaea, and some Eukarya, they may be unable to recognize these forms of life as common ancestors, just as many of us are still unable to see our evolutionary relationship to unicellular forms of life or to other primates for that matter! The future life forms would be less to blame, because they do not even share the same physicochemical properties of genetic material.

Bingham, P. M., M. G. Kidwell, and G. M. Rubin. 1982. The molecular basis of P-M hybrid dysgenesis: the role of the P element, a P-strain-specific transposon family. Cell 29:995–1004.
Blackburn, E. H. 1991. Telomeres. Trends in Biochemical Sciences 16:378–381.
Blackmore, S. 1999. The meme machine. Oxford University Press, Oxford.
Boeke, J. D. 2003. The unusual phylogenetic distribution of retrotransposons: a hypothesis. Genome Research 13:1975–1983.
Brenner, S. 1998. Refuge of spandrels. Current Biology 8:R669.
Britten, R. J. 1996. DNA sequence insertion and evolutionary variation in gene regulation. Proceedings of the National Academy of Sciences USA 93:9374–9377.
———. 1997. Mobile elements inserted in the distant past have taken on important functions. Gene 205:177–182.
Britten, R. J., and E. H. Davidson. 1969. Gene regulation for higher cells: a theory. Science 165:349–357.
———. 1971. Repetitive and non-repetitive DNA sequences and a speculation on the origins of evolutionary novelty. Quarterly Review of Biology 46:111–138.
Brosius, J. 1991. Retroposons—seeds of evolution. Science 251: 753.
———. 1999a. Genomes were forged by massive bombardments with retroelements and retrosequences. Genetica 107:209–238.
———. 1999b. Many G-protein-coupled receptors are encoded by retrogenes. Trends in Genetics 15:304–305.
———. 1999c. RNAs from all categories generate retrosequences that may be exapted as novel genes or regulatory elements. Gene 238:115–134.
———. 1999d. Transmutation of tRNA over time. Nature Genetics 22:8–9.
———. 2001. tRNAs in the spotlight during protein biosynthesis. Trends in Biochemical Sciences 26:653–656.
———. 2003a. The contribution of RNAs and retroposition to evolutionary novelties. Genetica 118:99–116.
———. 2003b. From Eden to a hell of uniformity? Directed evolution in humans. Bioessays 25:815–821.
———. 2003c. Gene duplication and other evolutionary strategies: from the RNA world to the future. Journal of Structural and Functional Genomics 3:1–17.
———. 2003d. How significant is 98.5% 'junk' in mammalian genomes? Bioinformatics 19(Suppl. 2):35.
———. 2005a. Echoes from the past—are we still in an RNP world? Cytogenetic and Genome Research (in press).
———. 2005b. Waste not, want not—transcript excess in multicellular Eukaryotes. Trends in Genetics 21:287–288.
Brosius, J., and S. J. Gould. 1992. On "genomenclature": a comprehensive (and respectful) taxonomy for pseudogenes and other "junk DNA." Proceedings of the National Academy of Sciences USA 89:10706–10710.
Brosius, J., and M. Kreitman. 2000. Eugenics—evolutionary nonsense? Nature Genetics 25:253.
Brosius, J., and H. Tiedge. 1995. Reverse transcriptase: mediator of genomic plasticity. Virus Genes 11:163–179.
———. 2004. RNomenclature. RNA Biology 1:81–83.
Cairns-Smith, A. G. 1982. Genetic takeover and the mineral origins of life. Cambridge University Press, Cambridge.
Cold Spring Harbor Laboratory Press. 1987. Cold Spring Harbor symposia on quantitative biology, Vol. 52. Evolution of catalytic function. Cold Spring Harbor Laboratory, Cold Spring Harbor, N.Y.
Conway Morris, S. 1998. The crucible of creation. Oxford University Press, Oxford.
———. 2003. Life, the universe and everything: inevitable humans in a lonely universe. Cambridge University Press, Cambridge.
Dagan, T., R. Sorek, E. Sharon, G. Ast, and D. Graur. 2004. AluGene: a database of Alu elements incorporated within protein-coding genes. Nucleic Acids Research 32(Database issue):D489–492.
Dahl, H. H., R. M. Brown, W. M. Hutchison, C. Maragos, and G. K. Brown. 1990. A testis-specific form of the human pyruvate dehydrogenase E1 alpha subunit is coded for by an intronless gene on chromosome 4. Genomics 8:225–232.
Darnell, J. E., and W. F. Doolittle. 1986. Speculations on the early course of evolution. Proceedings of the National Academy of Sciences USA 83:1271–1275.
Darwin, C. 1859. On the origin of species by means of natural selection, or the preservation of favoured races in the struggle for life. John Murray, London.
Dawkins, R. 1976. The selfish gene. Oxford University Press, Oxford.
———. 1982. The extended phenotype. W. H. Freeman, San Francisco.
———. 2003. A devil's chaplain. Weidenfeld and Nicolson, London.
DeChiara, T. M., and J. Brosius. 1987. Neural BC1 RNA: cDNA clones reveal nonrepetitive sequence content. Proceedings of the National Academy of Sciences USA 84:2624–2628.
Deininger, P. L., J. V. Moran, M. A. Batzer, and H. H. Kazazian Jr. 2003. Mobile elements and mammalian genome evolution. Current Opinion in Genetics and Development 13:651–658.
D'Erchia, A. M., C. Gissi, G. Pesole, C. Saccone, and U. Arnason. 1996. The guinea-pig is not a rodent. Nature 381:597–600.
Dermitzakis, E. T., A. Reymond, N. Scamuffa, C. Ucla, E. Kirkness, C. Rossier, and S. E. Antonarakis. 2003. Evolutionary discrimination of mammalian conserved non-genic sequences (CNGs). Science 302:1033–1035.
Devos, K. M., J. K. M. Brown, and J. L. Bennetzen. 2002. Genome size reduction through illegitimate recombination counteracts genome expansion in Arabidopsis. Genome Research 12: 1075–1079.
Dewannieux, M., C. Esnault, and T. Heidmann. 2003. LINE-mediated retrotransposition of marked Alu sequences. Nature Genetics 35:41–48.
Dietrich, M. R. 2003. Richard Goldschmidt: hopeful monsters and other 'heresies.' Nature Reviews Genetics 4:68–74.
Dobzhansky, T. G. 1937. Genetics and the origin of species. Columbia University Press, New York.
Doolittle, W. F., and C. Sapienza. 1980. Selfish genes, the phenotype paradigm and genome evolution. Nature 284:601–603.
Dworkin, J. P., A. Lazcano, and S. L. Miller. 2003. The roads to and from the RNA world. Journal of Theoretical Biology 222: 127–134.
Ellington, A. D., and J. W. Szostak. 1990. In vitro selection of RNA molecules that bind specific ligands. Nature 346:818–822.
Famulok, M., G. Mayer, and M. Blind. 2000. Nucleic acid aptamers-from selection in vitro to applications in vivo. Accounts of Chemical Research 33:591–599.
Franchini, L. F., E. W. Ganko, and J. F. McDonald. 2004. Retrotransposon-gene associations are widespread among *D. melanogaster* populations. Molecular Biology and Evolution 21: 1323–1331.
Gehring, W. J., and K. Ikeo. 1999. Pax 6: mastering eye morphogenesis and eye evolution. Trends in Genetics 15:371–377.
Gesteland, R. F., T. R. Cech, and J. F. Atkins. 1999. The RNA world. Cold Spring Harbor Laboratory Press, Cold Spring Harbor, N.Y.
Ghiselin, M. T. 1974. A radical solution to the species problem. Systematic Zoology 23:554–556.
Gilbert, W. 1978. Why genes in pieces? Nature 271:501.
Ginzburg, L. R., P. M. Bingham, and S. Yoo. 1984. On the theory

of speciation induced by transposable elements. Genetics 107: 331–341.
Goldschmidt, R. B. 1940. Material basis of evolution. Yale University Press, New Haven, Conn.
Gould, S. J. 1977a. Ever since Darwin: reflections in natural history. Norton, New York.
———. 1977b. Ontogeny and phylogeny. Belknap Press of Harvard University Press, Cambridge.
———. 1980. The panda's thumb. Norton, New York.
———. 1981. What happens to bodies if genes act for themselves? Natural History. November.
———. 1983. Hen's teeth and horse's toes. Norton, New York.
———. 1989. Wonderful life: the Burgess Shale and the nature of history. Norton, New York.
———. 2002. The structure of evolutionary theory. Belknap Press of Harvard University Press, Cambridge.
Gould, S. J., and E. A. Lloyd. 1999. Individuality and adaptation across levels of selection: how shall we name and generalize the unit of Darwinism? Proceedings of the National Academy of Sciences USA 96:11904–11909.
Gould, S. J., and E. S. Vrba. 1982. Exaptation—a missing term in the science of form. Paleobiology 8:4–15.
Graur, D. 1993. Molecular deconstructivism. Nature 363:490.
Graur, D., W. A. Hide, and W. H. Li. 1991. Is the guinea-pig a rodent? Nature 351:649–652.
Gregory, T. R. 2001. Coincidence, coevolution, or causation? DNA content, cell size, and the C-value enigma. Biological Reviews of the Cambridge Philosophical Society 76:65–101.
Gregory, T. R., and P. D. Hebert. 1999. The modulation of DNA content: proximate causes and ultimate consequences. Genome Research 9:317–324.
Guerrier-Takada, C., K. Gardiner, T. Marsh, N. Pace, and S. Altman. 1983. The RNA moiety of ribonuclease P is the catalytic subunit of the enzyme. Cell 35:849–857.
Hagan, C. R., R. F. Sheffield, and C. M. Rudin. 2003. Human Alu element retrotransposition induced by genotoxic stress. Nature Genetics 35:219–220.
Hanczyc, M. M., and R. L. Dorit. 1998. Experimental evolution of complexity: in vitro emergence of intermolecular ribozyme interactions. RNA 4:268–275.
Herbert, A. 2004. The four Rs of RNA-directed evolution. Nature Genetics 36:19–25.
Hurst, G. D., and J. H. Werren. 2001. The role of selfish genetic elements in eukaryotic evolution. Nature Reviews Genetics 2: 597–606.
Hüttenhofer, A., M. Kiefmann, S. Meier-Ewert, J. O'Brien, H. Lehrach, J.-P. Bachellerie, and J. Brosius. 2001. RNomics: an experimental approach that identifies 201 candidates for novel, small, non-messenger RNAs in mouse. EMBO Journal 20: 2943–2953.
Huxley, J. S. 1942. Evolution: the modern synthesis. Allen and Unwin, London.
Jaanusson, V. 1981. Functional thresholds in evolutionary progress. Lethaia 14:251–260.
Jepsen, G. L., G. G. Simpson, and E. Mayr, eds. 1949. Genetics, paleontology, and evolution. Princeton University Press, Princeton, N.J.
Jordan, I. K., I. B. Rogozin, G. V. Glazko, and E. V. Koonin. 2003. Origin of a substantial fraction of human regulatory sequences from transposable elements. Trends in Genetics 19:68–72.
Jurka, J. 1998. Repeats in genomic DNA: mining and meaning. Current Opinion in Structural Biology 8:333–337.
Kajikawa, M., and N. Okada. 2002. LINEs mobilize SINEs in the eel through a shared 3′ sequence. Cell 111:433–444.
Keller, L., ed. 1999. Levels of selection in evolution. Princeton University Press, Princeton, N.J.
Kidwell, M. G., and D. Lisch. 1997. Transposable elements as sources of variation in animals and plants. Proceedings of the National Academy of Sciences USA 94:7704–7711.
King, M. C., and A. C. Wilson. 1975. Evolution at two levels in humans and chimpanzees. Science 188:107–116.
Kirkness, E. F., V. Bafna, A. L. Halpern, S. Levy, K. Remington, D. B. Rusch, A. L. Delcher, M. Pop, W. Wang, C. M. Fraser, and J. C. Venter. 2003. The dog genome: survey sequencing and comparative analysis. Science 301:1898–1903.
Kreahling, J., and B. R. Graveley. 2004. The origins and implications of Aluternative splicing. Trends in Genetics 20:1–4.
Kruger, K., P. J. Grabowski, A. J. Zaug, J. Sands, D. E. Gottschling, and T. R. Cech. 1982. Self-splicing RNA: autoexcision and autocyclization of the ribosomal RNA intervening sequence of Tetrahymena. Cell 31:147–157.
Kurychev, V. Y., B. V. Skryabin, J. Kremerskothen, J. Jurka, and J. Brosius. 2001. Birth of a gene: locus of neuronal BC200 snmRNA in three prosimians and human BC200 pseudogenes as archives of change in the Anthropoidea lineage. Journal of Molecular Biology 309:1049–1066.
Lander, E. S., L. M. Linton, B. Birren, C. Nusbaum, M. C. Zody, J. Baldwin, K. Devon, K. Dewar, M. Doyle, W. FitzHugh, et al. 2001. Initial sequencing and analysis of the human genome. Nature 409:860–921.
Lev-Maor, G., R. Sorek, N. Shomron, and G. Ast. 2003. The birth of an alternatively spliced exon: 3′ splice-site selection in Alu exons. Science 300:1288–1291.
Li, T., J. Spearow, C. M. Rubin, and C. W. Schmid. 1999. Physiological stresses increase mouse short interspersed element (SINE) RNA expression in vivo. Gene 239:367–372.
Makalowski, W. 2000. Genomic scrap yard: how genomes utilize all that junk. Gene 259:61–67.
———. 2003. Genomics: not junk after all. Science 300:1246–1247.
Makalowski, W., G. A. Mitchell, and D. Labuda. 1994. Alu sequences in the coding regions of mRNA: a source of protein variability. Trends in Genetics 10:188–193.
Margulies, E. H., M. Blanchette, N. C. S. Program, D. Haussler, and E. D. Green. 2003. Identification and characterization of multi-species conserved sequences. Genome Research 13: 2507–2518.
Martignetti, J. A., and J. Brosius. 1993. BC200 RNA: a neural RNA polymerase III product encoded by a monomeric Alu element. Proceedings of the National Academy of Sciences USA 90:11563–11567.
Mattick, J. S. 2003. Challenging the dogma: the hidden layer of non-protein-coding RNAs in complex organisms. Bioessays 25:930–939.
Maxam, A. M., and W. Gilbert. 1977. A new method for sequencing DNA. Proceedings of the National Academy of Sciences USA 74:560–564.
Maynard Smith, J. 1989. Did Darwin get it right? Essays on games, sex and evolution. Chapman and Hall, New York.
Maynard Smith, J., and E. Szathmáry. 1995. The major transitions in evolution. Oxford University Press, Oxford.
Mayr, E. 1942. Systematics and the origin of species from the viewpoint of a zoologist. Columbia University Press, New York.
———. 2001. What evolution is. Basic Books, New York.
Mazlish, B. 1993. The fourth discontinuity: the co-evolution of humans and machines. Yale University Press, New Haven, Conn.
McCarrey, J. R., and K. Thomas. 1987. Human testis-specific PGK gene lacks introns and possesses characteristics of a processed gene. Nature 326:501–505.
McDonald, J. F. 1990. Macroevolution and retroviral elements. Bioscience 40:183–191.
McLaughlin, P. J., and M. O. Dayhoff. 1973. Eukaryote evolution:

a view based on cytochrome c sequence data. Journal of Molecular Evolution 2:99–116.

Mendel, G. 1866. Versuche über Pflanzenhybriden. Verhandlungen des Naturforschenden Vereins in Brünn 4:3–47.

———. 1870. Über einige aus künstlicher Befruchtung gewonnene Hieraciumbastarde. Verhandlungen des naturforschenden Vereins in Brünn 8:26–31.

Murphy, W. J., E. Eizirik, W. E. Johnson, Y. P. Zhang, O. A. Ryder, and S. J. O'Brien. 2001. Molecular phylogenetics and the origins of placental mammals. Nature 409:614–618.

Nekrutenko, A., and W. H. Li. 2001. Transposable elements are found in a large number of human protein-coding genes. Trends in Genetics 17:619–621.

Nelson, K. E., M. Levy, and S. L. Miller. 2000. Peptide nucleic acids rather than RNA may have been the first genetic molecule. Proceedings of the National Academy of Sciences USA 97:3868–3871.

Nelson, P., M. Kiriakidou, A. Sharma, E. Maniataki, and Z. Mourelatos. 2003. The microRNA world: small is mighty. Trends in Biochemical Sciences 28:534–540.

Nevo, E. 2001. Evolution of genome-phenome diversity under environmental stress. Proceedings of the National Academy of Sciences USA 98:6233–6240.

Novina, C. D., and P. A. Sharp. 2004. The RNAi revolution. Nature 430:161–164.

Numata, K., A. Kanai, R. Saito, S. Kondo, J. Adachi, L. G. Wilming, D. A. Hume, Y. Hayashizaki, and M. Tomita. 2003. Identification of putative noncoding RNAs among the RIKEN mouse full-length cDNA collection. Genome Research 13: 1301–1306.

O'Brien, S. J., and W. J. Murphy. 2003. Genomics: a dog's breakfast? Science 301:1854–1855.

Oei, S. L., V. S. Babich, V. I. Kazakov, N. M. Usmanova, A. V. Kropotov, and N. V. Tomilin. 2004. Clusters of regulatory signals for RNA polymerase II transcription associated with Alu family repeats and CpG islands in human promoters. Genomics 83:873–882.

Okada, N. 1991. SINEs. Current Opinion in Genetics and Development 1:498–504.

Orgel, L. E., and F. H. C. Crick. 1980. Selfish DNA: the ultimate parasite. Nature 284:604–607.

Pasquinelli, A. E., and G. Ruvkun. 2002. Control of developmental timing by micro RNAs and their targets. Annual Review of Cell and Developmental Biology 18:495–513.

Piatigorsky, J. 1998. Gene sharing in lens and cornea: facts and implications. Progress in Retinal and Eye Research 17:145–174.

Raup, D. M. 1991. Extinction: bad genes or bad luck? Norton, New York.

Rensch, B. 1947. Neuere Probleme der Abstammungslehre (Die Transspezifische Evolution). Ferdinand Enke, Stuttgart.

Rokas, A., and P. W. Holland. 2000. Rare genomic changes as a tool for phylogenetics. Trends in Ecology and Evolution 15: 454–459.

Ronshaugen, M., N. McGinnis, and W. McGinnis. 2002. Hox protein mutation and macroevolution of the insect body plan. Nature 415:914–917.

Rose, M. R., and W. F. Doolittle. 1983. Molecular biological mechanisms of speciation. Science 220:157–162.

Runnegar, B. 1987. Rates and modes of evolution in the Mollusca. Pp. 39–60 *in* K. S. W. Campbell and M. F. Day, eds. Rates of evolution. Allen and Unwin, London.

Ryan, S. C., and A. Dugaiczyk. 1989. Newly arisen DNA repeats in primate phylogeny. Proceedings of the National Academy of Sciences USA 86:9360–9364.

Sadegh-Zadeh, K. 2000. Als der Mensch das Denken Verlernte. Die Entstehung der Machina sapiens. [When man forgot thinking: the emergence of Machina sapiens.] Burgverlag, Tecklenburg, Germany.

Sakharkar, M. K., P. Kangueane, D. A. Petrov, A. S. Kolaskar, and S. Subbiah. 2002. SEGE: a database on 'intron less/single exonic' genes from eukaryotes. Bioinformatics 18:1266–1267.

Sanger, F., and A. R. Coulson. 1975. A rapid method for determining sequences in DNA by primed synthesis with DNA polymerase. Journal of Molecular Biology 94:441–448.

Sanger, F., S. Nicklen, and A. R. Coulson. 1977. DNA sequencing with chain-terminating inhibitors. Proceedings of the National Academy of Sciences USA 74:5463–5467.

Schmalhausen, I. I. 1949. Factors of evolution: the theory of stabilizing selection. Blakiston, Philadelphia.

Schmitz, J., M. Ohme, and H. Zischler. 2001. SINE insertions in cladistic analyses and the phylogenetic affiliations of *Tarsius bancanus* to other primates. Genetics 157:777–784.

Schmitz, J., M. Ohme, B. Suryobroto, and H. Zischler. 2002. The colugo (*Cynocephalus variegatus*, Dermoptera): the primates' gliding sister? Molecular Biology and Evolution 19:2308–2312.

Shedlock, A., K. Takahashi, and N. Okada. 2004. SINEs of speciation: tracking lineages with retroposons. Trends in Ecology and Evolution 19:545–553.

Shimamura, M., H. Yasue, K. Ohshima, H. Abe, H. Kato, T. Kishiro, M. Goto, I. Munechika, and N. Okada. 1997. Molecular evidence from retroposons that whales form a clade within even-toed ungulates. Nature 388:666–670.

Shippen-Lentz, D., and E. H. Blackburn. 1990. Functional evidence for an RNA template in telomerase. Science 247:546–552.

Simpson, G. G. 1944. Tempo and mode in evolution. Columbia University Press, New York.

Singer, M. F. 1982. SINEs and LINEs: highly repeated short and long interspersed sequences in mammalian genomes. Cell 28: 433–434.

Singer, S. S., D. N. Männel, T. Hehlgans, J. Brosius, and J. Schmitz. 2004. From "junk" to gene: curriculum vitae of a primate receptor isoform gene. Journal of Molecular Biology 341: 883–886.

Sober, E. 1984. The nature of selection: evolutionary theory in philosophical focus. MIT Press, Cambridge.

Sober, E., and M. Steel. 2002. Testing the hypothesis of common ancestry. Journal of Theoretical Biology 218:395–408.

Sorek, R., G. Ast, and D. Graur. 2002. Alu-containing exons are alternatively spliced. Genome Research 12:1060–1067.

Stebbins, G. L. 1950. Variation and evolution in plants. Columbia University Press, New York.

Sterelny, K. 2001. Dawkins vs. Gould: survival of the fittest. Icon Books, Cambridge, U.K.

Stock, G. 1993. Metaman. The merging of humans and machines into a global superorganism. Simon and Schuster, New York.

Sturtevant, A. H., and T. Dobzhansky. 1936. Inversions in the third chromosome of wild races of drosophila pseudoobscura, and their use in the study of the history of the species. Proceedings of the National Academy of Sciences USA 22:448–450.

Szathmáry, E. 1990. Towards the evolution of ribozymes. Nature 344:115.

Szathmáry, E., and J. M. Smith. 1995. The major evolutionary transitions. Nature 374:227–232.

Szostak, J. W., D. P. Bartel, and P. L. Luisi. 2001. Synthesizing life. Nature 409(Suppl.):387–390.

Thomas, J. W., J. W. Touchman, R. W. Blakesley, G. G. Bouffard, S. M. Beckstrom-Sternberg, E. H. Margulies, M. Blanchette, A. C. Siepel, P. J. Thomas, J. C. McDowell, et al. 2003. Comparative analyses of multi-species sequences from targeted genomic regions. Nature 424:788–793.

Tiedge, H., W. Chen, and J. Brosius. 1993. Primary structure, neural-specific expression, and dendritic location of human BC200 RNA. Journal of Neuroscience 13:2382–2390.

Tuerk, C., and L. Gold. 1990. Systematic evolution of ligands by exponential enrichment: RNA ligands to bacteriophage T4 DNA polymerase. Science 249:505–510.

van de Lagemaat, L. N., J. R. Landry, D. L. Mager, and P. Medstrand. 2003. Transposable elements in mammals promote regulatory variation and diversification of genes with specialized functions. Trends in Genetics 19:530–536.

Wächtershauser, G. 1992. Groundworks for an evolutionary biochemistry: the iron-sulphur world. Progress in Biophysics and Molecular Biology 58:85–201.

Wagner, G. P., C. Amemiya, and F. Ruddle. 2003. Hox cluster duplications and the opportunity for evolutionary novelties. Proceedings of the National Academy of Sciences USA 100: 14603–14606.

Ware, T. L., H. Wang, and E. H. Blackburn. 2000. Three telomerases with completely non-telomeric template replacements are catalytically active. EMBO Journal 19:3119–3131.

Watson, J. D., and F. H. C. Crick. 1953. Molecular structure of nucleic acids. Nature 171:737–738.

Weiner, A. M., P. L. Deininger, and A. Efstratiadis. 1986. Nonviral retroposons: genes, pseudogenes, and transposable elements generated by the reverse flow of genetic information. Annual Review of Biochemistry 55:631–661.

Williams, G. C. 1966. Adaptation and natural selection: a critique of some current evolutionary thought. Princeton University Press, Princeton, N.J.

Wilson, D. S. 1975. A theory of group selection. Proceedings of the National Academy of Sciences USA 72:143–146.

———. 1980. The natural selection of populations and communities. Benjamin/Cummings, Menlo Park, Calif.

Wilson, D. S., and E. Sober. 1989. Reviving the superorganism. Journal of Theoretical Biology 136:337–356.

Wilson, D. S., and J. W. Szostak. 1999. In vitro selection of functional nucleic acids. Annual Review of Biochemistry 68:611–647.

Wilson, E. O. 1971. The insect societies. Belknap Press of Harvard University Press, Cambridge.

Woese, C. R. 2001. Translation: in retrospect and prospect. RNA 7:1055–1067.

Woese, C. R., and G. E. Fox. 1977. Phylogenetic structure of the prokaryotic domain: the primary kingdoms. Proceedings of the National Academy of Sciences USA 74:5088–5090.

Wolfe, K. H., and W. H. Li. 2003. Molecular evolution meets the genomics revolution. Nature Genetics 33(Suppl.):5–265.

Zuckerkandl, E. 1975. The appearance of new structures and functions in proteins during evolution. Journal of Molecular Evolution 7:1–57.

Paleobiology, 31(2), 2005, pp. 17–26

Heterochrony, disparity, and macroevolution

Kenneth J. McNamara and Michael L. McKinney

Abstract.—The concept of heterochrony has long had a central place in evolutionary theory. During their long history, heterochrony and several associated concepts such as paedomorphosis and neoteny have often been contentious and they continue to be criticized. Despite these criticisms, we review many examples showing that heterochrony and its associated concepts are increasingly cited and used in many areas of evolutionary study. Furthermore, major strides are being made in our understanding of the underlying genetic and developmental mechanisms of heterochrony, and in the methods used to describe heterochronic changes. A general theme of this accumulating research is that some of the simplistic notions of heterochrony, such as terminal addition, simple rate genes, and "pure" heterochronic categories are invalid. However, this research also shows that a more sophisticated view of the hierarchical nature of heterochrony provides many useful insights and improves our understanding of how ontogenetic changes are translated into phylogenetic changes.

Kenneth J. McNamara. Department of Earth and Planetary Sciences, Western Australian Museum, Francis Street, Perth, Western Australia 6000, Australia. E-mail: ken.mcnamara@museum.wa.gov.au
Michael L. McKinney. Department of Earth and Planetary Sciences, University of Tennessee, Knoxville, Tennessee 37996. E-mail: mmckinney@utk.edu

Accepted: 22 July 2004

Introduction

Of all of the books that Stephen J. Gould wrote in his illustrious career, the one with the most impact on evolutionary studies is probably *Ontogeny and Phylogeny,* published in 1977. To say that this work is a hallmark in this area of evolutionary theory would be an understatement. It proved to be the catalyst for much of the future work in the field, and to a large degree was the inspiration for the modern field of "evolutionary developmental biology." Gould's hope was to show that the relationship between ontogeny and phylogeny is fundamental to evolution, and at its heart is a simple premise—that variations to the timing and rate of development provide the raw material upon which natural selection can operate. He restated this argument, at length, in his final magnum opus on evolutionary theory, *The Structure of Evolutionary Theory* (Gould 2002).

Before addressing the interrelationships between heterochrony, disparity, and macroevolution we will briefly discuss the fundamental question of what "heterochrony" means. This is no simple task, and there has been increasing debate about whether heterochrony plays a minor or a major role in evolution. Much of the confusion over this issue is, we believe, largely semantic in origin, depending largely on how broadly one defines heterochrony (McKinney 1999; McNamara 2002a). Given that we have been criticized for taking an extreme view that developmental evolution is "all heterochrony" (Raff 1996; Zelditch 2001), we discuss below that this is a "straw man" debate. Although we argue for a broad definition of heterochrony, we strongly agree that there are many kinds of developmental evolutionary changes that are *not* usefully understood as heterochronic processes. In fact, we noted this in our book, which instigated much of this debate: "Heterochrony is no panacea and not all developmental events can be fruitfully viewed in rate and timing terms" (McKinney and McNamara 1991: p. 332).

The relationship between heterochrony (however defined) and disparity has, surprisingly, received little attention, apart from work by Eble (1998, 2000, 2002) and Zelditch et al. (2003), and we will examine this important relationship further. Furthermore, we investigate a theme that was of particular interest to Gould, namely the extent to which heterochrony plays a significant role in macroevolution. The answer to this question is of

0094-8373/05/3102-0002/$1.00

course linked to how heterochrony is defined. If one accepts a very narrow definition, the answer is likely to be "not much." Adopting a much broader interpretation opens up many possibilities for changes to the rate and timing of development playing a central role in macroevolution. Although Gould sought to "rescue" heterochrony from the proliferation of jargon and criticized some of our own broader interpretations (Gould 2000), much of his career was indeed spent arguing for a central role for heterochrony in macroevolution (Gould 1977, 2002).

What Is Heterochrony?

To Gould (1977) heterochrony was the phenomenon that produces parallels between ontogeny and phylogeny. The crucial conclusion that he came to in this book was that both paedomorphosis and the opposite, recapitulatory patterns, were equally common phenomena. Acceleration, he argued, produced recapitulatory patterns, whereas retardation leads to paedomorphosis. As Gould (1977: p. 215) observed, "ontogeny of the most remote ancestor goes through the same stages as a phylogeny of adult stages read in reverse order," producing what is essentially a reverse biogenetic law. Haeckel's biogenetic law stated that "ontogeny is a recapitulation of phylogeny." Gould (1977) used the term "heterochrony" in a similar way to de Beer (1930), as a descriptor for a mechanism that produces recapitulation and paedomorphosis.

Gould's (1977) use is actually quite different from Haeckel's original use of the term: "Heterochronism [sic], or variation in time . . . consists in the fact that the series of forms in which the organs successively appear is different in embryology from what the stem history leads us to expect" (see Haeckel 1905: p. 10). Haeckel viewed heterochrony as the exception to recapitulation, when ontogeny does not parallel phylogeny. He regarded the temporal or spatial displacement of organs as occurring either by locality (which he called heterotopism) or time (heterochronism), specifically accelerations or delays in the "rise of an organ." In this sense heterochrony was used to describe the condition when the ancestral ontogenetic sequence of events failed to recapitulate the sequence of events in phylogeny—in other words no parallelism.

Although Gould stressed parallelism between ontogeny and phylogeny in his 1977 book, and therefore adopted a relatively narrow view of heterochrony, just two years later, in an article by Alberch et al. (1979) parallelism was not an issue; indeed it was not even addressed. And one of the coauthors on this seminal paper was Stephen J. Gould. Alberch et al. regarded heterochrony in the broad sense of de Beer (1930), namely displacement of a feature relative to the time that this same feature appeared in the ancestral form. They argued that heterochrony could occur either by changes to the timing of appearance of a trait, or when it ceased to grow. Moreover, they also included changes in the rate of growth of traits as mechanisms for heterochrony. In doing so, they effectively negated parallelism as a factor in heterochrony: the consequence of changes to the rates of development of traits was to change allometries, thus eliminating any direct parallelism between ontogeny and phylogeny. Altering ontogenetic trajectories in this manner results in a descendant's ontogenetic pathway being different from that of its ancestor. There is no parallelism.

The model proposed by Alberch et al. (1979), in which heterochrony is defined as "change in timing or rate of developmental events, relative to the same events in the ancestor," became the cornerstone for studies, particularly paleontological, in heterochrony through much of the remainder of the twentieth century, and has been the basis for our own studies in the subject (McKinney 1988; McKinney and McNamara 1991; McNamara 1995, 1997; Minugh-Purvis and McNamara 2002). In this model, paedomorphosis ("underdevelopment") is produced by the processes of neoteny, postdisplacement, and progenesis, whereas peramorphosis ("overdevelopment") occurs by acceleration, predisplacement, and hypermorphosis.

McKinney and McNamara (1991) subsequently developed the Alberch et al. model by arguing for a hierarchical structure to heterochrony. Not only does it operate from the cellular to phenotypic level, but even in the sphere of behavior (Parker and McKinney (1999). Significantly, heterochrony is viewed not as confined to global processes but also as operating locally, targeting specific traits, as Alberch et al. (1979) indicated. Recently there has been some criticism of this broad view of heterochrony (Hall 1998). Thus those who have followed the model of Alberch et al. have been labeled "panheterochronists" and their work is considered as being "virtually synonymous with evolutionary change of ontogeny" (Zelditch 2001: p. xiv).

This alternative interpretation, argues that heterochrony should be restricted to "conservation of a shared trajectory between ancestor and descendant" (Webster et al. 2001). Moreover, Webster et al. argued that heterochrony can be invoked only when there is parallelism in the ontogenetic trajectories between presumed ancestors and descendants. Any rate change, and allometric repatterning, is called "heterotopy" (another Haeckelian term meaning, literally, "different position"). Indeed, Hall (2001) went so far as to contend that "some see vastly more evolutionary potential in heterotopy than in heterochrony."

Similarly, Nehm (2001) believed that ontogeny paralleling phylogeny is "pure heterochrony" (not withstanding Haeckel's original definition). According to Nehm, heterochrony occurs when there is change in onset or offset time and "no rate change in bivariate allometric relationships." Nehm's definition of heterochrony thus becomes essentially the same as Haeckel's pangenesis (his term for embryonic recapitulation) if there is more development in the descendant ontogeny, or reverse recapitulation if this is less. In this interpretation of heterochrony, any localized rate changes resulting in "spatial patterning" become heterotopy and not heterochrony; only by heterotopy can evolutionary novelties arise (Zelditch and Fink 1996).

This view is at odds with Alberch et al. (1979), who specifically pointed out that many evolutionary novelties can be explained by heterochrony: "slight perturbation(s) in the ontogenetic trajectory of an organ (that) can be amplified through time, by the dynamics of growth and tissue interactions, to produce an adult phenotype drastically different from that of the ancestor" (Alberch et al. 1979: p.315). Indeed, Gould himself recently reinforced this view by arguing that the "mixing and matching" of dissociated heterochronic processes can be a powerful influence on the evolution of novelties (Gould 2002). In addition, heterochronic changes in developmental sequences can produce evolutionary novelties. We agree with Smith (2002) that heterochronic discussions should not be limited to changes in size and shape but expanded to include changes in developmental sequences. Such changes can produce cell and tissue juxtapositions in development that can produce much novelty.

A main problem with arguments about the relative importance of heterochrony versus heterotopy is that like is not being compared with like. Organisms possess shape. A change in shape between ancestor and descendant could be called heterotopy. But our contention is that heterochrony involves changes to the rate and timing of development of whatever part of a developing organism, so that as rates change differentially within a growing structure between species, relative positions or points on these structures will change. Heterochrony is thus the cause, whereas heterotopy is the effect. It then becomes more appropriate to regard the many heterotopic changes as the consequence of changes to rates of growth (McNamara 2002a). A heterochronic process (acceleration or neoteny/deceleration) will therefore induce a pattern change (a spatial repatterning), when acting on a local growth field.

Some critics of a broad interpretation of heterochrony have argued against expanding the traditional view of it too far. Gould (2000) lamented the proliferation of jargon as a kind of "terminal addition" and argued for a return to using heterochrony mainly for the description of size and shape change. Similarly, in a chapter entitled "It's Not All Heterochrony" Raff (1996) cautioned that extending heterochronic terms to include developmental rate and timing changes at smaller scales, including genes, tissue interactions, and local growth fields, will cause many developmental changes to be labeled as heterochrony, without any important gain in our understanding of ontogeny and phylogeny. Finally, Alberch and Blanco (1996) also cautioned against "reductionistic" attempts to apply classical morphological heterochronic terms to cellular, genetic, and molecular scales of development. They argued that this would be appropriate only if events at these finer scales were isomorphic with those at coarser scales.

While we reiterate our agreement that much developmental evolution is not usefully seen as heterochronic (McKinney and McNamara 1991: p. 332), we stand by our original argument that there are good reasons to expand the definition of heterochrony. First, we note that the history of science is replete with terms that have evolved and now have a much different and often broader meaning than they originally had. The language of science is much like that of the legal system, where interpretations of the laws evolve through time, often expanding to include new concepts and social changes. In the case of heterochrony, the original coiners of the term had as yet no knowledge of the molecular, genetic, and cellular processes that we now have. So, we must respectfully disagree with Gould's (2000) assertion that we should "rescue" the terminology of heterochrony because of historical precedence alone.

Criticisms questioning the practical utility of heterochronic terms at many scales are more convincing to us. However, as more papers are published every year on heterochrony, an increasing number of these papers incorporate references to heterochronic patterns at scales below the level of whole organisms. For example, Kim et al. (2000) discussed molecular heterochrony in *Drosophila*, and Skaer and others (2002) examined transcriptional heterochrony of scute and bristle pattern in blowfly species. A much larger number of papers describe heterochronic shifts that affect some tissues and organs but not others. Examples include acceleration of spinal cord development in frogs (Schlosser 2003), early maturation of the stomata in plants (Snir and Sachs 2002), and paedo- and peramorphosis in late Triassic rhyncosaur dentition (Langer et al. 2000).

The fact that heterochrony is rarely "global" (affecting the whole organism) is in fact what has created much of the debate over how to define heterochrony. The idea that a descendant species could be fully characterized as a juvenilized (or overdeveloped) version of its ancestor species is an attractive idea. But of course evolution is rarely that simple and so we must wrestle with the mosaic of changes that typify most developmental evolutionary changes. Jaecks and Carlson (2001) discussed the example of Mesozoic-Cenozoic brachiopods, where the peramorphic evolution of a few traits is superimposed against the background of a largely paedomorphic pattern of evolution for most traits. That increasing numbers of researchers are applying heterochronic terms and concepts to other scales indicates that they are useful. Indeed, we will conclude this paper with several new directions being taken by heterochronic research.

Heterochrony and Disparity

Disparity is the degree of morphological differentiation among taxa within groups (Foote 1999; Eble 2000; Ciampaglio et al. 2001). The concept of disparity grew out of the notion of distance in state space (Sneath & Sokal 1973; Van Valen 1974; Foote 1997). The aim was to represent the average spread and spacing of forms in morphospace (Eble 2002). Most studies of disparity have been concerned solely with the consideration of adult morphological characters. However, given that adult morphological variety is a consequence of the differences in the developmental programs of individual taxa, then it can be argued that disparity is a corollary of heterochrony. For instance, Fortey et al. (1996) suggested that heterochronic genes played a role in the rapid expansion of disparity during the Cambrian explosion.

The only studies that have directly addressed this relationship are those of Eble (1998, 2000, 2002) and Zelditch et al. (2003), who have extended the concept of morphospace and disparity to include developmental morphospace. As Eble (2002) has pointed out, understanding developmental morphospace is a prerequisite for identifying phenomena such as heterochrony. In introducing the concept of developmental disparity, Eble has argued that it can be quantified between taxa or within taxa. As he pointed out, "Instead of changes in amount of variation in evolutionary time, the focus is shifted to changes in amount of variation in ontogenetic time" (Eble 2002: p.59).

From a developmental perspective Eble (2000) considered that disparity could be viewed in three basic ways: (1) how it changes among taxa across ontogenetic stages; (2) how disparate ontogenetic end points are; (3) how this degree of disparity changes during evolution by developmental changes. If the usual measures of evolutionary disparity are the differences between ontogenetic end points, then given that these differences arise from variation in the rate and timing of development, the extent of disparity between taxa is a direct consequence of heterochrony (sensu Alberch et al. 1979). This type of developmental disparity can be considered as cross-sectional developmental disparity.

However, developmental disparity can also be viewed longitudinally, along a single ontogenetic trajectory—this we call *ontogenetic disparity*. This is particularly evident in taxa that show episodic growth, or episodic manifestations of growth, such as arthropods, although comparative ontogenetic disparity can also be successfully analyzed in organisms with continuous development (Zelditch et al. 2003). The morphospace occupied by each ontogenetic stage in organisms that show episodic growth will be slightly different from the preceding one. Moreover, in most ontogenetic trajectories ontogenetic disparity between successive stages will be highest earliest in development, steadily reducing through ontogeny. Consequently, if selection targets early developmental changes, then cross-sectional evolutionary disparity can be quite significant.

Von Baer's second law suggests that disparity between closely related taxa should be smaller in earlier growth stages than in later ones. In a study of hominid evolution, Eble (2002) showed, on the basis of dental characteristics, that this was indeed the case. However, other groups do not follow von Baer's law in this regard. For instance, in the echinoid *Heliocidaris* the adults of the two species, *Heliocidaris erythrogramma* and *H. tuberculata* from southern Australia, show a very low degree of disparity as adults. Moreover, they exhibit similar behavioral characteristics (Wray and Raff 1991; Wray 1995). However, their earliest embryonic stages undergo quite different developmental pathways and show a high degree of disparity in the early stages of development. This arises from profound heterochronic changes. In *H. erythrogramma* one developmental stage, the pluteus larval stage, has undergone such an extreme reduction that it has disappeared altogether. Not only is such a timing change expressed in very different morphological characteristics early in ontogeny, but the larval life histories of the two species, and therefore their larval behaviors, are quite different—one has a free-swimming, feeding larval stage that persists for weeks, whereas the other has a nonfeeding, nonplanktic larval stage that metamorphoses into a juvenile after two days. Developmental disparity at this stage is extremely high. In their study of the ontogenetic variability in disparity in nine species of living piranha fishes Zelditch et al. (2003) likewise showed that during ontogeny disparity decreases. Juveniles of the different species track along ontogenetic trajectories that converge later in development to produce adult morphotypes displaying low disparity.

In terms of the three basic heterochronic factors—changes in growth rate, and changes in offset and onset timing—the extent to which each of these varies between descendant and ancestor will affect both developmental and evolutionary disparity. The greater the changes in growth rate of traits between related taxa, then the greater the resultant evolutionary disparity between adults (assuming the onset and offset times are the same). Conversely, the smaller the rate changes, the less the disparity. Changes in offset times will have an effect on disparity only if the allometric changes (arising from variable growth rates) are sufficiently different. Small variability in growth rates, or allometries between related taxa, can result in high disparity if there is also an appreciable difference in offset times. If there are no rate differences, and no differences in allometry (which would be most unusual), then offset changes would not change disparity. The most potent mix inducing the highest disparity levels would be when there are both major rate changes and offset changes. However, both would have to be in the same heterochronic "direction"; i.e., both would have to be either peramorphic or paedomorphic. Thus, acceleration combined with hypermorphosis will produce extreme disparity; likewise, neoteny and progenesis. Dissociated heterochronies affecting the same trait—i.e., traits are predominantly neotenic but the organism is hypermorphic—could result in reduced disparity. The same applies with changes in onset times; they are dependent on the extent of allometric change.

High degrees of disparity can be induced also by changes to the timing and rate of very early developmental events, in particular the timing of limb condensation in vertebrates. Unlike *Heliocidaris* species, in most instances such differences can have a profound effect on the adult phenotype and the degree of disparity. As Richardson (1999) has pointed out, differences in adult morphology are often related to differences in gene regulation in early embryonic development. Thus in different vertebrates the timing of onset of development of various organs, such as the heart tube, aortic arch, liver bud, and optic vessel, relative to one another, may vary. Very small variations in onset times at very early developmental stages can have a major impact on the adult phenotype, resulting in a high degree of disparity. For example, Kordikova (2002) showed that the morphological disparity of turtles can be explained by heterochronic shifts in the early stages of ontogeny

Moreover, disparate limb sizes in adults may ensue from size of initial limb bud and subsequent growth rates, relative to the rest of the body. For example, in the flightless kiwi, *Apteryx australis*, hind limb growth rate is relatively high in early embryonic stages. By comparison, in the bat *Rousettus amplexicaudatus* hind limb growth rate is relatively much less, but forelimb growth rate is appreciably higher (Richardson 1999: Fig. 2). Disparity between kiwis and bats is extremely high.

Heterochrony and Macroevolution

Heterochrony has long been invoked as an important agent in macroevolution (McKinney and McNamara 1991). Jablonski (2000) provided an excellent overview of the role of heterochrony in macroevolution. Hanken and Wake (1993) reviewed the role of miniaturization in macroevolution. Specific examples of heterochrony in producing major evolutionary innovations include seed plants (Friedman and Carmichael 1998), grasses (Kellogg 2000), colonial invertebrates and social insects (Harvell 1994), and deep-sea fishes (Miya and Nishida 1996). Bininda-Emonds and others (2003) provided evidence that timing changes between developmental modules may be an important mechanism giving rise to the diversity of vertebrates.

Gould (1977) argued that of all the heterochronic processes, progenesis was potentially the most important. However, his argument centered not just on the pronounced disparity that can ensue from a major shift in earlier offset of growth, but in selection primarily targeting precocious maturation as a life history strategy. This highlights why it is inadvisable to take the simplistic approach of linking high disparity with macroevolutionary outcomes. As we have argued before, in support of Gould (McKinney and McNamara 1991), selection may often be targeting life history strategies rather than morphology per se. As disparity sensu stricto is concerned with morphospace, it cannot be considered exclusively as an agent for macroevolution.

Although Gould (1977) focused on progenesis as the dominant agent in macroevolution, particularly for major evolutionary novelties, he also discussed the role of neoteny, and of hypermorphosis. The latter he believed offered fewer opportunities for macroevolution on the grand scale he envisaged for progenesis, as it led to overspecialization and "blind alleys" (such as the "Irish Elk," *Megaloceros*). Whether or not this giant red deer became a victim of its perceived "overspecialization," it is undeniable that its success, both temporally and geographically, arose because of the operation of hypermorphosis. It might not have had the potential for the evolution of a major new bauplan, but it still had limited macroevolutionary success.

Likewise, a recent study of the largest known terrestrial lizard, the Pleistocene varanid *Megalania prisca*, has shown that it evolved by hypermorphosis (Erickson et al. 2003). More than twice the length of any living varanid this dominant Australian carnivore achieved its enormous size by delaying onset of maturity. In the next largest varanid, *V. komodoensis*, this is at about ten years, whereas in *M. prisca* it was between 13 and 14 years. By continuing the preadult rate of growth it was able to double its size.

The role of heterochrony as an agent in macroevolution can also be seen in sequential heterochrony (McKinney and McNamara 1991; McNamara 2002b). Here, changes to the timing of onset or offset of various ontogenetic stages can have very significant macroevolutionary consequences and be an agent for macroevolution. This is because a consequence of sequential heterochrony can be the generation of high disparity within the ontogenetic trajectory, not just at the end. Extending the duration of particular ontogenetic phases is sequential hypermorphosis. Contraction of the phases is sequential progenesis.

Sequential heterochrony affecting, in particular, the timing of metamorphosis from one growth stage to another can clearly also have important effects on life history. This is particularly apparent in lamprey evolution (McNamara 1997). It has also been invoked as a principal factor in hominid evolution (McNamara 2002b). However, sequential heterochrony can also affect local growth fields. This can have profound macroevolutionary implications and result in the evolution of macroevolutionary novelties and the establishment of new evolutionary pathways. One particularly significant example is the evolution of tetrapod digits.

In their examination of the embryological development of the teleost zebrafish *Danio rerio*, Sordino et al. (1995) found that the first condensation of cells of the fin bud occurs by a thickening and growth of patches of mesodermal cells. These are surrounded by an ectoderm layer, a configuration comparable with that observed in tetrapod limb buds. As development proceeds, the ectoderm very rapidly protrudes and folds in on itself. The fold moves distally and dermal skeleton appears inside it. There is a sudden concurrent reduction in production of mesenchymal cells. The time of transition determines the relative amount of endodermal, compared with dermal, skeleton. Fin rays are constructed from dermal skeleton, whereas digits are formed from endodermal skeleton.

In tetrapods, folding of the ectoderm fails to occur, and the limb develops entirely by the production of endodermal skeleton. Each element of the limb forms sequentially, with continued proliferation of endodermal cells generating digits as the terminal expression of the endoskeleton limb. Thus digits are morphological novelties, arising from sequential hypermorphosis. The transition from endodermal to dermal skeletal elements is delayed to such an extent that it fails to occur.

Lobe-finned fishes, such as the Late Devonian *Eusthenopteron*, the group from which tetrapods are thought to have evolved (Long 1995), display an intermediate condition, a result of an intermediate transition time from endoderm to dermal expression. Compared with the ancestral state in fishes that possess only fins, initiation of the folding of the ectoderm is delayed in lobe-finned fishes, allowing a longer period for more endoskeleton production. The result is that, like tetrapods, lobe-finned fishes possess a humerus, radius, and ulna, as well as a femur, tibia, and fibula. However, because of the late onset of this folding, fin rays sprout from the endodermal bones—an intermediate transition time produces an intermediate morphology.

Many groups of vertebrates, such as whales, dolphins, ichthyosaurs, and plesiosaurs, show hyperphalangy (numerous finger bones). By analyzing growth of the dolphin embryo flipper, Richardson and Oelschläger (2002) have argued that the developmental basis for this phenomenon may be linked to timing shifts, in particular hypermorphic delays in cessation of phalange production. Conversely, limb reduction, such as in cetacean hind limbs, might arise from early cessation of production of phalanges. The underlying mechanism for extending or contracting phalange development might lie in the period of activity of the apical ectodermal ridge. By extending its duration more cells can be added to the limb tip and then generate extra skeletal elements (Holder 1983).

Variation in ontogenetic trajectories of individual digits is a feature that characterizes different groups of tetrapods. Recent studies have revealed that in early embryogenesis birds have Anlagen for five digits (see Galis et al. 2003 for review), even though as adults they have only three digits. Experimental work on chicken and ostrich digits supports the hypothesis of evolutionary digit reduction in bird wings due to so-called arrested development of digit I, followed by its degeneration.

How does arrested development fit into Alberch et al.'s (1979) model? Digit I can be viewed as having undergone progenesis very early in development. This has implications for arguments over whether birds evolved from dinosaurs. Wing digits in birds consist of II–IV, whereas in theropod dinosaurs, birds' presumed ancestors, they are I–III. This seemingly fundamental difference has been used to argue against direct evolution of birds from dinosaurs (Feduccia 2003). Although the three digits of each group are presumably not homologous, there is no reason why birds could not have evolved from an ancestor with digits I–III.

Although some have argued for a homeotic shift in identity of the digits (Wagner and Gauthier 1999) and others that birds could have had a hypothetical theropod ancestor that had digits II–IV (Galis et al. 2003), heterochrony targeting specific digits provides a simpler, alternative explanation. To derive a II–IV pattern from a I–III one, the following two simple heterochronic changes would need to occur: early progenesis of digit I, combined with hypermorphosis in digit IV.

Future Directions

Despite the many controversies that seem constantly to pervade the study of heterochrony, it remains a rapidly growing facet of evolutionary investigation. Even many critics of heterochronic ideas point out that the debate has more to do with terminological disputes than with the importance of the process itself. Gould himself was well aware of this, and he often noted that vigorous debate was the sign of a healthy science (Gould 2000, 2002).

As further evidence for the growing interest in heterochrony, we note some of the many areas of heterochronic study that promise to make major advances in our understanding. In some cases, these advances are in methods used to study heterochrony. An example is the application of isotopic data to resolve the ontogenetic age of fossil individuals (Jones and Gould 1999; Ivany et al. 2003). This is an essential part of describing whether, for example, paedomorphic species matured earlier, grew more slowly, or perhaps both. Bininda-Emonds and others (2002) described a means of quantifying changes in the sequence of developmental events (sensu Smith 2002) through analysis of event-pairing in a phylogenetic context (see also Jeffery et al. 2002). Schlosser (2001) compared developmental sequences by introducing heterochrony plots as a new graphic method to detect temporal shifts in the development of characters in pairwise species comparisons. These plot the timing of character development in one species against the timing of character development in another species. Such plots can be embedded into comparative phylogenetic analysis and can detect whether suites of characters are dissociated from one another. However, as a point of caution, Bininda-Emonds et al. (2002) have noted that even when comparing identical sequences, heterochrony plots can sometimes give misleading results. Finally, Cubo and others (2002) described a method of detecting heterochronic change by the study of growth curves in ontogenetic bone growth.

Many other advances include the expansion of heterochronic analyses into the evolution of life history and behavior. Life history studies that incorporate heterochrony span a wide variety of groups including fishes (Takeshi and Yoshino 2002), newts (Denoel and Joly 2000), salamanders (Ryan and Semlitsch 1998), and plants (Zopfi 1998). Heterochrony also influences behavioral evolution, as shown in newts (Denoel 2002), birds (McDonald and Smith 1994), and mice (Gariepy et al. 2001). The last study is especially interesting for its use of experimental selection to alter the ontogenetic timing of aggressive behavior.

Yet another rapidly growing exciting area of research relates to the underlying genetic and molecular mechanisms of heterochronic patterns. Much of this work is outside the scope of most paleobiological research, and this literature is so large that we cannot hope to do it justice in this short essay. However, this is a fascinating area for those who would like to connect developmental mechanisms to larger-scale phenomena. We have already noted, for example, discussions of molecular heterochrony in *Drosophila* (Kim et al. 2000) and transcriptional heterochrony of scute and bristle pattern in blowfly species (Skaer et al. 2002). Another example is that of microRNAs, which may mediate transitions on a variety of time-scales to pattern the activities of particular target protein-coding genes and in turn generate sets of cells over a period of time. Plasticity in these microRNA genes or their targets may lead to changes in relative developmental timing between related species (Pasquinelli and Ruvkun 2002).

Literature Cited

Alberch, P., and M. J. Blanco. 1996. Evolutionary patterns in ontogenetic transformation: from laws to regularities. International Journal of Developmental Biology 40:845–858.

Alberch, P., S. J. Gould, G. F. Oster, and D. B. Wake. 1979. Size and shape in ontogeny and phylogeny. Paleobiology 5:296–317.

Bininda-Emonds, O. R. P., J. E. Jeffery, M. I. Coates, and M. K. Richardson. 2002. From Haeckel to event-pairing: the evolution of developmental sequences. Theory in Biosciences 121: 297–320.

Bininda-Emonds, O. R. P., J. E. Jeffery, and M. K. Richardson. 2003. Inverting the hourglass: quantitative evidence against the phylotypic stage in vertebrate development. Proceedings of the Royal Society of London B 270:341–346.

Ciampaglio, C. N., M. Kemp, and D. W. McShea. 2001. Detecting changes in morphospace occupation patterns in the fossil record: characterizations and analysis of measures of disparity. Paleobiology 27:695–715.

Cubo, J., D. Azagra, A. Casinos, and J. Castanet. 2002. Heterochronic detection through a function for the ontogenetic variation of bone shape. Journal of Theoretical Biology 215:57–66.

De Beer, G. R. 1930. Embryology and evolution. Clarendon, Oxford.

Denoel, M. 2002. Paedomorphosis in the Alpine newt (*Triturus alpestris*): decoupling behavioural and morphological change. Behavioral Ecology and Sociobiology 52:394–399.

Denoel, M., and P. Joly. 2000. Neoteny and progenesis as two heterochronic processes involved in paedomorphosis in *Triturus alpestris* (Amphibia: Caudata). Proceedings of the Royal Society of London B 267:1481–1485.

Eble, G. J. 1998. The role of development in evolutionary radiations. Pp. 132–161 *in* M. L. McKinney and J. A. Drake, eds. Biodiversity dynamics: turnover of populations, taxa, and communities. Columbia University Press, New York.

———. 2000. Contrasting evolutionary flexibility in sister groups: disparity and diversity in Mesozoic atelostomate echinoids. Paleobiology 26:56–79.

———. 2002. Multivariate approaches to development and evolution. Pp. 51–78 *in* N. Minugh-Purvis and K. J. McNamara, eds. Human evolution through developmental change. Johns Hopkins University Press, Baltimore.

Erikson, G. M., A. de Ricqlès, V. de Buffrénil, R. E. Molnar, and M. K. Bayless. 2003. Vermiform bones and the evolution of gigantism in *Megalania*—how a reptilian fox became a lion. Journal of Vertebrate Paleontology 23:966–970.

Feduccia, A. 2003. Bird origins: problem solved, but the debate continues. Trends in Ecology and Evolution 18:9–10.

Foote, M. 1997. The evolution of morphological diversity. Annual Review of Ecology and Systematics 28:129–152.

———. 1999. Morphological diversity in the evolutionary radiation of Paleozoic and Post-Paleozoic crinoids. Paleobiology Memoirs No. 1. Paleobiology 25(Suppl. to No. 2).

Fortey, R. A., D. E. G. Briggs, and M. A. Wills. 1996. The Cambrian evolutionary "explosion": decoupling cladogenesis from morphological disparity. Biological Journal of the Linnean Society 57:13–33.

Friedman, W. E., and J. S. Carmichael. 1998. Heterochrony and developmental innovation: evolution of female gametophyte ontogeny in Gnetum, a highly apomorphic seed plant. Evolution 52:1016–1030.

Galis, F., M. Kundrát, and B. Sinervo. 2003. An old controversy solved: bird embryos have five fingers. Trends in Ecology and Evolution 18:7–9.

Gariepy, J. L., D. J. Bauer, and R. B. Cairns. 2001. Selective breeding for differential aggression in mice provides evidence for heterochrony in social behaviours. Animal Behaviour 61:933–947.

Gould, S. J. 1977. Ontogeny and phylogeny. Belknap Press of Harvard University Press, Cambridge.

———. 2000. Of coiled oysters and big brains: how to rescue the terminology of heterochrony, now gone astray. Evolution and Development 2:241–248.

———. 2002. The structure of evolutionary theory. Belknap Press of Harvard University Press, Cambridge.

Haeckel, E. 1905. The evolution of man, 5th ed. Watts, London.

Hall, B. K. 1998. Evolutionary developmental biology, 2d ed. Kluwer Academic, Dordrecht, The Netherlands.

———. 2001. Foreword. Pp. vii–ix *in* Zelditch 2001.

Hanken, J., and D. B. Wake. 1993. Miniaturization of body size: organismal consequences and evolutionary significance. Annual Review of Ecology and Systematics 24:501–519.

Harvell, C. D. 1994. The evolution of polymorphism in colonial invertebrates and social insects. Quarterly Review of Biology 69:155–185.

Holder, N. 1983. The vertebrate limb: patterns and constraints in development and evolution. Pp. 399–425 *in* B. C. Goodwin, N. Holder, and C. C. Wylie, eds. Development and evolution. Cambridge University Press, Cambridge.

Ivany, L. C., B. H. Wilkinson, and D. S. Jones. 2003. Using stable isotopic data to resolve rate and duration of growth throughout ontogeny: an example from the surf clam, *Spisula solidissima*. Palaios 18:126–137.

Jablonski, D. 2000. Micro- and macroevolution: scale and hierarchy in evolutionary biology and paleobiology. Paleobiology 26:15–52.

Jaecks, G. S., and S. J. Carlson. 2001. How phylogenetic inference can shape our view of heterochrony: examples from thecideide brachiopods. Paleobiology 27:205–225.

Jeffery, J. E., M. K. Richardson, M. I. Coates, and O. R. P. Bininda-

Emonds. 2002. Analyzing developmental sequences within a phylogenetic framework. Systematic Biology 51:478–491.

Jones, D. S., and S. J. Gould. 1999. Direct measurement of age in fossil *Gryphaea*: the solution to a classic problem in heterochrony. Paleobiology 25:158–187.

Kellogg, E. A. 2000. The grasses: a case study in macroevolution. Annual Review of Ecology and Systematics 31:217–238.

Kim, J., J. Q. Kerr, and G. S. Min. 2000. Molecular heterochrony in the early development of *Drosophila*. Proceedings of the National Academy of Sciences USA 97:212–216.

Kordikova, E. G. 2002. Heterochrony in the evolution of the shell of Chelonia. Neues Jahrbuch für Geologie und Paläontologie, Abhandlungen. 226:343–417.

Langer, M. C., J. Ferigolo, and J. Schultz. 2000. Heterochrony and tooth evolution in hyperodapedontine rhynchosaurs (Reptilia, Diapsida). Lethaia 33:119–128.

Long, J. A. 1995. The rise of fishes. Johns Hopkins University Press, Baltimore.

McDonald, M. A., and M. H. Smith. 1994. Behavioral and morphological correlates of heterochrony in Hispaniolan palm tanagers. Condor 96:433–446.

McKinney, M. L., ed. 1988. Heterochrony in evolution: a multidisciplinary approach. Plenum, New York.

———. 1999. Heterochrony: beyond words. Paleobiology 25: 149–153.

McKinney, M. L., and K. J. McNamara. 1991. Heterochrony: the evolution of ontogeny. Plenum, New York.

McNamara, K. J., ed. 1995. Evolutionary change and heterochrony. Wiley, Chichester, U.K.

———. 1997. Shapes of time: the evolution of growth and development. Johns Hopkins University Press, Baltimore.

———. 2002a. Changing times, changing places: heterochrony and heterotopy. Paleobiology 28:551–558.

———. 2002b. Sequential hypermorphosis: stretching ontogeny to the limit. Pp. 102–121 *in* N. Minugh-Purvis and K. J. McNamara, eds. Human evolution through developmental change. Johns Hopkins University Press, Baltimore.

Minugh-Purvis, N., and K. J. McNamara, eds. 2002. Human evolution through developmental change. Johns Hopkins University Press, Baltimore.

Miya, M., and M. Nishida. 1996. Molecular phylogenetic perspective on the evolution of the deep-sea fish genus *Cyclothone* (Stomiiformes: Gonostomatidae). Ichthyological Research 43: 375–398.

Nehm, R. H. 2001. The developmental basis of morphological disarmament in *Prunum* (Neogastropoda: Marginellidae). Pp. 1–26 *in* Zelditch 2001

Parker, S. T., and M. L. McKinney. 1999. Origins of intelligence: the evolution of cognitive development in monkey, apes, and humans. Johns Hopkins University Press, Baltimore.

Pasquinelli, A. E., and G. Ruvkun. 2002. Control of developmental timing by microRNAs and their targets. Annual Review of Cell and Developmental Biology 18:495–513.

Raff, R. A. 1996. The shape of life. University of Chicago, Chicago.

Richardson, M. K. 1999. Vertebrate evolution: the developmental origins of adult variation. Bioessays 21:604–613.

Richardson, M. K., and H. H. A. Oelschläger. 2002. Time, pattern, and heterochrony: a study of hyperphalangy in the dolphin embryo flipper. Evolution and Development 4:435–444.

Ryan, T. J., and R. D. Semlitsch. 1998. Intraspecific heterochrony and life history evolution: decoupling somatic and sexual development in a facultatively paedomorphic salamander. Proceedings of the National Academy of Sciences USA 95:5643–5648.

Schlosser, G. 2001. Using heterochrony plots to detect the dissociated coevolution of characters. Journal of Experimental Zoology 291:282–304.

———. 2003. Mosaic evolution of neural development in anurans: acceleration of spinal cord development in the direct developing frog *Eleutherodactylus coqui*. Anatomy and Embryology 206:215–227.

Skaer, N., D. Pistillo, and P. Simpson. 2002. Transcriptional heterochrony of scute and changes in bristle pattern between two closely related species of blowfly. Developmental Biology 252: 31–45.

Smith, K. K. 2002. Sequence heterochrony and the evolution of development. Journal of Morphology 252:82–97.

Sneath, P. H. A., and R. R. Sokal. 1973. Numerical taxonomy. W. H. Freeman, San Francisco.

Snir, S., and T. Sachs. 2002. The evolution of epidermal development: examples from the Fabaceae. Israel Journal of Plant Sciences 50(Suppl.):S129–S139.

Sordino, P., F. van der Hoeven, and D. Duboule. 1995. Hox gene expression in teleost fins and the origin of vertebrate digits. Nature 375:678–681.

Takeshi, K., and T. Yoshino. 2002. Diversity and evolution of life histories of gobioid fishes from the viewpoint of heterochrony. Marine and Freshwater Research 53:377–402.

Van Valen, L. 1974. Multivariate structural statistics in natural history. Journal of Theoretical Biology 45:235–247.

Wagner, G. P., and J. A. Gauthier. 1999. 1,2,3 = 2,3,4: a solution to the problem of the homology of the digits in the avian hand. Proceedings of the National Academy of Sciences USA 96:5111–5116.

Webster, M., H. D. Sheets, and N. C. Hughes. 2001. Allometric patterning in trilobite ontogeny: testing for heterochrony in *Nephrolenellus*. Pp. 105–144 *in* Zelditch 2001.

Wray, G. A. 1995. Causes and consequences of heterochrony in early echinoderm development. Pp. 197–223 *in* K. J. McNamara, ed. Evolutionary change and heterochrony. Wiley, Chichester, U.K.

Wray, G. A., and R. A. Raff. 1991. The evolution of developmental strategy in marine invertebrates. Trends in Ecology and Evolution 6:45–50.

Zelditch, M. L. 2001. Beyond heterochrony: the evolution of development. Wiley-Liss, New York.

Zelditch, M. L., and W. L. Fink. 1996. Heterochrony and heterotopy: stability and innovation in the evolution of form. Paleobiology 22:241–254.

Zelditch, M. L., H. D. Sheets, and W. L. Fink. 2003. The ontogenetic dynamics of shape disparity. Paleobiology 29:139–156.

Zopfi, H. J. 1998. Life-history variation among populations of *Euphrasia rostkoviana* Hayne (Scrophulariaceae) in relation to grassland management. Biological Journal of the Linnean Society 64:179–205.

Paleobiology, 31(2), 2005, pp. 27–35

Whale barnacles: exaptational access to a forbidden paradise

Adolf Seilacher

Abstract.—Of all sessile filtrators, only some species of acorn barnacles managed to permanently settle on whales. Their key exaptation was probably a kind of biochemical cleaning process, which could be modified to penetrate into the host's dead cutis. Anchorage was further increased by coring prongs out of the whale skin (*Coronula*) or by transforming the wall into a cylindrical tube that added new rings at the base, while old ones flaked off at the surface in tandem with skin shedding (*Tubicinella*). *Xenobalanus* even everted its naked body into a stalked structure and reduced the wall plates to a minute, but highly efficient, anchor. *Cryptolepas* combines the strategies of *Tubicinella* and *Coronula*, but with a different structure of the radial folds. Because of a shared exaptational inventory, it is impossible to unravel phylogenetic relationships within the Coronulida from skeletal morphology alone.

A. Seilacher. *Department of Geology and Geophysics, Yale University, Post Office Box 208109, New Haven, Connecticut 06520. E-mail: geodolf@tuebingen.netsurf.de*

Accepted: 17 August 2004

Riding on a whale! This should be the dream not only for kids, but also for all sessile filter feeders, such as bryozoans, serpulids, mussels, oysters, and tunicates, to name just a few. As any sailor knows, these foulers hitchhike happily on ships. Yet, except for acorn barnacles (the gooseneck barnacle *Conchoderma* can settle there only after its balanid cousins have provided a hard substrate), none of them made it to whales. What should fence them off from an environment that has been around for millions of years, is big enough for large populations, is virtually free from predators, and provides a reliable filter current? One might argue that whale skin is not hard enough for cementation; yet kelp, just as soft and flexible, is readily colonized (Gutmann 1960). The morphology of whale barnacles tells us the real problem: whale skin, as any mammalian cutis, flakes gradually off as the animal grows. Thus an organism simply attaching to the skin will soon be shed with it. Whale barnacles, in contrast, managed to get anchored *within* the skin and to constantly deepen their anchorage ahead of the shedding front. What relates in particular to Steve Gould is the way this crucial innovation came about: by a serendipitous *exaptation* (Gould and Vrba 1982), or predisposition.

Cementation under water is not a trivial task. Aquatic organisms can't dry the substrate, as is instructed for commercial glues, but they must *clean* it. In removing the mucilage covering all submarine surfaces, barnacles have a particular problem: their cementation front is around the basal plate, way out of reach for the appendages. So the cleaning job has to be done by a narrow ring of sutural tissue that emerges at the contact between the basal plate and the wall plates. Along this suture, the mucilage is presumably removed not mechanically, but biochemically, i.e., by a kind of external digestion. This was probably the *crucial exaptation* that allowed barnacles, and only them, to cope with the whale-skin problem: in an organic substrate their cleaning mechanism could also be used for *penetration*, provided that the basal plate was not mineralized. In fact, there are several groups of balanids other than whale barnacles in which the basal plate remains unmineralized.

For those who may be frightened by such a seemingly speculative story, there is enough experimental evidence for it, though not in the usual sense (experimenting with whales may be tricky). What I mean are the multiple experiments made by evolution itself, because the exaptational pathway to the promised (whale) land was probably used more than once. So let us turn to modern whale barnacles and their skeletal morphologies to learn what constructional tricks they used to further improve life in the new paradise.

0094-8373/05/3102-0003/$1.00

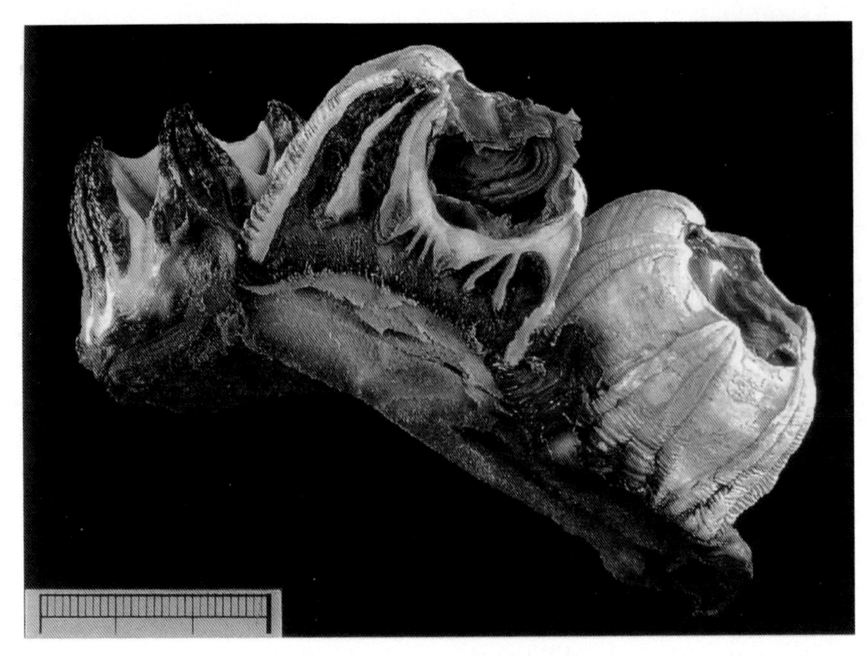

FIGURE 1. Colony of *Coronula diadema* preserved in alcohol with a piece of whale skin. Right specimen intact, middle one sectioned, and left one etched away with hydrochloric acid to show the anchoring prongs of black whale cutis. These prongs did not form by tissue growth, but by the coring effect of the barnacle's mural edges growing into the dead cutis of the host's skin (YPM 24494). The juvenile (Fig. 3B) also comes from this colony.

Coronula

This genus has been named after its habit to settle on the forehead of the host. Although the white spots may look to an imaginative viewer like the whale's diadem, the barnacles prefer to settle in this region because it directly faces the current needed for filtration.

In stranded whale carcasses, the barnacles may still be alive; but when trying to collect specimens, we would find that they cannot be broken off by any force. For this reason, most pickled museum specimens are still embedded in chunks of whale blubber cut out with a large knife. Barnacles are also the only indications of whales in the shell middens of Tierra del Fuego, because the ancient Yamana people cut pieces of meat from stranded carcasses, but not the heavy bones (E. Piana, Ushuaya, personal communication 2004). If a wet sample of *Coronula* is put into hydrochloric acid for removal of the calcitic barnacle skeleton, a most unexpected structure is revealed: prongs of solid whale skin protruding well above the surrounding skin surface (Fig. 1). For a biochemist studying the biography of the particular whale, these cores preserve skin layers otherwise lost. For the barnacle, however, these tough, rubberlike structures provide an attachment strong enough to elevate the filter well above the substrate.

The life history of *Coronula* reflects the origin of this structure. Juveniles a few millimeters in diameter still look like ordinary balanids, except that their bases are already sunk in the substrate, with radial cracks reflecting the force of penetration (Fig. 3B). This detail reminds us of another serendipitous exaptation that enhanced the initial anchorage: because it widens by accretion at the immersed base, the conical shell of acorn barnacles turns out to be an ideal screw anchor.

In the second stage, four radial (i.e., growth-normal) folds develop on each mural plate. As they get more accentuated, they become T- and L-shaped in cross-section and eventually fuse with their neighbors. The result is three coring devices, whose inner walls are part of the external surface and whose penetration is effected by resorption of the whale skin along the growing edge. In an evolutionary sense, this transformation had the advantage that it required no stepping-stone function, because

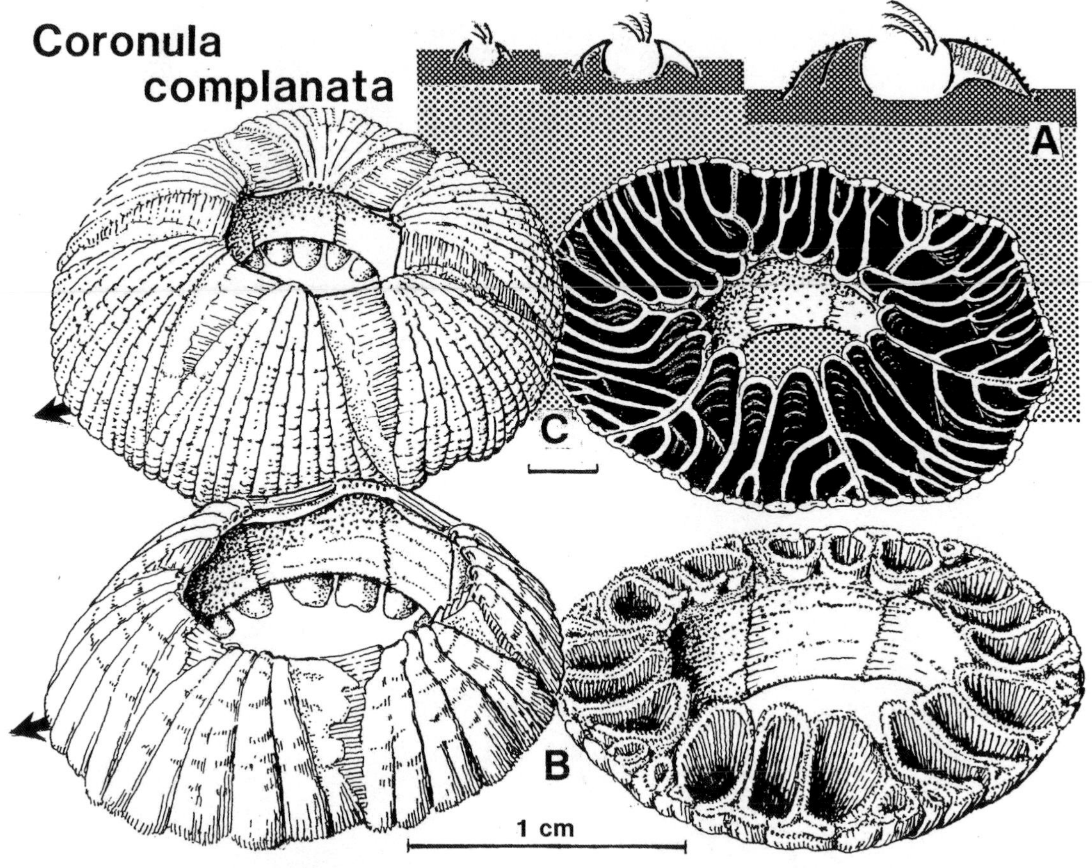

FIGURE 2. *Coronula complanata*. A, The whale cutis (black) thickens by addition of new layers from below, while older layers flake off at the surface. The barnacle adjusts to this shift by penetrative growth at the basal margins of the wall plates. At the same time, prongs of old, but still elastic, whale skin are preserved within these plates for anchorage. B, Juvenile specimen (YPM 2369A). Its basal view shows how in each plate three primary "skin corers" formed by four ribs that fuse marginally with neighbors. Note that the resulting cavities are lined by the outer cuticle of the barnacle. Arrow in anterior direction, facing current. C, In the adult specimen (YPM 2731), anchorage has become improved by bifurcation of the primary ribs and introduction of new ones at the sutures between plates. (From Seilacher 1992.)

the increasing rib profile improved anchorage at every step even before the folds fused.

In the third stage, improvement went in different directions in the two studied species. The mural plates of *Coronula complanata* continue to grow wider at the base but increase the number of cores by introducing new fold pockets (Fig. 2B). Consequently the cutting edges at the base of the adult barnacle (Fig. 2C) have a treelike pattern. In contrast, the more common *Coronula diadema* (Fig. 3), maintains the original three corers in each wall plate and improves anchorage by basal constriction. Accordingly, adult skeletons have a bulbous shape and a deeply sunken basal surface (Fig. 3C). In reality, the basal edges of the six plates do not have to shorten, because the bulge is due to accretion along the *radial* plate boundaries. In ordinary balanids, lateral plate growth is expressed by triangular wings. They shingle with the wings of the neighbors and glide past each other (as expressed by their horizontal striation) in tandem with marginal accretion. In *Coronula diadema*, however, the shingled relationship between wall plates is replaced by broad sutures that are almost perpendicular to the mural surface (horizontal section in Fig. 3D). Combined with the increased curvature of the wall plates, growth along these sutures would produce pillars of solid calcite that exceed the width of the coring sectors. Instead, the hermaphroditic bar-

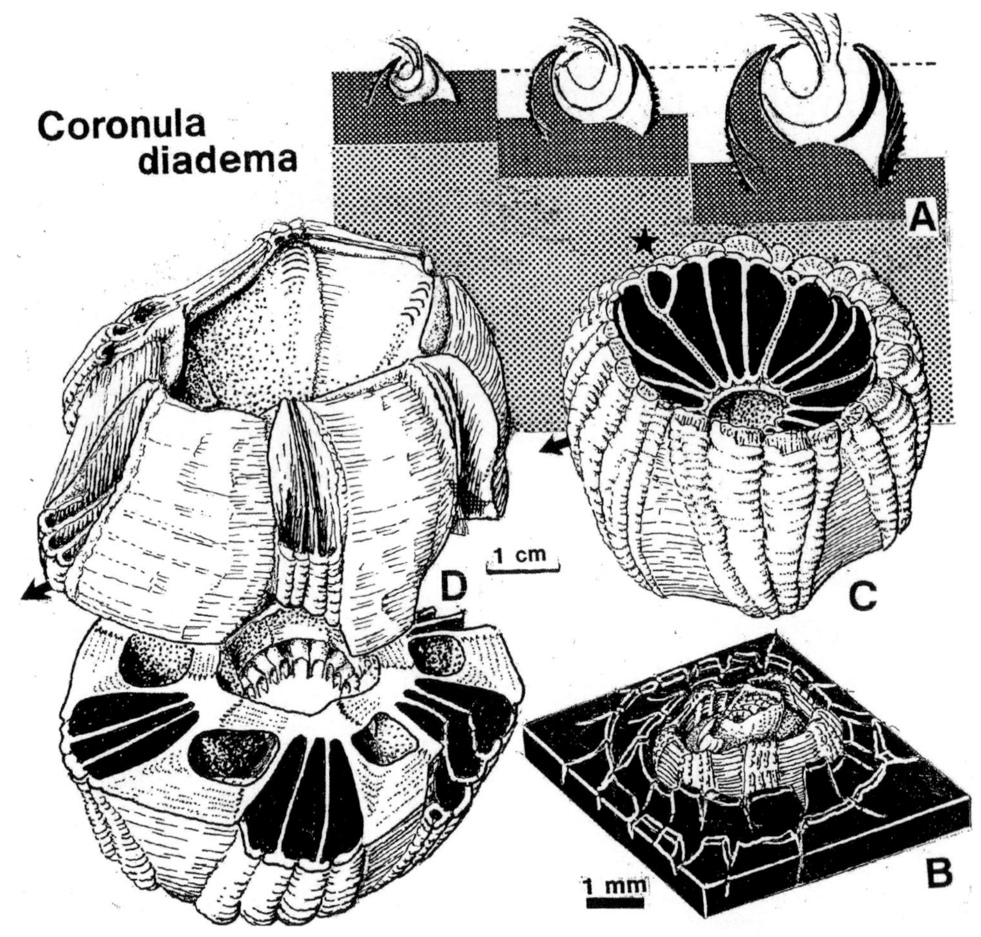

FIGURE 3. *Coronula diadema*. A, Scheme of penetration. B, Juvenile individual penetrating like a screw anchor produced radial cracks in the whale skin. Four ridges on each plate are already present, but as yet without coring function. Interplate sutures are still shingled (YPM 28494). C, Base of older specimen (YPM 2368A). The three original coring chambers are maintained, except for one adventitious chamber (asterisk). D, Fully grown individual (YPM 28495). As seen in the equatorial section, lateral plate growth along abutting sutures (dotted lines) has separated the three coring chambers by sectors of solid calcite, whose cavities house the ovaries. At the same time, the penetrating basal margins turn inward. This further enhances the anchorage of the whole structure, in which the body chamber occupies only a small volume (From Seilacher 1992.)

nacles use this space for housing the ovaries. Because the ovarial cavities are crossed by the sutures, their diameters can increase during growth without resorption; but their walls are internal surfaces, in contrast to the external surfaces of the coring cavities.

Tubicinella

Although it uses the same mechanism of penetration, *Tubicinella* (Fig. 4) is so different from *Coronula* that it probably represents an independent entry into the guild of whale riders. After initial penetration (probably in the screw-anchor mode) and after the animal has reached a diameter of 1–2 cm, the accreting basal edges of the wall plates cease to widen. As a result, the adult *Tubicinella* skeleton has become a cylindrical tube that grows at the base while being eroded at the top, just as the surrounding whale skin. Because of lateral growth along still shingled sutures, the tube even widens somewhat toward the top. In other words, plate growth continues only to compensate for whale skin growth. As long as the height of the tubular wall (or what remains of it) does not exceed the thickness of the whale cutis, this is an effective way to get firmly anchored. But as the top of the tube becomes

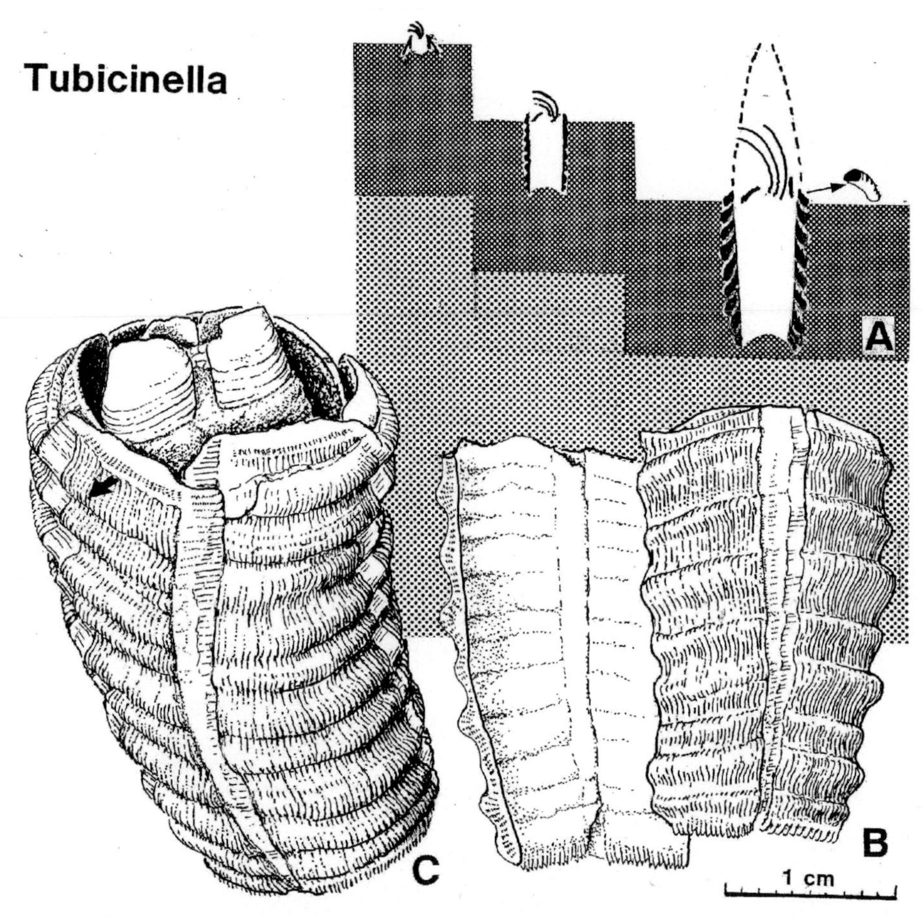

FIGURE 4. *Tubicinella major*. As in *Coronula*, shedding of outer whale skin layers is compensated by growth penetration of the basal plate margins (A). Ratcheted transversal corrugation (B; YPM 2734A) improves anchorage and provides preformed fracture lines, along which ring sectors break off at the surface. This implies (1) that the soft body and its opercular plates (C; YPM 258A) slide down in the mural tube during subsequent moltings and (2) that growth of the tube must continue throughout life, even after the soft body has stopped to grow. Note that sutures between mural plates are still shingled, but grow very little. (From Seilacher 1992.)

eroded, adjustment of the body itself becomes a problem. In fact, the body and filter apparatus of *Tubicinella* move down in their own tube like an elevator, but in steps related to molting. So the animal must continue to penetrate and to molt, after adult size has been reached, lest it would get out of step with whale-skin shedding. As G. Eble suggested (personal communication 2004), this could be called a correlative exaptation.

Tubicinella tubes show another important innovation: molting steps are expressed by strong annular *corrugations*. Because they are ratcheted (Fig. 4B), these rings enhance anchorage; but in addition they provide zones of weakness, along which ring segments break off on top while new ones are added at the base and the body retreats deeper into the tube in the rhythm of molting.

Xenobalanus

The penetration mechanism of *Tubicinella* (adding new ring segments at the base of the mural wall while old ones are shed at the upper edge) is maintained in its much smaller cousin *Xenobalanus* (Fig. 5A). On the other hand, this genus has broken with all balanid traditions. In an ordinary barnacle, the body is nested within the wall of its mural plates; but in *Xenobalanus* it elevates with a long flexible stalk so that it looks like a naked gooseneck barnacle that has also lost the opercular plates. This startling transformation, possible only in the predator-free environment of the

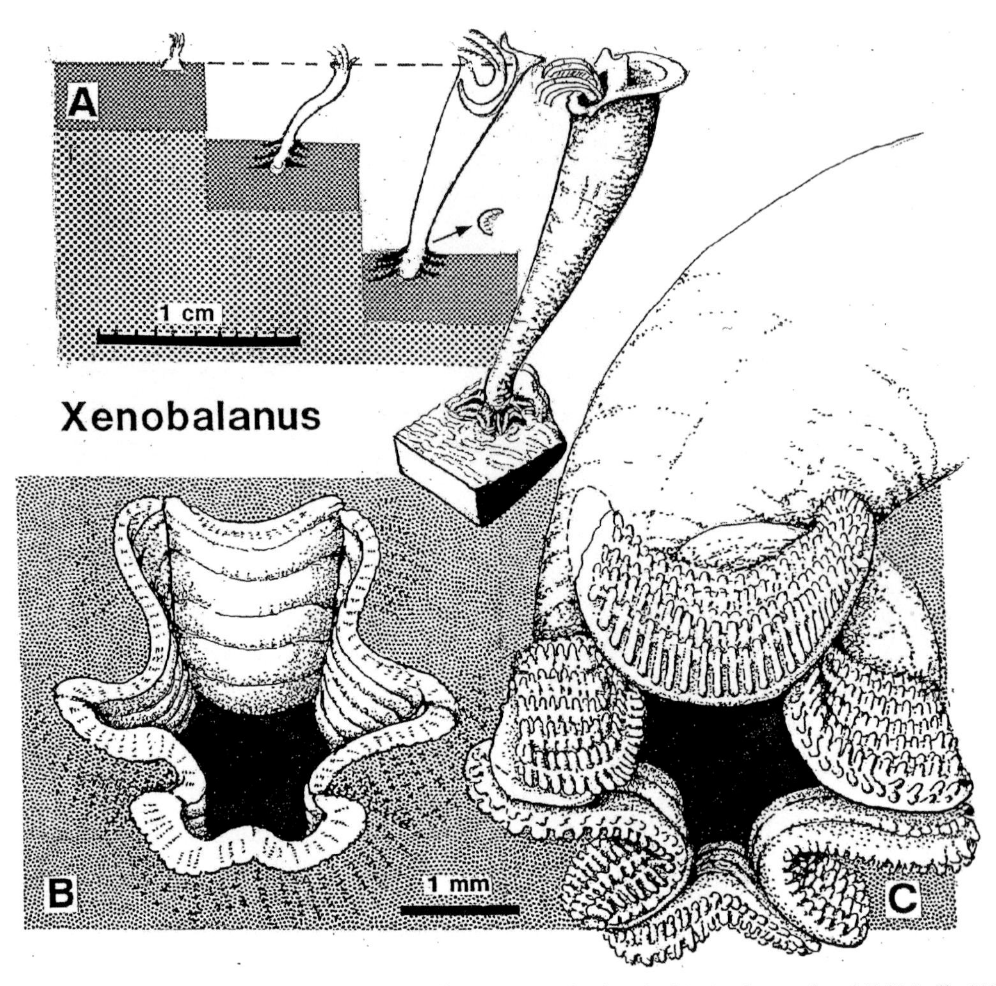

FIGURE 5. *Xenobalanus globicibites*. Although it still penetrates the host's skin in the mode of *Tubicinella* (A), the mural tube of the minute *Xenobalanus* has become profoundly modified by the eversion of the body into a stalked structure. This change reduced the tube to a small, star-shaped anchoring device. With the hydrostatic body pressure being replaced by tensional forces, the wall plates reversed their convexity and loosened their sutural connections (B, basal view). The seleniform ring-elements also improved anchorage by delicate spine patterns on their concave exterior surfaces (C). Along with this transformation went a change in the settling behavior of the larvae. Instead of selecting areas with the strongest tractional currents, they prefer the rear edges of dolphin flukes, where food can be collected from chains of vortices. (From Seilacher 1992.)

whale skin, went along with a change in the settling behavior of the larvae. Instead of choosing surfaces exposed to tractional current, *Xenobalanus* is typically attached to the rear edges of the dolphin's flukes, where the water separates from the wing structure in a tail of vortices.

Even more interesting is the effect that body eversion had on the *morphogenesis* of the mural wall. As long as the wall remained part of the animal's protective carapace, the mural plates were convex-out in horizontal section. With the hydrostatic body pressure being replaced by pulling forces from the everted body, however, the stress regime became reversed. As a result, the mural plates are concave on the outside and their sutures stick out to form a star-shaped anchor. Initially, this new feature was mere fabricational noise (Seilacher 1973), or *non-adaptation* (Gould and Vrba 1982): an outcome of physical rules without biological relevance. But selection soon "discovered" the utility of the new configuration and worked on its optimization toward a novel and very effective anchoring system.

In its present form, the anchor of *Xenobala-*

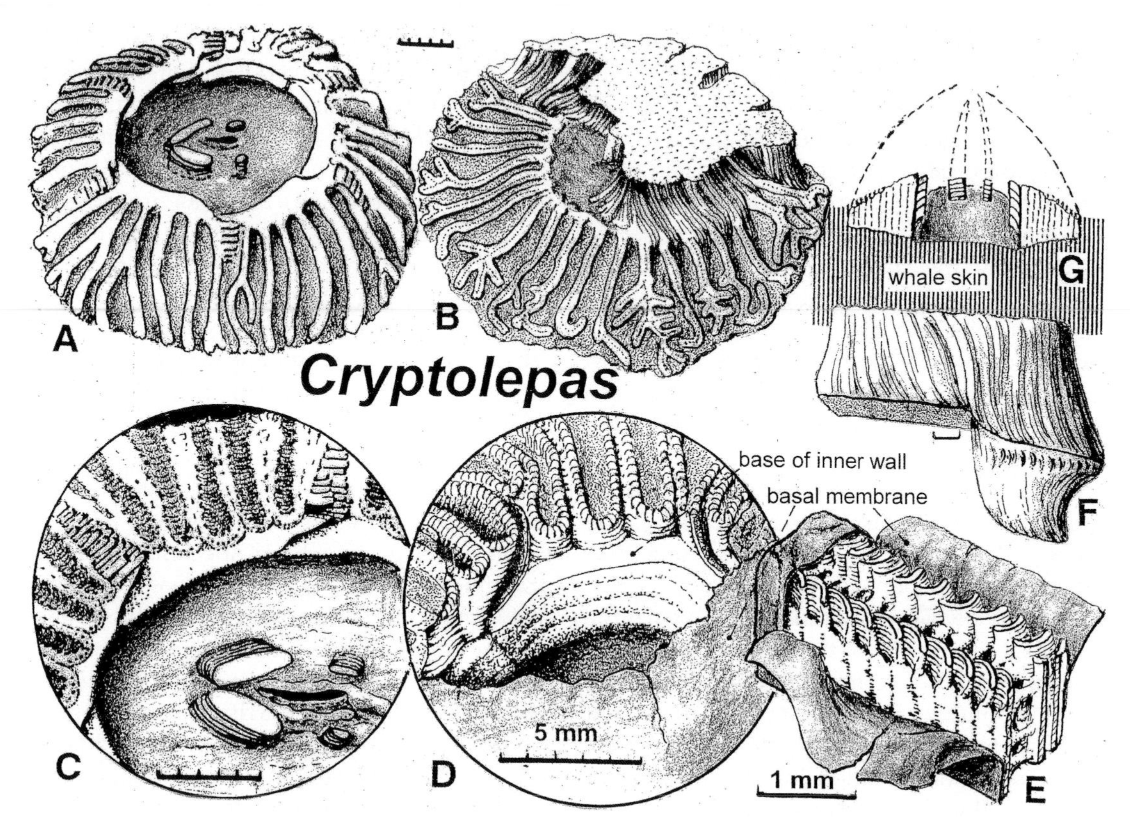

FIGURE 6. *Cryptolepas rhachianecti.* This genus uses the tubicinellid mode of shedding growth rings of the inner wall on top. It also developed folds of the outer wall, like *Coronula*, but in a very different mode. Note the transformation of the opercular plates into self-sharpening mouth parts. A, B, Top and bottom views of YPM 8305. C–E, Details of same specimen. F, Prong of whale skin extracted from between mural folds. G, Diagrammatic section.

nus bears no resemblance to the original balanid wall. Its concave-out crescent-shaped elements fall apart as soon as one tries to free them from the matrix. But not only are the sutural connections between the six plates gone; the growth rings, or corrugations, in the wall plates transformed as well into stacks of free ossicles. Moreover, these seleniform elements are ornamented with delicate knobs and spines (Fig. 5C), which enhance anchorage at the small scale and hardly reveal their derivation from the growth-normal ribs of the *Tubicinella* plates.

Cryptolepas

This genus combines the strategies of *Tubicinella* and *Coronula* As in the previous genus, the inner wall of *Cryptolepas* (Fig. 6) penetrates steeply into the whale skin and flakes off former growth rings at the surface. At the same time, the outer wall develops radial folds, as in *Coronula*, but with a different effect. Whereas the four folds of *Coronula* plates fuse at the tips to form a solid outer wall around prongs of whale skin, those of *Cryptolepas* never fuse. Not only are they more numerous, but they also enhance their contact with the whale skin by forming secondary folds. In a formal sense, these branches could be compared to the secondary folds in *Coronula complanata* (Fig. 2). The complex structure of the *Cryptolepas* folds, however, points to a different origin.

In *Coronula complanata*, vertical riblets protrude at the cutting edges of the folds as minute toothlets that merge into a solid wall behind the marginal growth zone. In *Cryptolepas*, however, the riblets remain but loosely connected with their neighbors, as well as with previous growth rings. The resulting skeletal elements (Fig. 6E) are morphologically complex and able to transfer radial pressure, but they crumble readily as the surrounding whale skin becomes eroded. This is why the upper edges of the folds are always flush with the skin surface and hard to see, as expressed by the genus name. Another effect is that it is almost impossible to free this barnacle from the skin matrix mechanically. In trying to do this, one makes another important observation: below the barnacle, the fibers of the whale skin bend radially out before they enter the spaces between the folds (Fig. 6B,F). This indicates that digestive *penetration* of the substrate combines with radial *deformation* due to the widening of the inner wall by basal and sutural growth.

In reality, of course, it is not the denticulate edge of the mineralized skeleton that penetrates, but the *cuticle* covering it. In *Cryptolepas*, this cuticle can be easily peeled off, as if it were a mere wrapping of the fold. Toward the center, it passes into the uncalcified basal membrane, which is not indented but flat or slightly bulging. So one may speculate that the basal membrane participates in the digestive penetration, assisted by the pressure of the body fluid and activated by glands between cuticle and skeleton.

The *opercular plates* of *Cryptolepas* pose another problem. Together with the soft body, they retreat into the frame of wall plates in the rhythm of moltings. One would therefore expect that these plates retained their protective function, as in any other acorn barnacle except the everted *Xenobalanus*. Yet the four opercular plates of *Cryptolepas* deviate from the norm (1) by being too small to cover the whole aperture, (2) by growing higher than wide through addition of new tablets on the inner surface, and (3) by shedding old tablets on top, just like growths rings are shed off the inner wall. This phenomenon could be explained in two ways: *either* the shedding mode of growth affected the opercular plates as pure fabricational noise without functional significance, *or* the new growth mode led to a *novel function* (for instance as self-sharpening mouth parts) in an environment in which protection was no more at a premium.

Conclusions

By their sessile mode of life, cirripeds are far removed from the ancestral design of an arthropod (Gutmann 1960; Newman et al. 1969). The history of the stalked lepadomorphs goes possibly back to the Cambrian (Collins and Rutkin 1981).The stalkless balanomorphs evolved from them only in the Eocene. Whale barnacles (Coronulina) are known since the late Miocene. They all used a serendipitous crucial exaptation (probably related to cleaning the substrate) to successfully colonize a substrate that grows from the base and wears off on top. To improve anchorage, they also modified the basic balanid design by using other (secondary) exaptations. From our present viewpoint, their roads may appear as dead-ends. In fact, some species may already have become, or soon will be, extinct with the whale species they have specialized on. In principle, however, evolution does not know dead ends. For example, tubicinellids penetrating deep enough to reach the living layers of the skin could potentially become real parasites, deriving their nutrition directly from the host. On the other hand, one could speculate: On what hosts might acorn barnacles have hitchhiked in the past?

Acorn barnacles evolved too late to make it for the giant marine reptiles of Mesozoic times. But they are regularly found on sea snakes and sea turtles. Because reptile skin is much thinner than whale skin, barnacles that attach to the necks of sea turtles (*Platylepas*) show very different aptational structures. This is also true for species (*Chelonibia*) living on the horny carapace, whose vibration during swimming poses a different set of problems (Seilacher 1992). One thing, however, is sure: barnacles cannot settle on furred skins. Therefore, it would be worthwhile to search the sediment around the skeletons of ancestral whales for barnacle plates to answer the question of when whales got naked. As the constructional problems to be solved are known, isolated mural plates of whale dwellers could be singled out even if they belong to taxa now extinct. We should also know more about the *settling behavior* of the larvae. For instance, wouldn't it be a good idea for a whale barnacle to coordinate its reproduction cycle with that of the host in order to allow larval transfer to whale babies while they are still with the mother?

Whale barnacles also illustrate the difficulty of analysis at lower taxonomic levels. As members of a group that share the same exaptational inventory, identical pathways into a specific new function may have been used repeatedly (homoplasy by parallel evolution). Thus, the present study does not allow us to decide whether whale barnacles as a whole (Coronulida) are monophyletic. Nor is it certain that novelties, such as the folding of the outer wall (*Coronula*, *Cryptolepas*), the shedding of growth rings in the inner wall (*Tubicinella*, *Xenobalanus*, *Cryptolepas*), or the transformation of the opercular plates, were invented only once. Data other than skeletal morphology will be necessary to unravel the true phylogenetic relationships between the members of what is presently called Coronulida.

Afterword

Dear Steve, you know that it was not planned this way. When I decided years ago that the constructional history of whale barnacles was too interesting to remain buried in the report of an architectural research group (Seilacher 1992), you were the obvious partner to approach. First, this story is very much in your line of evolutionary thinking. Second, the greatest barnacle specialist of all times was your hero, Charles Darwin (1854). He already knew all the strange morphological details I thought I had discovered. Yet he hardly ever mentioned barnacles in his evolutionary writings. Why? You thought that he did the barnacle studies as an antidote to his heretic evolutionary ideas, whereas I suspected the tendency of all specialists to see the trees rather than the forest. In any case I was very happy that you agreed to join in. Apart from the theoretical and historical aspects you would certainly have added, nobody could tell this story better than you, Steve, in your unique style. Unfortunately you had other priorities, as testified by your last grand book. So whenever I touched the issue, you answered evasively that the project was not forgotten. At our last visit in your and Rhonda's New York home, I asked ironically whether you wanted to save this theme for my own obituary. At that time none of us expected the fatal die to fall to the side of the younger partner and so soon!

Steve, we have been friends since your student times. On several occasions, you supported my provocative ideas. In turn, I admired you as I saw you mature, change your views (remember your initial reaction to our first symposium on constructional morphology in the Journal of Paleontology [Gould 1971]), and master the crises in your private life. But most of all I admired your language in talks as well as your writings. Now the readers have to do with my own humble wording. Nevertheless, the paper should convey your basic message: that life owes its seeming directionality to chance and contingency, combined with the ability to cope with anything short of extinction and death. As an obituary in a leading German newspaper put it, you have been blessed with enormous success, but not with these few extra decades after retirement. We miss you!

Acknowledgments

Specimens from the zoology collection of the Yale Peabody Museum (YPM) were kindly made available by E. Lazem-Wasem. G. Eble and an anonymous reviewer made helpful comments.

Literature Cited

Collins, D., and D. M. Rutkin. 1981. *Priacansermarinus barnetti*, a probable lepadomorph barnacle from the Middle Cambrian Burgess Shale of British Columbia. Journal of Paleontology 55:1006–1015.

Darwin, C. 1854. A monograph on the sub-class Cirripedia with figures of all the species. Royal Society, London.

Gould, S. J. 1971. Tübingen meeting on form. Journal of Paleontology 45:1042–1043.

Gould, S. J., and E. Vrba. 1982. Exaptation—a missing term in the science of form. Paleobiology 8:4–15.

Gutmann, W. F. 1960. Funktionelle Morphologie von *Balanus balanoides*. Abhandlungen der Senckenbergischen Naturforschenden Gesellschaft 500. Frankfurt am Main.

Newman, W. A., V. A. Zullo, and T. H. Withers. 1969. Cirripedia. Pp. R206–R 295 *in* H. K. Brooks et al. Arthropoda 4, Vol. 1. Part R of R.C. Moore, ed. Treatise on invertebrate paleontology. Geological Society of America, Boulder, Colo., and University of Kansas, Lawrence.

Seilacher, A. 1973. Fabricational noise in adaptive morphology. Systematic Zoology 22:451–465.

———. 1992. Whale barnacles: how an evolutionary dream could become true. Mitteilungen des Sonderforschungsbereichs 230(8):131–136. Universität Stuttgart.

Paleobiology, 31(2), 2005, pp. 36–55

Tempo and mode of early animal evolution: inferences from rocks, Hox, and molecular clocks

Kevin J. Peterson, Mark A. McPeek, and David A. D. Evans

Abstract.—One of the enduring puzzles to Stephen Jay Gould about life on Earth was the cause or causes of the fantastic diversity of animals that exploded in the fossil record starting around 530 Ma—the Cambrian explosion. In this contribution, we first review recent phylogenetic and molecular clock studies that estimate dates for high-level metazoan diversifications, in particular the origin of the major lineages of the bilaterally-symmetrical animals (Bilateria) including cnidarians. We next review possible "internal" triggers for the Cambrian explosion, and argue that pattern formation, those processes that delay the specification of cells and thereby allow for growth, was one major innovation that allowed for the evolution of distinct macroscopic body plans by the end of the Precambrian. Of potential "external" triggers there is no lack of candidates, including snowball earth episodes and a general increase in the oxygenation state of the world's oceans; the former could affect animal evolution by a mass extinction followed by ecological recovery, whereas the latter could affect the evolution of benthic animals through the transfer of reduced carbon from the pelagos to the benthos via fecal pellets. We argue that the most likely cause of the Cambrian explosion was the evolution of macrophagy, which resulted in the evolution of larger body sizes and eventually skeletons in response to increased benthic predation pressures. Benthic predation pressures also resulted in the evolution of mesozooplankton, which irrevocably linked the pelagos with the benthos, effectively establishing the Phanerozoic ocean. Hence, we suggest that the Cambrian explosion was the inevitable outcome of the evolution of macrophagy near the end of the Marinoan glacial interval.

Kevin J. Peterson and Mark A. McPeek. Department of Biological Sciences, Dartmouth College, Hanover, New Hampshire 03755. E-mail: kevin.peterson@dartmouth.edu*
David A. D. Evans. Department of Geology and Geophysics, Yale University, New Haven, Connecticut 06520-8109
*Corresponding author

Accepted: 4 August 2004

> The Cambrian explosion ranks as such a definitive episode in the history of animals that we cannot possibly grasp the basic tale of our own kingdom until we achieve better resolution for both the antecedents and the unfolding of this cardinal geological moment.
>
> [Gould 1998]

Introduction

The Cambrian explosion stands out, both in the history of animal evolution and in the writings of Stephen Jay Gould, as the pivotal event in animal evolution. Although multifarious like no other episode in the history of life, Gould saw three particularly important evolutionary issues associated with the Cambrian explosion: (1) the rapidity of morphological evolution in the Early Cambrian, and its independence from genealogy; (2) the cause of this rapidity, whether triggered environmentally or genetically; and (3) the notion of disparity, or the stability of animal body plans over the ensuing 530 million years. Although this stability is, according to some (e.g., Levinton 2001), the single most important fact the fossil record has contributed to the science of evolutionary biology, here we will restrict our discussion to the first two themes, whether the anatomical innovations characterizing numerous groups were separate from the events generating the clades, and the nature of the triggers causing the innovations because these two issues are obviously entwined. As Gould (2002) argued: if only a single Early Cambrian lineage generated all Cambrian diversity, then an internal trigger based upon some genetic or developmental "invention" is plausible; if, alternatively, most phylum-level lineages were already established well before the Precambrian/Cambrian boundary, then their transformation into the larger and well-differ-

0094-8373/05/3102-0004/$1.00

entiated body plans, apparent by the Atdabanian, suggests some sort of external trigger. Our goals here are to (1) place the evolutionary history of animals into a proper phylogenetic and temporal context; (2) address the notion of internal versus external triggers to the Cambrian explosion; and (3) summarize early animal evolution in the context of the late Precambrian to Cambrian transition.

Clocks: How Many Worms Crawled across the Precambrian/Cambrian Boundary?

> The Cambrian explosion, as paleontologists propose and understand the concept, marks an anatomical transition in the overt phenotypes of bilaterian organisms—that is, a geologically abrupt origin of the major *Baupläne* of bilaterian phyla and classes—not a claim about times of initial phyletic branching. (Gould 2002)

Of considerable interest to Gould was the conflict between two competing views of the Cambrian explosion: whether the Cambrian explosion reflects the rapid appearance of fossils with animals having a deep, but cryptic, Precambrian history, as suggested by most molecular clock studies; or whether it reflects the true sudden appearance and diversification of animals, as suggested by a literal reading of the fossil record (see Runnegar 1982). Although Gould stressed that the Cambrian explosion stands as one of the most important evolutionary events in natural history irrespective of when the lineages actually diverged from one another, a deeper understanding of the mechanisms underlying the Cambrian explosion requires knowledge about how these lineages are interrelated, and when in time the lineages themselves came into existence.

Metazoan Phylogeny.—With the advent of molecular systematics, much progress has been made in our understanding of how animals are related to one another. Figure 1 shows a maximum parsimony analysis of 28 metazoan species, plus five non-metazoan taxa, based on a concatenation of 2039 amino acid positions derived from seven different housekeeping genes, 228 amino acid positions from the cytochrome oxidase I gene, 1747 nucleotide characters from the 18S rDNA gene, and 155 morphological characters coded for the genus where possible (Peterson unpublished). This "total evidence" tree supports the monophyly of both well-and newly-established nodes including all recognized phyla, Metazoa, Deuterostomia (represented here by echinoderms and hemichordates), Trochozoa (nemerteans, annelids, and molluscs), Platyzoa (the rotifer and the flatworm), Spiralia (the trochozoans and the platyzoans), Ecdysozoa (the priapulid and the insects), and Protostomia (the spiralians and the ecdysozoans) (e.g., Halanych et al. 1995; Aguinaldo et al. 1997; Zrzavy et al. 1998; Giribet et al. 2000; Peterson and Eernisse 2001; Mallatt and Winchell 2002; reviewed in Eernisse and Peterson 2004).

As expected, we find strong support for the monophyly of triploblasts, those bilaterians possessing true mesoderm. Recent studies on cnidarians have demonstrated that mesoderm is a triploblast apomorphy because "mesoderm" genes are expressed in the endoderm of the anthozoan *Nematostella* (Scholtz and Technau 2003; Martindale et al. 2004) and are not absent as would be expected if mesoderm was secondarily lost. We also find strong support for the clade Cnidaria + Triploblastica, a clade we will refer to, somewhat unconventionally, as "Bilateria," because it appears that bilateral symmetry is primitive for Cnidaria and Triploblastica and is still present in anthozoans, as assessed by both morphological and genetic criteria (Hayward et al. 2002; Finnerty et al. 2004). Finally, we find weak support for the monophyly of Eumetazoa (Ctenophora + Bilateria), which suggests that tissues, the nervous system, and importantly a true gut each evolved once, contra the hypothesis of Cavalier-Smith et al. (1996).

Two phylogenetic results are particularly important for what is to follow. First, we do not find a monophyletic Porifera, consistent with most recent ribosomal analyses (Borchiellini et al. 2001; Medina et al. 2001; Peterson and Eernisse 2001; but see Manuel et al. 2003). In particular, those sponges whose skeletons are composed of calcareous spicules (Calcispongia) are the sister taxon of Eumetazoa, whereas the remaining sponges, whose

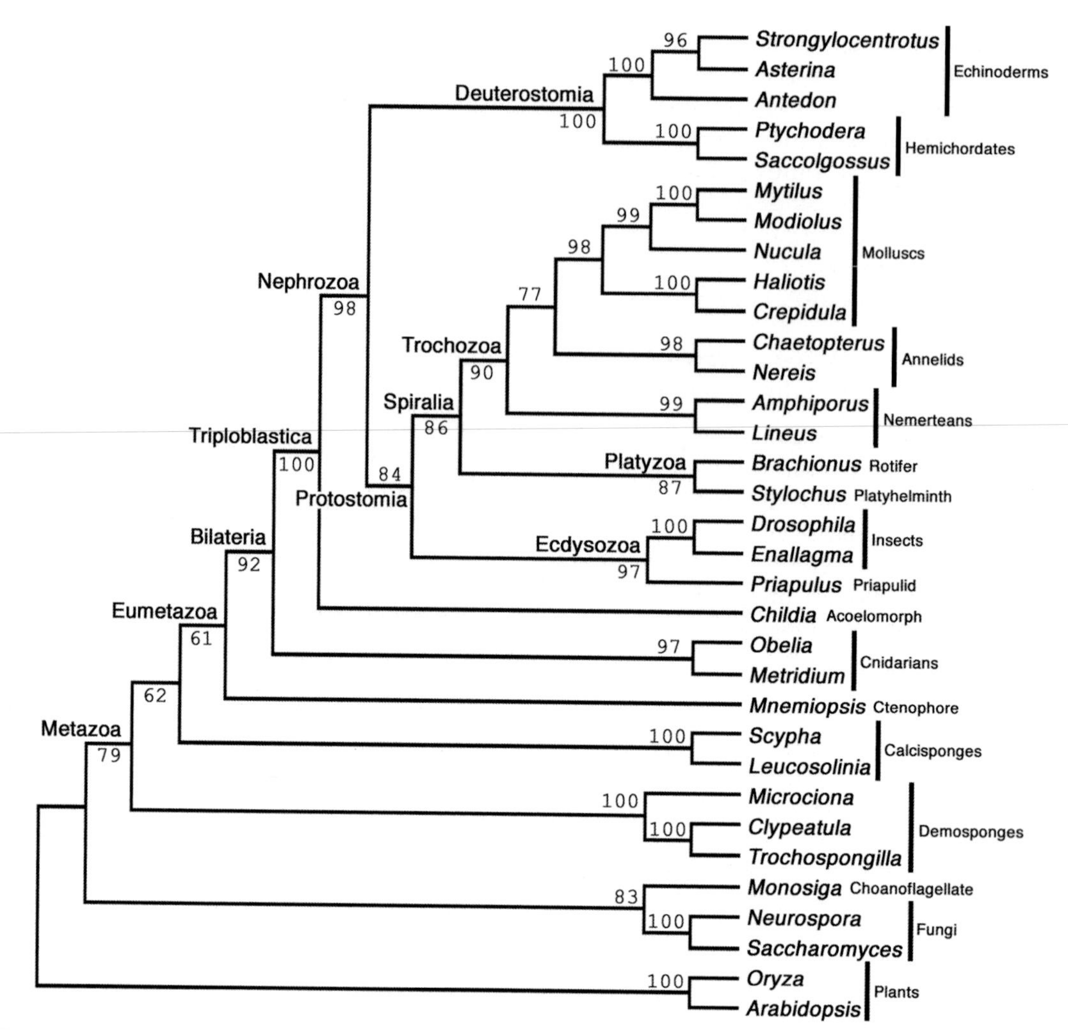

FIGURE 1. Maximum parsimony analysis (PAUP v. 4.0b10 for Macintosh [Swofford 2002]) of a total-evidence data set consisting 2039 amino acid positions derived from seven different housekeeping genes (Peterson et al. 2004), 228 amino acid positions from the cytochrome oxidase I gene, 1747 nucleotide characters from the 18S rDNA gene, and 155 morphological characters coded for the genus where possible, for 28 metazoans, the choanoflagellate *Monosiga*, two fungi, and two plant taxa. The tree is 13,266 steps long (number of parsimony informative characters = 1989); C.I. = 0.50; R.I. = 0.48; R.C. = 0.24. Bootstrap percentages are derived from 1000 replicates. Of particular importance is the paraphyly of Porifera with calcisponges more closely related to eumetazoans than to demosponges, and the basal position of the acoel flatworm *Childia* with respect to the other triploblasts. (From Peterson unpublished.)

skeletons are composed of siliceous spicules (Silicispongia = Demospongia + Hexactinellida), are the sister taxon of the Calcispongia + Eumetazoa clade. This result is not driven solely by the ribosomal sequences, as the paraphyly of Porifera is also found when the amino acid sequences of the seven different housekeeping genes are analyzed by Minimum Evolution alone (Peterson and Butterfield unpublished). An obvious implication of these results is that both the last common ancestor of Metazoa, and the last common ancestor of Calcispongia + Eumetazoa were constructed like a modern sponge, an obligate benthic organism with a water-canal system (Eernisse and Peterson 2004).

Second, the acoel flatworm *Childia* falls at the base of the triploblasts with high precision, again consistent with virtually all recent phylogenetic inquiries (Ruiz-Trillo et al. 1999,

2002; Jondelius et al. 2002; Pasquinelli et al. 2003; Telford et al. 2003). Acoels are direct-developing, micrometazoan acoelomates, suggesting that the earliest triploblasts were also small, benthic, directly developing animals with internal fertilization and a uniphasic life cycle (i.e., eggs laid on the benthos, not released into the pelagos) (Baguñà et al. 2001; Jondelius et al. 2002). An important implication is that both coeloms (contra Budd and Jensen 2000) and complex life cycles (contra Peterson et al. 2000, see below) arose multiple times within the clade Nephrozoa (all triploblasts exclusive of acoelomorphs, and which primitively possess nephridia [Jondelius et al. 2002]).

Our analysis does not consider the phylogenetic position of nematodes. This is a contentious issue because nematodes are placed within the Ecdysozoa on the basis of ribosomal sequence analysis (Aguinaldo et al. 1997; Peterson and Eernisse 2001; Mallatt and Winchell 2002; Mallatt et al. 2004), the possession of ecdysozoan-specific posterior Hox genes (de Rosa et al. 1999), and a nervous system with HRP immunoreactivity (Haase et al. 2001), whereas genome-wide phylogenetic analyses support a basal triploblast position (Blair et al. 2002; Wolf et al. 2004). Nonetheless, given the amount of gene loss in the *C. elegans* system, it is difficult to say where exactly nematodes fall within Triploblastica, although an affinity with the fruit fly *Drosophila* is not ruled out (Copley et al. 2004). Despite this controversy, it makes little difference to our arguments; in fact, a basal triploblast position would only reinforce the conclusions reached herein.

The Geologic Time Frame.—In order to relate animal evolution to terminal Neoproterozoic geology and the metazoan fossil record, accurate and precise geochronological dates are required. Probably the most important date of all is the formally defined Proterozoic/Cambrian boundary (Landing 1994), which was calibrated to 543–542 Ma by precise U-Pb zircon geochronology on correlative sections in Siberia (Bowring et al. 1993), Namibia (Grotzinger et al. 1995), and Oman (Amthor et al. 2003). The spectacular embryos entombed in chert and phosphate of the Doushantuo Formation in southern China (Xiao et al. 1998; Li et al. 1998), which represent the oldest known unequivocal metazoan remains, are suggested to be less than 580 Myr old. The first appearance of macroscopic Ediacara fossils—including giant frondose specimens up to 2 m long—is within the 575 Ma Drook Formation in southeast Newfoundland (Narbonne and Gehling 2003). These ancient Ediacara lineages continue upsection to the Mistaken Point deposits dated at 565 Ma (Benus 1988). Finally, macroscopic bilaterians make their first appearance at 555 Ma (Martin et al. 2000). These ages are shown on Figure 2 and are listed in Table 1.

A New Metazoan Time Frame.—Using the topology in Figure 1 as our phylogenetic framework, and the high-precision geochronological dates reviewed above, we can now discuss the time frame for bilaterian evolution. The molecular clock is the tool of choice to test hypotheses of metazoan originations independent of the fossil record (Runnegar 1982). Although problems exist with certain aspects of molecular clock analyses (Smith and Peterson 2002; Benton and Ayala 2003), two recent analyses came to the same conclusion: bilaterians arose ca. 600–630 Ma, and nephrozoans arose ca. 560–580 Ma (Aris-Brosou and Yang 2002, 2003; Peterson et al. 2004; see each individual paper for details of the analyses and confidence intervals). The estimates from an updated analysis of Peterson et al. (2004; Peterson unpublished) are shown on Figure 2.

These dates differ from almost every previous molecular clock study, many of which argue that the last common ancestor of Nephrozoa originated ca. 1000 Ma (e.g., Wray et al. 1996; Bromham et al. 1998; Wang et al. 1999; Nei et al. 2001), because the rate of molecular evolution across taxa was addressed in detail. Peterson et al. (2004) examined seven new protein sequences from over 20 new taxa, which allowed for the use of multiple calibration points scattered across bilaterian phylogeny and through time, whereas Aris-Brosou and Yang (2002, 2003) undertook sophisticated analyses of published sequences using a Bayesian posterior-probability approach to account for rate heterogeneity. Importantly, both groups of authors demonstrated the ex-

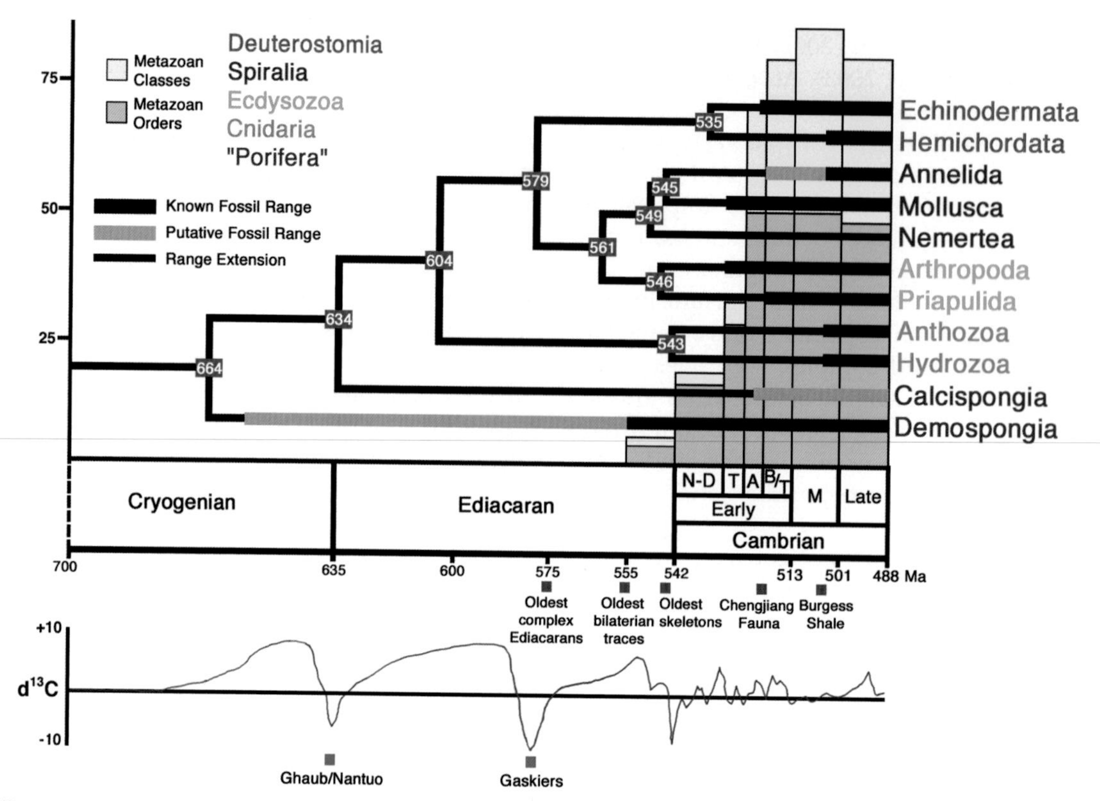

FIGURE 2. Tempo of early animal evolution placed into the geologic context of the Neoproterozoic/Cambrian transition. Tree nodes are positioned according to the updated molecular clock estimates of Peterson et al. (2004; Peterson unpublished). The generalized Precambrian carbon isotope curve is from Knoll 2000. Boundary ages are from the 2003 ICS International Stratigraphic Chart except for the Cryogenian/Ediacaran boundary (Knoll et al. 2004), which is placed at 635 Ma on the basis of Hoffman et al. 2004 and Condon et al. 2005. Note that according to Calver et al. (2004) this boundary could be as young as 580 Ma and thus the Marinoan and Gaskiers glaciations could be synchronous; see the text for further discussion. All other ages are listed in Table 1. Abbreviations: N-D; Nemakit-Daldynian; T; Tommotian; A; Atdabanian; B/T; Botomian/Toyonian; M; Middle. (Adapted from Knoll and Carroll 1999.)

istence of significant rate heterogeneity associated with the vertebrate sequences when compared with most invertebrate sequences (see also Ayala et al. 1998). Indeed, Peterson et al. (2004) showed that the rate of molecular evolution decreased in vertebrates about twofold with respect to three different invertebrate lineages (echinoderms, bivalves, and insects). This is significant because vertebrates are usually used to calibrate the molecular clock, and thus a twofold decrease in the rate of vertebrate sequence evolution causes about a two-fold overestimate of the vertebrate-dipteran divergence (see also Benton and Ayala 2003). Studies that either exclude vertebrates (Peterson et al. 2004) or allow substitution rates to vary among taxa (Aris-Brosou and Yang 2002, 2003) give estimates much closer to the those derived from a synoptic reading of the fossil record (Fig. 2).

In addition, Peterson et al. (2004) and Peterson (unpublished) tested the accuracy of their clock by estimating a divergence when both a minimum and a maximum are known from the fossil record. The minimum age for the split between crinozoan and eleutherozoan echinoderms is 485 Ma (Landing et al. 2000) and is derived from the first occurrence of unequivocal crown-group echinoderms (stem-group crinoids from the Tremadocian [Guensburg and Sprinkle 2001]). The maximum age is 525 Ma (Landing et al. 1998) and is derived from the first occurrence of stereom in the fossil record (stem-group echinoderms from the

TABLE 1. High-precision ages from late Neoproterozoic stratified successions.

Location	Formation (Group)	Age (Ma)	Relation to stratigraphically important events	Ref.
Terminal Proterozoic cratonic successions				
Oman	(Ara)	542 ± 0	Ediacaran/Cambrian boundary (−)$\delta^{13}C$ excursion	1
Namibia	Urusis	543 ± 1	Immediately below Cambrian; Cloudina	2
Oman	Fara (Ara)	545 ± 3	Immediately below Cambrian	3
Namibia	Zaris	549 ± 1	(+)δ^{13} C excursion; diverse Ediacara fossils	2
Poland	(Volhyn)	551 ± 4	In borehole 200 m above diamictite	4
White Sea	Ust-Pinega	555 ± 0	*Kimberella*-bearing assemblage	5
Avalonian-Cadomian arc				
England	Beacon Hill	559 ± 2	Immediately underlies Ediacara-bearing strata	6
Newfound.	Mistaken Point	565 ± 3	Diverse frondose Ediacara assemblage	7*
Newfound.	Drook	ca. 575	Simple frondose Ediacara assemblage	8*
Newfound.	Gaskiers	ca. 580	Ashes below, within, and above glaciogenic unit	8*
NE U.S.A.	Mattapan	595 ± 2	Older than Squantum glaciogenic deposit	9
Laurentian late-stage rifts				
SW Canada	(Hamill/Gog)	570 ± 5	Above marker dolostone = Icebrook cap equiv?	10
E U.S.A.	(Catoctin)	ca. 570	Closely associated with Fauquier Gp diamictite	11
Scotland	Tayvallich	601 ± 4	12 km above Port Askaig "Tillite"	12
Proto-Gondwanaland terminal Proterozoic successions				
Tasmania	(Grassy)	ca. 580	Pre-(?) and postglacial volcanic rocks	13
S China	Doushantuo	551 ± 1	Minimum age for metazoan embryos	14
		635 ± 1	Ash bed within post-Nantuo cap carbonate	14
N Namibia	Ghaub	636 ± 1	Ash bed within glaciogenic formation	15
S Australia	Marino	ca. 650	Two detrital zircons in arkose below Elatina level	16
S China	Datangpo	663 ± 4	Older than the Nantuo glaciogenic deposit	17

References: (1) Amthor et al. 2003; (2) Grotzinger et al. 1995; (3) Brasier et al. 2000; (4) Compston et al. 1995; (5) Martin et al. 2000; (6) Compston et al. 2002; (7) Benus 1988; (8) Bowring et al. 2003; Narbonne and Gehling 2003; (9) Thompson and Bowring 2000; (10) Colpron et al. 2002; (11) Aleinikoff et al. 1995; (12) Dempster et al. 2002; (13) Calver et al. 2004; (14) Condon et al. 2005; (15) Hoffmann et al. 2004; (16) Ireland et al. 1998; (17) Zhou et al. 2004.

* Age not fully published with analytical data.

Nevadella trilobite zone [Smith 1988]). Peterson et al. (2004) estimated this divergence at 508 Ma, well within the range of dates established by the fossil record. The second point is the divergence between the vetigastropod *Haliotis* and the sorbeconch *Crepidula*, which the fossil record estimates at 501–488 Ma (Lindberg and Guralnick 2003) and the molecular clock estimates at 500 Ma (Peterson unpublished). These two checks suggest that the molecular-clock estimates summarized here are accurate.

Summary.—The most recent clock studies, coupled with a new understanding of metazoan relationships and precise geochronological tie points, estimate that many different lineages of triploblastic "worms" crawled across the boundary, in addition to the other extant metazoan lineages that were present in the Ediacaran (calcisponges, silicisponges, ctenophores, and cnidarians). Thus, the Ediacaran/Cambrian boundary seems to correspond to the time when many metazoan phylum- and class-level divergences were occurring (Fig. 2). Hence, the cladogenic events underlying the origin of modern phylum-level lineages and the gradual acquisition of each of the phylum-level body plans would seem to go hand in hand: Unique bilaterian body plans evolved in the context of a diversification event that started around 600 Ma and culminated with the phylum-level splits around 540 Ma, dates consistent with the known fossil record of metazoans (Xiao et al. 1998; Budd and Jensen 2000; Jensen 2003; Condon et al. 2005), and with the Proterozoic fossil record in general (Butterfield 2004).

Hox: Internal Trigger(s)

> The question of one vs. ten ["worms" crawling across the Precambrian/Cambrian boundary] does, however, bear strongly upon the important question of internal vs. external triggers for the explosion. If only one lineage generated all Cambrian diversity, then an internal trigger based upon some genetic or developmental "invention" becomes plausible. (Gould 2002)

Our thesis below is that the spectacular radiation of animals during the late Precambrian and early Cambrian was inevitable once macrophagous predators evolved near the end of the Neoproterozoic. For the first time in earth history, predation pressures drove macroevolution of not just animals, but other eukaryotic lineages as well. Hence, understanding how and why bilaterians achieved macroscopic size within a few tens of millions of years after their origination at ca. 600 Ma is the first key to unraveling the mystery of the Cambrian explosion.

Macrophagy and the Cambrian Explosion.—The first two essays Gould wrote on the Cambrian explosion were a perspective on Stanley's (1973) paper on cropping (Gould 1979a: pp. 126–133), and an argument that the explosion itself follows a general law of growth (Gould 1979: 119–125). Both essays convey the general idea that the Cambrian explosion was inevitable: "Perhaps the explosion itself was merely the predictable outcome of a process inexorably set in motion by an earlier Precambrian event" (Gould 1979). We agree with this view and contend that the Cambrian explosion is the paleontological manifestation of the evolution of predation, specifically the origin of and defense against macrophagy, within the phylogenetic context of the bilaterian radiation.

Ignoring the highly unusual form of macrophagy found in cladorhizid sponges (Vacelet and Boury-Esnault 1995), macrophagy is a derived state within Metazoa because the paraphyly of sponges (Fig. 1) implies that the last common ancestor of metazoans, and also calcareous sponges + eumetazoans, had a water-canal system, and thus must have fed intracellularly on free organic matter and demersal bacteria, as both groups of sponges still do today (Brusca and Brusca 2002). This reliance upon intracellular digestion restricts the upper limit of their prey size to approximately 5 μm (Brusca and Brusca 2002), which excludes most forms of eukaryotic phytoplankton and, of course, other animals as prey items. Thus, planktonic eukaryotic algae would have been invulnerable to metazoan predators during this time interval, and, not surprisingly, acritarch taxa are morphologically monotonous from ca. 2 Ga until just after the 653 Ma Nanto glacial deposit (Butterfield 1997; Peterson and Butterfield unpublished), a morphology characteristic of phytoplankton that do not experience herbivory (Reynolds 1984; Sommer et al. 1984; Leibold 1989).

Macrophagy must have evolved at the base of Eumetazoa because all three major groups—ctenophores, cnidarians, and triploblasts—each have a mouth and a functional gut to digest food extracellularly. None use a water-canal system, which was lost by the time eumetazoans arose, and thus no necessary restrictions of prey size like that of sponges would exist. We hypothesize that benthic grazing is the primitive mode of feeding for eumetazoans, which means that these animals should have been able to feed upon anything living in the bacterial mats that was smaller than themselves, including bacteria, algae, and presumably other animals.

Bearing in mind the obvious taphonomic window through which we view the Burgess Shale biota (Butterfield 2003), this hypothesis does conform to the known distribution of feeding modes such that an overwhelming majority of the described Cambrian eumetazoans are epifaunal or infaunal vagrants, dominated by arthropods and priapulids, respectively (Conway Morris 1979, 1986). In contrast, the epifaunal suspension feeders are primarily sponges, with only brachiopods and a few echinoderms and other enigmatic forms (e.g., *Dinomischus*) feeding upon eukaryotic plankton (see also Signor and Vermeij 1994). Nonetheless, the trophic mode of both cnidarians and ctenophores, as well as ecological considerations of Ediacara assemblages (e.g., Conway Morris 1986; Clapham and Narbonne 2002; Clapham et al. 2003), would seem to suggest that suspension feeding is primitive for Eumetazoa. However, whether or not Ediacaran organisms were metazoans is a contentious issue, and several recent treatments find more similarities with modern fungi than with modern animals (Steiner and Reitner 2001; Peterson et al. 2003). More importantly, the fractal design of many Ediacara organisms (Narbonne 2004) and the lack of any obvious openings, coupled with their ecological tiering (Clapham and Narbonne 2002; Clapham et al. 2003), suggest that they could only feed on dissolved organic manner in a unique fashion when compared with either sponges or eumetazoans, irrespective of their actual phylogenetic position, and thus are irrelevant for polarization.

Suspension feeding is unlikely to be primitive for cnidarians because the earliest cnidarians could not have yet evolved nematocysts. We hypothesize that the earliest cnidarians grazed upon benthic organisms in a manner similar to that of triploblasts. Indeed, it may be better to think about cnidarians as originating from acoelomorph-like animals, as opposed to acoelomorph-like animals evolving from planula larva (e.g., Jondelius et al. 2002), especially given the presence of both an anterior/posterior and a cryptic dorsal/ventral axis in anthozoans (Hayward et al. 2002; Finnerty et al. 2004). The evolution of cnidarian polyps with nematocyst-laden tentacles could only have occurred after the evolution of an appropriate food source, namely mesozooplankton, which is probably no older than about ca. 560 Ma (Peterson unpublished). Given that total-group Cnidaria originated ca. 600 Ma, there must have been stem-group cnidarians that fed on other food sources. With respect to ctenophores, not only are their colloblast-laden tentacles of unique construction (Brusca and Brusca 2002), but they are also absent from all known Cambrian ctenophores (Conway Morris and Collins 1996; Chen and Zhou 1997), making suspension feeding a derived condition here as well.

The Evolution of Complex Life Cycles.—The evolution of complex life cycles is coupled with the relative size increase in animals in response to predation. The idea that direct development with a benthic adult stage is primitive for Triploblastica contrasts sharply with the hypotheses of Nielsen (1998) and Peterson et al. (2000), who suggested that the earliest triploblasts were ciliated planktotrophic larval-like forms. This hypothesis is problematic for numerous reasons (see also Conway Morris 1998; Budd and Jensen 2000; Valentine and Collins 2000; Sly et al. 2003). First, as discussed above, the basal position of acoels suggests that the primitive condition for Triploblastica is a uniphasic life cycle that lacks a pelagic larval stage (Baguñà et al. 2001; Jondelius et al. 2002). Second, most metazoan phyla, including triploblasts, are adapted to live on the benthos, not in the pelagos (Valentine et al. 1999; Collins and Valentine 2001). Third, the paraphyly of Porifera (Fig. 1), and the primitiveness of the cnidarian polyp stage (e.g., Bridge et al. 1995; Collins 2002), implies that the last common ancestor of metazoans, the last common ancestor of calcisponges and eumetazoans, and the last common ancestor of bilaterians all had an obligate benthic adult stage. Fourth, the general direction of evolution among both metazoans and non-metazoans is from the benthos to the pelagos and not the reverse (Rigby and Milsom 1996, 2000). Fifth, because internal fertilization is primitive for both Triploblastica and Nephrozoa (Buckland-Nicks and Scheltema 1995), small size combined with direct development is most likely primitive as well (Olive 1985; Chaffee and Lindberg 1986). Finally, the known fossil record suggests that the benthos is the primitive site of animal evolution given that the earliest known bilaterian macrofossil, *Kimberella*, is a benthic animal (Fedonkin and Waggoner 1997), and pelagic forms are extremely rare, if present at all, in the Ediacara biotas (Fedonkin 2003). Therefore, pelagic larvae are not primitive; instead larvae were independently intercalated into an existing direct-developing strategy (Sly et al. 2003) multiple times (Hadfield 2000; Hadfield et al. 2001).

This convergent evolution of maximal indirect development (Davidson et al. 1995) from direct development is difficult to understand because in the modern world larvae offer few, if any, evolutionary advantages (Pechenik 1999). In addition, natural selection should not favor the lengthening of a pre-adult stage or the addition of pre-adult stages; selection should predominantly favor the evolution of rapid development (Williams 1966). To us, the independent evolution of larvae, as well as other forms of mesozooplankton, is a solution to an ecological problem, namely the problem of benthic predation (Signor and Vermeij 1994). This hypothesis is consistent with four key features of the evolutionary history of the rise of larvae themselves. First, the evolution of positive phototaxy and locomotive ciliary bands allow for negatively buoyant eggs and embryos to stay within the pelagos and away from the predator-rich benthos (Pechenik 1979). Second, benthic predation would have disproportionately affected the smallest size classes of each species; this accounts for the migration of eggs and initial free-living forms into the pelagic zone first, instead of older, and thus larger, life stages, which occurred in a few cases later in geological time (Rigby and Milsom 1996, 2000). Third, the rise of larvae is not due solely to the exploitation of novel resources (e.g., phytoplankton) because phylogenetic considerations suggest that the many of the earliest larval forms were non-feeding (McHugh and Rouse 1998; Rouse 2000). And fourth, predation pressures from macrozooplankton (e.g., cnidarian medusae, ctenophores, chaetognaths, and vertebrates [Young and Chia 1987]) was virtually nonexistent until the late Early Cambrian (Vannier and Chen 2000). In addition, although some larval forms show macroevolutionary responses to predation (Nützel and Fryda 2003), even today, thanks to their small size and optical clarity, larvae are not heavily preyed upon (Johnson and Shanks 2003). Hence, planktonic predation pressures have played a minor role in the evolution of larvae, consistent with the relatively long periods they spend in the pelagos.

Benthic predation pressures may have also selected for the evolution of "set-aside" cells (Davidson et al. 1995; Peterson et al. 1997). It appears that the principal function of a larval stage is not dispersal, but growth (Strathmann 1987, 1993). Growth is of fundamental importance because benthic suspension feeders ingest particles within a size range that includes most marine larval forms (Young and Chia 1987), and thus larvae cannot escape predation until they have grown larger than their typical size at metamorphosis (Pechenik 1999). Therefore, benthic predation pressures would select for the rapid growth of juveniles. Rapid growth on the benthos independent of juvenile feeding requires that much of the juvenile body plan already be "prepackaged" within the larvae; this could be accomplished if larvae began adult body plan construction while still in the plankton. Consistent with prediction, then (Williams 1966), the evolution of set-aside cells makes obvious ecological and evolutionary sense: they allow for maximal growth (Strathmann 1987; Hadfield 2000) during the time of maximal predation (MacGinitie 1934) and thus are a convergent solution to a ubiquitous problem (Bishop and Brandhorst 2003; Sly et al. 2003).

Molecular Control of Body Size.—The independent evolution of both coeloms and larvae strongly suggests that the direction of nephrozoan evolution was from small direct-developing acoelomates to large indirectly developing coelomates, and not the reverse. Consistent with this hypothesis, the dearth of body fossils and, more importantly, trace fossils before 555 Ma (Martin et al. 2000; Jensen 2003) suggests that the earliest nephrozoans were under 1 cm, and probably closer to 1 mm, in length (Erwin and Davidson 2002). This predicts, then, that the evolutionary increase in body size (or, more appropriately, mass [see Conlon and Raff 1999]), occurred numerous times independently. Growth is regulated by both extrinsic and intrinsic inputs that are linked in an unknown manner (Johnston and Gallant 2002) and must work at the appropriate scale (Nijhout 2003). These inputs can be classified into three types: (1) growth factors that stimulate protein synthesis, resulting in an increase in cell size; (2) mitogens that increase cell number by stimulating cell division; and (3) survival factors that prevent programmed cell death (Conlon and Raff 1999). These inputs are primitive for at least Bilateria, because an insulin receptor was characterized from the cnidarian *Hydra* (Steele et al. 1996) and the mitogen *myc* appears in the *Hydra* EST database (CN550970).

Importantly, defects in the mitogen pathway result in a smaller body size in both flies and mice, but the pattern of the body plan appears unaffected (Trumpp et al. 2001). Hence, interrelated with growth is the ability to pattern groups of cells during development (Davidson et al. 1995). Pattern formation, as defined by Davidson et al. (1995), consists of the regulatory processes required to partition undifferentiated regions of an embryo into areas of specific morphogenetic fate, and it is the depth of the gene regulatory networks responsible for pattern formation that delays the specification of cells and thereby allows for growth (Davidson 2001). Arguably the best-known pattern formation genes are the Hox genes, which play a crucial role in controlling cell proliferation in a position-dependent manner (Duboule 1995). Although a link between Hox genes and proliferation was apparent from studies on carcinomas (Carè et al. 1996; Naora et al. 2001; see Abate-Shen 2002 for review), it was not apparent how a developmental regulatory gene interacts with the cell cycle. Recently, Luo et al. (2004) showed that one link is the protein geminin. Geminin controls cell replication by interacting with an essential component of the replication complex, Cdt1: binding with Cdt1 inhibits DNA replication and hence cell division, whereas the degradation of geminin allows Cdt1 to assemble with several other proteins and results in replication initiation, and ultimately cell proliferation (Li and Rosenfeld 2004). Luo et al. (2004) also demonstrated that geminin interacts with Hox genes as well, which displaces it from Cdt1 and thus promotes cell division. In addition, geminin can antagonize Hox gene function by displacing it from target sites of downstream genes. Therefore, important developmental regulators can control the cell cycle by preventing DNA replication, and a cell cycle regulator can affect development by inhibiting the DNA binding of transcription factors, and all of this can be mediated by the competitive regulation of a single protein.

Using the presence of a Hox complex as a proxy for the evolution of pattern formation suggests that pattern formation is primitive for at least Bilateria, given that the planula stage of the anthozoan *Nematostella* expresses Hox genes in a colinear manner along the oral-aboral axis of the larva (Finnerty et al. 2004). Curiously, to date no authentic Hox genes have been reported from ctenophores or sponges (Martindale et al. 2002) despite the presence of MetaHox genes such as *Tlx, NK2,* and *Msx* in sponges (Coutinho et al. 2003), which were primitively linked to Hox genes in a "MegaHox" cluster (Coulier et al. 2000; Pollard and Holland 2000; Holland 2001). Whether this means that the tandem duplications that gave rise to Hox genes from a MetaHox gene had not yet occurred, or instead that sponges lost all Hox genes, or simply that they are not easily amplified under standard PCR conditions, is unknown and will remain so at least until a sponge genome is sequenced. Nonetheless, their apparent absence in ctenophores as well suggests that pattern formation, as assessed by a cluster of collinearly expressed Hox genes, is restricted to bilaterians, and might be the innovation largely responsible for the rapid diversification of the bewildering array of bilaterian body plans during the late Precambrian and Early Cambrian (Davidson et al. 1995).

Summary.—All available data suggest that the earliest eumetazoans were direct developing, micrometazoan benthic predators (e.g., Boaden 1989; Fortey et al. 1996, 1997; Jondelius et al. 2002; Valentine and Collins 2000; Sly et al. 2003). The invention of pattern formation as a form of growth control during development is the key innovation that ultimately allowed for the evolution of macroscopic bilaterian body plans during the Ediacaran Period, and it is our contention that the evolution of macrophagy is what drove this increase in body size. It is inevitable, then, that large, indirectly developing skeletonized animals would appear within a few million years after the evolution of macrophagy, and it is no surprise that the first appearances of skeletons at 545 Ma (Amthor et al. 2003) follow, rather than precede, the large, but simple, traces first recorded at 555 Ma (Martin et al. 2000), as predicted by Stanley (1976a,b).

Rocks: External Trigger(s)

> But . . . the transformation of ten tiny worms into the larger and well differentiated *Baupläne* of Cambrian phyla suggests an external trigger . . . [and] the melting of a "snowball earth" sometime before the Cambrian transition may well represent [a] plausible environmental trigger. (Gould 2002)

Thus far, we have argued that the critical temporal window for understanding early bilaterian evolution is between 650 and 550 Ma. Furthermore, we identified a potential candidate for an "internal trigger" necessary for the rapid rise of macrophagic predators by the late Early Cambrian, namely pattern formation processes (Davidson et al. 1995). Of course, one cannot divorce genetic innovations from the environmental problems those innovations are developed to solve, and there is no lack of dramatic, global environmental changes in the terminal Neoproterozoic interval. In addition to the possibility that our planet was engulfed in ice for 5–10 million years at a time, perhaps several times during the latter half of the Neoproterozoic Era (Hoffman et al. 1998), one of the largest recognized bolide impact events in the entire geological record occurred within this same time period as well (Grey et al. 2003; Williams and Wallace 2003). Furthermore, paleomagnetic data suggest very rapid changes in global paleogeography, possibly involving episodes of true polar wander (e.g., Kirschvink et al. 1997; Evans 2003), and there appears to have been a significant increase in the oxidation state of the ocean system (Knoll 2003). The following discussion considers the evolutionary consequences of "snowball Earth" and the rise of atmospheric oxygen, because these two factors can be most easily implicated in generating high-level taxonomic innovations, although clearly a catastrophic impact and dramatic changes in ocean circulation must have had some effect upon Earth's late Precambrian and Cambrian biota as well (e.g., Kirschvink and Raub 2003).

Snowball Earth.—Distinctive hallmarks of glaciation pepper the late Neoproterozoic stratigraphic record of nearly every craton, among all present continents. Direct juxtapositions of glaciogenic diamictites with carbonate rocks, as well as numerous paleomagnetic data indicating near-equatorial paleolatitudes of deposition (reviewed by Evans 2000), suggest that the Neoproterozoic ice ages were global affairs (reviewed by Hoffman and Schrag 2002). Direct age constraints on glaciogenic deposits have only recently begun to provide precise relationships between the number of glaciations and the deduced timing of metazoan divergences as assessed by the molecular clock (see above). The number of Neoproterozoic ice ages has been considered to be as few as two and as many as five (e.g., Kaufman et al. 1997; Kennedy et al. 1998; Evans 2000; Hoffman and Schrag 2002), but this paper will simplify the debate by discussing only those with likely ages younger than 650 Ma, and thus relevant with respect to bilaterian evolution as assessed by the molecular clock (Fig. 2).

Of particular relevance for this discussion is the age of the true Marinoan glaciation interval, that is, defined by rocks of the near-equatorial Elatina Formation (Sohl et al. 1999, and references therein) and correlative units within the Marinoan chronostratigraphic series that is restricted to Australia. This ice age occurred between ca. 650 and 580 Ma, according to detrital zircons from an underlying arkose in South Australia and igneous zircons from subsequent volcanism on King Island, Tasmania (Ireland et al. 1998; Calver et al. 2004). It is important to emphasize the exact age constraints on this particular group of deposits, free of extrabasinal correlations, because they show the best evidence for deep tropical paleolatitudes (reviewed by Evans 2000)—other paleomagnetic determinations of tropical glaciogenic deposits (e.g., Macouin et al. 2004) suffer from the absence of tightly constrained field stability tests that are needed to prove a primary origin of the magnetizations. However, several non-Australian deposits that are generally correlated to the Marinoan ice age are directly dated at 635 Ma in central Namibia (Hoffmann et al. 2004) and South China (Condon et al. 2005).

Glaciogenic rocks of the Gaskiers Formation, in Newfoundland, are constrained to an age of 580 Ma (Bowring et al. 2003). However, it is uncertain whether the Gaskiers and related units in the Avalonian-Cadomian volcanic arc (Thompson and Bowring 2000) correlate with the Marinoan ice age, or if they instead represent a younger glaciation of merely regional extent (see Myrow and Kaufman 1999; Evans 2000; Xiao et al. 2004; Condon et al. 2005). Regardless of the specific correlations, most of the world's youngest Proterozoic glaciogenic deposits can be grouped into two episodes, dated at 635 and 580 Ma.

In addition to the uncertainty surrounding both the number of glacial episodes and their absolute ages, the effort to characterize a snowball event stretches the notion of climate change to the limits of imagination. The most complete model thus far advanced includes a "freeze-fry" scenario of extreme warming following the buildup of greenhouse gases needed to escape the snowball state (Hoffman et al. 1998; but see Pierrehumbert 2004). Syn-snowball accumulation of those greenhouse gases (mainly CO_2) would have been achieved via volcanic outgassing and highly attenuated chemical weathering of silicate rocks, the atmosphere's major long-term (and strongly temperature-dependent) CO_2 sink (Walker et al. 1981; Kirschvink 1992). A more conservative view of the Neoproterozoic ice ages in terms of Pleistocene-like glacial-interglacial cycles, but with recognition that the deposits encroached deep within the Tropics, has been dubbed the "slushball" model. Although supported by the glacial-interglacial sedimentary sequences (Condon et al. 2002; Leather et al. 2002), and some numerical climate models (e.g., Hyde et al. 2000), this perspective cannot explain the reappearance of banded-iron formations and the existence of cap carbonates with their ubiquitous carbon-isotopic excursions (Schrag and Hoffman 2001).

Each model hypothesizes different effects upon the biota. Biological refugia within a "slushball" ice age are easy to imagine given that the deep Tropics would have included vast expanses of open seawater (Hyde et al. 2000). Full-blown snowball ice ages would harbor more subtle refugia. Submarine hydrothermal vents at the mid-ocean ridges would provide a continuous source of energy and nutrients throughout the glaciation, permitting survival of that ecosystem at nearly modern conditions. Within the ice, ablation from the surface would be balanced by a continuous flux of basal freezing to maintain steady-state thickness (modified by equatorward flow; see below). In this way, nutrients such as hydrothermal iron could reach the photic zone where bacteria, algae, and metazoans could survive within brine cavities. Above the ice, long-lived continental volcanic fields could maintain hydrothermal systems for thermophilic microbes throughout the cold spell. Recent considerations of marine ice dynamics, which indicate maintenance of super-equilibrium thickness in the Tropics due to ocean-scale sea glacier flow from high latitudes (Goodman and Pierrehumbert 2004), suggest that large regions of the tropical ocean could have remained ice covered even while the ice within more protected embayments melted away as greenhouse gases accumulated over the course of several million years. Such a process would create several geographically isolated regions of warm seawater enduring throughout the late stages of a prolonged snowball regime (Halverson et al. 2004).

In sum, as long as microbes and primitive metazoans could find a refuge during the early stages of a snowball episode, then the waning stages could foster evolutionary innovation among small populations of survivors within several geographically distinct regions. Given that a widespread Marinoan glaciation occurred at 635 Ma (Hoffman et al. 2004; Condon et al. 2005), then our estimate of the origin of eumetazoans somewhere between 634 and 604 Ma could have occurred on the heels of snowball Earth. In addition, at least five lineages of living animals (silicisponges, calcisponges, ctenophores, cnidarians, and triploblasts) must have survived the Gaskiers ice age (Fig. 2), casting doubt on the global import of this glaciation. Alternatively, if the definitively low-latitude Marinoan glaciation is as young as 580 Ma (Calver et al. 2004), then a much more severe environmental perturbation is implied, but one may question how it could be possible for these five animal lineages to withstand a "hard" snowball ice age (Hoffman et al. 1998) given the plethora of feeding modes and reproductive strategies these animals utilize. Thus, the phylogenetic constraint derived from molecular-clock analyses could aid our perceptions of a snowball world and shed deeper insight into its impact upon the biota, once the number of episodes, their absolute ages, and their paleogeographic distributions are better characterized.

Oxygen.—Suggestions that rising oxygen levels permitted bilaterians to attain large size have been voiced for nearly 50 years (reviewed in Knoll and Carroll 1999; Knoll 2003). Independent evidence for an increase in atmospheric oxygen concentration during the terminal Proterozoic to Cambrian interval comes from carbon and sulfur isotope records, which suggest that oxygen levels increased dramatically after the Sturtian interval (Fig. 2). Importantly, this rise in oxygen cannot explain the origin of large bilaterians, because the relative levels necessary for increased size and complexity of eukaryotes was already in place (Canfield and Teske 1996) before 570 Ma (Colpron et al. 2002).

Nonetheless, the distribution of oxygen in the water column might have played a fundamental role in the evolutionary history of bilaterians. On the basis of stable carbon isotopes of molecular biomarkers, Logan et al. (1995) proposed that the nearly complete (sulfate-reducing) bacterial degradation of primary algal products inhibited the transport of oxygen into the deep Proterozoic ocean. This stands in contrast to Phanerozoic geochemical cycling, whereby fecal-pellet transport of reduced carbon to the deep benthos limits midwater sulfate reduction, and thus oxygenates the benthos. This model predicts that Proterozoic oceans should have been more redox-stratified than at present and is supported by isotopic data from ^{13}C (Shields et al. 1997) and ^{34}S (Shen et al. 2003) across Proterozoic paleodepth gradients. Hence, the continued increase in size of benthic animals would depend upon the increase in oxygenation provided by the carbon flux from the pelagos. Although this model provides an attractive link between our model of zooplankton evolution and the subsequent appearance of large metazoan body fossils, the utility of the biomarker stable-isotope proxy for dating the onset of fecal-pellet generation by zooplankton depends on the age constraints of the sampled sedimentary rocks, which at the moment can only be constrained to between the Early Cambrian and the late Neoproterozoic (Schaefer and Burgess 2003).

Summary.—Poor numerical age control of late Neoproterozoic sedimentary successions is presently the major hindrance to unraveling the web of conceivable external influences on early metazoan evolution. As determined by a variety of measurements, the terminal Proterozoic interval was clearly a time of Earth-system agitation. If further geochronology constrains the definitively low-latitude glaciogenic strata to 635 Ma, global glaciation could have been a driving factor in the origin and subsequent evolution of eumetazoans by giving ample opportunity for an extinction and subsequent ecological recovery event (e.g., Hoffman et al. 1998; Hoffman and Schrag 2000; Narbonne and Gehling 2003). In addition, the oxygenation of the benthos via reduced carbon derived from fecal pellets would have allowed for the evolution of increased size, a necessary cofactor for the successful evolution of indirectly developing life cycles (Olive 1985).

Internal and External Triggers: Putting It All Together

> The Cambrian explosion still requires a[n external] trigger . . . but our understanding of the geological rapidity of this most puzzling and portentous event in the evolution of animals will certainly be facilitated if the developmental prerequisites already existed in an ancestral taxon. (Gould 2002)

A key insight into the nature of the Cambrian explosion was made by Butterfield (1997): The ecological ramifications resulting from the invasion of small animals into the plankton to avoid benthic predation (Signor and Vermeij 1994) changed the pelagos from a single trophic to a multitiered ecosystem (Stanley 1973, 1976a,b). The effect of evolving mesozooplankton by ca. 510 Ma (Butterfield 1994), but possibly earlier (see above), would have been immediate, with pelagic algae evolving antipredator defenses by the early Cambrian (Butterfield 1997, 2001); such structures would have come at a cost given that, at least among modern prey, spines and other antipredatory devices are quickly lost when the predator is removed from the system (Sommer et al. 1984; Leibold 1989; Reynolds 1984). Furthermore, the advent of planktic skeletons by the end of the Cambrian (e.g., Tolmacheva et al. 2001; Pawlowski et al. 2003), in response to mesozooplankton predation pressures (and this in response to benthic predation pressures), not only changed the potential for future global glacial episodes (Ridgwell et al. 2003), but also severely reduced the potential for subsequent Burgess Shale-type deposits (Gaines 2003). Thus, the window through which Stephen Jay Gould discerned patterns and processes in early animal evolution (Gould 1989) was both created and destroyed by the same causal factors: the advent of pelagic multi-tiered food webs constructed on the heels of snowball Earth.

The evolution of mesozooplankton would have had another very important effect; they would have served as a food source for macrozooplankton and nekton, and it is their evolution that would have changed the oxygenation potential of the benthos and thus increased benthic productivity. As mentioned above, the dramatic reorganization of biogeochemical cycles that occurred sometime between the latest Neoproterozoic and Early Cambrian (Logan et al. 1995) is attributed to the evolution of pelagic metazoans whose fecal pellets would remove organic matter from the pelagos and deliver it to the benthos. Importantly, the fecal pellets of meso- and microzooplankton make little or no contribution to the sedimentary flux (Butterfield 1997; Turner 2002), and, if anything, their evolution would have decreased the amount of organic rain to the benthos as the fecal pellets of small metazoans were processed within the water column. However, the fecal pellets of macrozooplankton and nekton are large enough to settle before they are consumed in the pelagos (Turner 2002). The evolution of macrozooplankton and nekton sometime around the Precambrian/Cambrian boundary, as deduced by the data of Logan et al. (1995), is consistent with the maximal date of 543 Ma for the evolution of the cnidarian medusae as estimated by the molecular clock (Fig. 2), and the 520 Ma date for bona fide fossil macrozooplankton and nekton (Chen and Zhou 1997; Vannier and Chen 2000).

The evolutionary ramifications of mesozooplankton are not limited to the pelagos, as they strongly affected the subsequent evolution of the animals on the benthos as well, given that pelagic predators evolved from benthic ancestors to take advantage of this new and very plentiful food source, and also that benthic suspension-feeding predators evolved to capitalize on these new prey. Thus, the evolution of mesozooplankton links the eukaryotic plankton to the eukaryotic benthos and effectively establishes the "modern" or Phanerozoic ocean (Butterfield 1997). But like large size and skeletons, the evolution of mesozooplankton is itself a response to benthic predation, and hence it is the evolution of eumetazoans themselves as mobile multicellular heterotrophs near the end of the Cryogenian—possibly in response to a selective event involving global perturbations and mass extinctions—that was the ultimate cause of the Cambrian explosion, as their spectacular radiation, and the radiations of their prey (and predators) were inevitable once these animals evolved pattern formation processes. Although not by any means the last word on the subject, it is our sincere hope that the ideas presented in this paper, inspired by the perspicacious writings of Stephen Jay Gould, have brought our understanding of this singular episode in the evolutionary history of animals a bit closer in reach.

Acknowledgments

We would like to thank E. Davidson and D. Erwin for their insightful comments, and G. Budd and N. Butterfield for their very helpful reviews. K.J.P. would also like to extend his thanks to the members of the laboratory, especially J. Lyons, for their technical assistance. K.J.P. is supported by NASA-Ames and the National Science Foundation (NSF); M.A.M. is supported by the NSF; D.E. acknowledges support from the David and Lucile Packard Foundation.

Literature Cited

Abate-Shen, C. 2002. Deregulated homeobox gene expression in cancer: cause or consequence? Nature Reviews Cancer 2:777–785.

Aguinaldo, A. M. A., J. M. Turbeville, L. S. Linford, M. C. Rivera, J. R. Garey, R. A. Raff, and J. A. Lake. 1997. Evidence for a clade of nematodes, arthropods and other molting animals. Nature 387:489–493.

Aleinikoff, J. N., R. E. Zartman, M. Walter, D. W. Rankin, P. T. Lyttle, and W. C. Burton. 1995. U-Pb ages of metarhyolites of the Catoctin and Mount Rogers formations, Central and Southern Appalachians: evidence for two pulses of Iapetan rifting. American Journal of Science 295:428–454.

Amthor, J. E., J. P. Grotzinger, S. Schröder, S. A. Bowring, J. Ramezani, M. W. Martin, and A. Matter. 2003. Extinction of *Cloudina* and *Namacalathus* at the precambrian-Cambrian boundary in Oman. Geology 31:431–434.

Aris-Brosou, S., and Z. Yang. 2002. Effects of models of rate evolution on estimation of divergence dates with special reference to the metazoan 18S ribosomal RNA phylogeny. Systematic Biology 51:703–714.

———. 2003. Bayesian models of episodic evolution support a late precambrian explosive diversification of the Metazoa. Molecular Biology and Evolution 20:1947–1954.

Ayala, F. J., A. Rzhetsky, and F. J. Ayala. 1998. Origin of the metazoan phyla: molecular clocks confirm paleontological estimates. Proceedings of the National Academy of Sciences USA 95:606–611.

Baguñà, J., I. Ruiz-Trillo, J. Paps, M. Loukota, C. Ribera, U. Jondelius, and M. Riutort. 2001. The first bilaterian organisms: simple or complex? New molecular evidence. International Journal of Developmental Biology 45:S133–S134.

Benton, M. J., and F. J. Ayala. 2003. Dating the tree of life. Science 300:1698–1700.

Benus, A. P. 1988. Sedimentological context of a deep-water Ediacaran fauna (Mistaken Point Formation, Avalon Zone, eastern Newfoundland). *In* E. Landing, G. M. Narbonne, and P. M. Myrow, eds. Trace fossils, small shelly fossils and the Precambrian/Cambrian boundary. Bulletin of the New York State Museum 463:8–9.

Bishop, C. D., and B. P. Brandhorst. 2003. On nitric oxide signaling, metamorphosis, and the evolution of biphasic life cycles. Evolution and Development 5:542–550.

Blair, J. E., K. Ikeo, T. Gojobori, and S. B. Hedges. 2002. The evolutionary position of nematodes. BMC Evolutionary Biology 2:1–7.

Boaden, P. J. S. 1989. Meiofauna and the origins of the Metazoa. Zoological Journal of the Linnean Society 96:217–227.

Borchiellini, C., M. Manuel, E. Alivon, N. Boury-Esnault, J. Vacelet, and Y. Le Parco. 2001. Sponge paraphyly and the origin of Metazoa. Journal of Evolutionary Biology 14:171–179.

Bowring, S. A., and D. H. Erwin. 1998. A new look at evolutionary rates in deep time: uniting paleontology and high-precision geochronology. GSA Today 8(9):1–8.

Bowring, S. A., J. P. Grotzinger, C. E. Isachsen, A. H. Knoll, S. M. Pelechaty, and P. Kolosov. 1993. Calibrating rates of Early Cambrian evolution. Science 261:1293–1298.

Bowring, S., P. Myrow, E. Landing, J. Ramezani, and J. Grotzinger. 2003. Geochronological constraints on terminal Neoproterzoic events and the rise of metazoans. Geophysical Research Abstracts 5:13219.

Brasier, M., G. McCarron, R. Tucker, J. Leather, P. Allen, and G. Shields. 2000. New U-Pb zircon dates for the Neoproterozoic Ghubrrah glaciation and for the top of the Huqf Supergroup, Oman. Geology 28:175–178.

Bridge, D., C. W. Cunningham, R. DeSalle, and L. W. Buss. 1995. Class-level relationships in the phylum Cnidaria: molecular and morphological evidence. Molecular Biology and Evolution 12:679–689.

Bromham, L., A. Rambaut, R. Fortey, A. Cooper, and D. Penny. 1998. Testing the Cambrian explosion hypothesis by using a molecular dating technique. Proceedings of the National Academy of Sciences USA 95:12386–12389.

Brusca, R. C., and G. J. Brusca. 2002. Invertebrates, 2d ed. Sinauer, Sunderland, Mass.

Buckland-Nicks, J., and A. H. Scheltema. 1995. Was internal fertilization an innovation of early Bilateria? Evidence from sperm structure of a mollusc. Proceedings of the Royal Society of London B 261:11–18.

Budd, G. E., and S. Jensen. 2000. A critical reappraisal of the fossil record of the bilaterian phyla. Biological Reviews of the Cambridge Philosophical Society 75:253–295.

Butterfield, N. J. 1994. Burgess Shale-type fossils from a Lower Cambrian shallow-shelf sequence in northwestern Canada. Nature 369:477–479.

———. 1997. Plankton ecology and the Proterozoic-Phanerozoic transition. Paleobiology 23:247–262.

———. 2001. Ecology and evolution of Cambrian plankton. Pp.

200–216 *in* A. Y. Zhuravlev and R. Riding, eds. The ecology of the Cambrian Radiation. Columbia University Press, New York.

———. 2003. Exceptional fossil preservation and the Cambrian explosion. Integrative and Comparative Biology 43:166–177.

———. 2004. A vaucheriacean alga from the middle Proterozoic of Spitsbergen: implications for the evolution of Proterozoic eukaryotes and the Cambrian explosion. Paleobiology 30: 231–252.

Calver, C. R., L. P. Black, J. L. Everard, and D. B. Seymour. 2004. U-Pb zircon age constraints on late Neoproterozoic glaciation in Tasmania. Geology 32:893–896.

Canfield, D. E., and A. P. Teske. 1996. Late Proterozoic rise in atmospheric oxygen concentration inferred from phylogenetic and sulphur-isotope studies. Nature 382:127–132.

Carè, A., A. Silvani, E. Meccia, G. Mattia, A. Stoppacciaro, G. Parmiani, C. Peschle, and M. P. Colombo. 1996. HOXB7 constitutively activates basic fibroblast growth factor in melanomas. Molecular and Cellular Biology 16:4842–4851.

Cavalier-Smith, T., M. T. E. P. Allsopp, E. E. Chao, N. Boury-Esnault, and J. Vacelet. 1996. Sponge phylogeny, animal monophyly, and the origin of the nervous system: 18S rRNA evidence. Canadian Journal of Zoology 74:2031–2045.

Chaffee, C., and D. R. Lindberg. 1986. Larval biology of Early Cambrian molluscs: the implications of small body size. Bulletin of Marine Science 39:536–549.

Chen, J., and G.-Q. Zhou. 1997. Biology of the Chenjiang fauna. Pp. 11–105 *in* J. Chen, Y. N. Cheng, and H. V. Iten, eds. The Cambrian Explosion and the fossil record (Bulletin of the National Museum of Natural Science, Vol. 10). National Museum of Natural Science, Taichung.

Clapham, M. E., and G. M. Narbonne. 2002. Ediacaran epifaunal tiering. Geology 30:627–630.

Clapham, M. E., G. M. Narbonne, and J. G. Gehling. 2003. Paleoecology of the oldest known animal communities: Ediacaran assemblages at Mistaken Point, Newfoundland. Paleobiology 29:527–544.

Collins, A. G. 2002. Phylogeny of Medusozoa and the evolution of cnidarian life cycles. Journal of Evolutionary Biology 15: 418–432.

Collins, A. G., and J. W. Valentine. 2001. Defining phyla: evolutionary pathways to metazoan body plans. Evolution and Development 3:432–442.

Colpron, M., J. M. Logan, and J. K. Mortensen. 2002. U-Pb zircon age constraint for late Neoproterozoic rifting and initiation of the lower Paleozoic passive margin of western Laurentia. Canadian Journal of Earth Sciences 39:133–143.

Compston, W., M. S. Sambridge, R. F. Reinfrank, M. Moczydlowska, G. Vidal, and S. Claesson. 1995. Numerical ages of volcanic rocks and the earliest faunal zone within the late Precambrian of East Poland. Journal of the Geological Society of London 152:599–611.

Comptson, W., A. E. Wright, and P. Toghill. 2002. Dating the late precambrian volcanicity of England and Wales. Journal of the Geological Society, London 159:323–339.

Condon, D. J., A. R. Prave, and D. I. Benn. 2002. Neoproterozoic glacial-rainout intervals: observations and implications. Geology 30:35–38.

Condon, D., M. Zhu, S. Bowring, W. Wang, A. Yang, and Y. Jin. 2005. U-Pb ages from the Neoproterozoic Doushantuo Formation, China. Science (in press).

Conlon, I., and M. Raff. 1999. Size control in animal development. Cell 96:235–244.

Conway Morris, S. 1979. The Burgess Shale (Middle Cambrian) Fauna. Annual Review of Ecology and Systematics 10:327–349.

———. 1986. The community structure of the Middle Cambrian phyllopod bed (Burgess Shale). Palaeontology 29:423–467.

———. 1998. Eggs and embryos from the Cambrian. Bioessays 20:676–682.

Conway Morris, S., and D. H. Collins. 1996. Middle Cambrian ctenophores from the Stephen Formation, British Columbia, Canada. Philosophical Transactions of the Royal Society of London B 351:279–308.

Copley, R. R., P. Aloy, R. B. Russell, and M. J. Telford. 2004. Systematic searches for molecular synapomorphies in model metazoan genomes give some support for Ecdysozoa after accounting for the idiosyncrasies of *Caenorhabditis elegans*. Evolution and Development 6:164–169.

Coulier, F., C. Popovici, R. Villet, and D. Birnbaum. 2000. Meta*Hox* gene clusters. Journal of Experimental Biology 288: 345–351.

Coutinho, C. C., R. N. Fonseca, J. J. C. Mansure, and R. Borojevic. 2003. Early steps in the evolution of multicellularity: deep structural and functional homologies among homeobox genes in sponges and higher metazoans. Mechanisms of Development 120:429–440.

Davidson, E. H. 2001. Genomic regulatory systems: development and evolution. Academic Press, San Diego.

Davidson, E. H., K. J. Peterson, and R. A. Cameron. 1995. Origin of adult bilaterian body plans: evolution of developmental regulatory mechanisms. Science 270:1319–1325.

Dempster, T. J., G. Rogers, P. W. G. Tanner, B. J. Bluck, R. J. Muir, S. D. Redwood, T. R. Ireland, and B. A. Paterson. 2002. Timing of deposition, orogenesis and glaciation within the Dalradian rocks of Scotland: constraints from U-Pb zircon ages. Journal of the Geological Society, London 159:83–94.

de Rosa, R., J. K. Grenier, T. Andreeva, C. E. Cook, A. Adoutte, M. Akam, S. B. Carroll, and G. Balavoine. 1999. Hox genes in brachiopods and priapulids and protostome evolution. Nature 399:772–776.

Duboule, D. 1995. Vertebrate *Hox* genes and proliferation—an alternative pathway to homeosis. Current Opinion in Genetics and Development 5:525–528.

Eernisse, D. J., and K. J. Peterson. 2004. The history of animals. Pp. 197–208 *in* J. Cracraft and M. J. Donoghue, eds. Assembling the tree of life. Oxford University Press, Oxford.

Erwin, D. H., and E. H. Davidson. 2002. The last common bilaterian ancestor. Development 129:3021–3032.

Evans, D. A. D. 2000. Stratigraphic, geochronological, and paleomagnetic constraints upon the Neoproterozoic climatic paradox. American Journal of Science 300:347–433.

———. 2003. True polar wander and supercontinents. Tectonophysics 362:303–320.

Fedonkin, M. A. 2003. The origin of the Metazoa in light of the Proterozoic fossil record. Paleontological Research 7:9–41.

Fedonkin, M. A., and B. M. Waggoner. 1997. The Late Precambrian fossil *Kimberella* is a mollusc-like bilaterian organism. Nature 388:868–871.

Finnerty, J. R., K. Pang, P. Burton, D. Paulson, and M. Q. Martindale. 2004. Origins of bilateral symmetry: *Hox* and *Dpp* expression in a sea anemone. Science 304:1335–1337.

Fortey, R. A., D. E. G. Briggs, and M. A. Wills. 1996. The Cambrian evolutionary 'explosion' decoupling cladogenesis from morphological disparity. Biological Journal of the Linnean Society 57:13–33.

———. 1997. The Cambrian evolutionary 'explosion' recalibrated. Bioessays 19:429–434.

Gaines, R. R. 2003. Understanding Burgess-Shale-type preservation: new insights from the Wheeler Shale, Utah. Geological Society of America Abstracts with Programs 35: 40:7.

Giribet, G., D. L. Distel, M. Polz, W. Sterrer, and W. C. Wheeler.

2000. Triploblastic relationships with emphasis on the acoelomates and the position of Gnathostomulida, Cycliophora, Plathelminthes, and Chaetognatha: a combined approach of 18S RNDA sequences and morphology. Systematic Biology 49:539–562.

Goodman, J. C., and R. T. Pierrehumbert. 2004. Glacial flow of floating marine ice in 'Snowball Earth.' Journal of Geophysical Research (in press).

Gould, S. J. 1979. Ever since Darwin. Norton, New York.

———. 1989. Wonderful life. Norton, New York.

———. 1998. On embryos and ancestors. Natural History 107(6): 20–22, 58–65.

———. 2002. The structure of evolutionary theory. Belknap Press of, Harvard University Press, Cambridge.

Grey, K., M. R. Walter, and C. R. Calver. 2003. Neoproterozoic biotic diversification: snowball earth or aftermath of the Acraman impact. Geology 31:459–462.

Grotzinger, J. P., S. A. Bowring, B. Z. Saylor, and A. J. Kaufman. 1995. Biostratigraphic and geochronologic constraints on early animal evolution. Science 270:598–604.

Guensburg, T. E., and J. Sprinkle. 2001. Earliest crinoids: new evidence for the origin of the dominant Paleozoic echinoderms. Geology 29:131–134.

Haase, A., M. Stern, K. Wächtler, and G. Bicker. 2001. A tissue-specific marker of Ecdysozoa. Development, Genes and Evolution 211:428–433.

Hadfield, M. G. 2000. Why and how marine-invertebrate larvae metamorphose so fast. Seminars in Cell and Developmental Biology 11:437–443.

Hadfield, M. G., E. J. Carpizo-Ituarte, K. Del Carmen, and B. T. Nedv ed. 2001. Metamorphic competence, a major adaptive convergence in marine invertebrate larvae. American Zoologist 41:1123–1131.

Halanych, K. M., J. D. Bacheller, A. M. A. Aguinaldo, S. M. Liva, D. M. Hillis, and J. A. Lake. 1995. Evidence from 18S Ribosomal DNA that the lophophorates are protostome animals. Science 267:1641–1643.

Halverson, G. P., A. C. Maloof, and P. F. Hoffman. 2004. The Marinoan glaciation (Neoproterozoic) in northeast Svalbard. Basin Research 16:297–324.

Hayward, D. C., G. Samuel, P. C. Pontynen, J. Catmull, R. Saint, D. J. Miller, and E. E. Ball. 2002. Localized expression of a *dpp/BMP2/4* ortholog in a coral embryo. Proceedings of the National Academy of Sciences USA 99:8106–8111.

Hoffman, P. F., and D. P. Schrag. 2000. Snowball earth. Scientific American 282:68–75.

———. 2002. The snowball Earth hypothesis: testing the limits of global change. Terra Nova 14:129–155.

Hoffman, P. F., A. J. Kaufman, G. P. Halverson, and D. P. Schrag. 1998. A Neoproterozoic snowball earth. Science 281:1342–1346.

Hoffmann, K. H., D. J. Condon, S. A. Bowring, and J. L. Crowley. 2004. U-Pb zircon date from the Neoproterozoic Ghaub formation, Namibia: constraints on Marinoan glaciation. Geology 32:817–820.

Holland, P. W. H. 2001. Beyond the Hox: how widespread is homeobox gene clustering? Journal of Anatomy 199:13–23.

Hyde, W. T., T. J. Crowley, S. K. Baum, and R. Peltier. 2000. Neoproterozoic 'snowball Earth' simulations with a coupled climate/ice-sheet model. Nature 405:425–429.

Ireland, T. R., T. Flöttman, C. M. Fanning, G. M. Gibson, and W. V. Preiss. 1998. Development of the early Paleozoic Pacific margin of Gondwana from detrital-zircon ages across the Delamerian orogeny. Geology 26:243–246.

Jensen, S. 2003. The Proterozoic and earliest Cambrian trace fossil record: patterns, problems and perspectives. Integrative and Comparative Biology 43:219–228.

Johnson, K. B., and A. L. Shanks. 2003. Low rates of predation on planktonic marine invertebrate larvae. Marine Ecology Progress Series 248:125–139.

Johnston, L. A., and P. Gallant. 2002. Control of growth and organ size in *Drosophila*. Bioessays 24:54–64.

Jondelius, U., I. Ruiz-Trillo, J. Baguñà, đ M. Riutort. 2002. The Nemertodermatida are basal bilaterians and not members of the Platyhelminthes. Zoologica Scripta 31:201–215.

Kaufman, A. J., and A. H. Knoll. 1995. Neoproterozoic variations in the C-isotopic composition of seawater: stratigraphic and biogeochemical implications. Precambrian Research 73: 27–49.

Kaufman, A. J., A. H. Knoll, and G. M. Narbonne. 1997. Isotopes, ice ages, and terminal Proterozoic earth history. Proceedings of the National Academy of Sciences USA 94:6600–6605.

Kennedy, M. J., B. Runnegar, A. R. Prave, K. H. Hoffmann, and M. A. Arthur. 1998. Two or four Neoproterozoic glaciations? Geology 26:1059–1063.

Kirschvink, J. L. 1992. Late Proterozoic low-latitude global glaciation: the snowball Earth. Pp. 51–52 *in* J. W. Schopf and C. C. Klein, eds. The Proterozoic biosphere: a multidisciplinary study. Cambridge University Press, Cambridge.

Kirschvink, J. L., and T. D. Raub. 2003. A methane fuse for the Cambrian explosion: carbon cycles and true polar wander. Comptes Rendus de l'Académie des Sciences (Géosciences) 335:65–78.

Kirschvink, J. L., R. L. Ripperdan, and D. A. Evans. 1997. Evidence for a large-scale reorganization of early Cambrian continental masses by inertial interchange true polar wander. Science 277:541–545.

Knoll, A. H. 2000. Learning to tell Neoproterozoic time. Precambrian Research 100:3–20.

———. 2003. Life on a young planet. Princeton University Press, Princeton.

Knoll, A. H., and S. B. Carroll. 1999. Early animal evolution: emerging views from comparative biology and geology. Science 284:2129–2137.

Knoll, A. H., M. R. Walter, G. M. Narbonne, and N. Christie-Blick. 2004. A new period for the geologic time scale. Science 305:621–622.

Landing, E. 1994. Precambrian-Cambrian boundary global stratotype ratified and a new perspective of Cambrian time. Geology 22:179–182.

Landing, E., S. A. Bowring, K. L. Davidek, S. R. Westrop, G. Geyer, and W. Heldmaier. 1998. Duration of the Early Cambrian: U-Pb ages of volcanic ashes from Avalon and Gondwana. Canadian Journal of Earth Science 35:329–338.

Landing, E., S. A. Bowring, K. L. Davidek, A. W. A. Rushton, R. A. Fortey, and A. P. Wibledon. 2000. Cambrian-Ordovician boundary age and duration of the lowest Ordovician Tremadoc series based on U-Pb zircon dates from Avalonian Wales. Geological Magazine 137:485–494.

Leather, J., P. A. Allen, M. D. Brasier, and A. Cozzi. 2002. Neoproterozoic snowball Earth under scrutiny: evidence from the Fiq glaciation of Oman. Geology 30:891–894.

Leibold, M. A. 1989. Resource edibility and the effects of predators and productivity on the outcome of trophic interactions. American Naturalist 134:922–949.

Levinton, J. S. 2001. Genetics, paleontology, and macroevolution, 2d ed. Cambridge University Press, Cambridge.

Li, C.-W., J. Chen, and T.-E. Hua. 1998. Precambrian sponges with cellular structures. Science 279:879–882.

Li, X., and M. G. Rosenfeld. 2004. Origins of licensing control. Nature 427:687–688.

Lindberg, D. R., and R. P. Guralnick. 2003. Phyletic patterns of early development in gastropod molluscs. Evolution and Development 5:494–507.

Logan, G. A., J. M. Hayes, G. B. Hieshima, and R. E. Summons. 1995. Terminal Proterozoic reorganization of biogeochemical cycles. Nature 376:53–56.

Luo, L., X. Yang, Y. Takihara, H. Knoetgen, and M. Kessel. 2004. The cell-cycle regulator geminin inhibits Hox function through direct and polycomb-mediated interactions. Nature 427:749–753.

MacGinitie, G. E. 1934. The egg-laying activities in the sea hare *Tethys californicus* (Cooper). Biological Bulletin 67:300–303.

Macouin, M., J. Besse, M. Ader, S. Gilder, Z. Yang, Z. Sun, and P. Agrinier. 2004. Combined paleomagnetic and isotopic data from the Doushantuo carbonates, South China: implications for the "snowball Earth" hypothesis. Earth and Planetary Science Letters 224:387–398.

Mallatt, J., and C. J. Winchell. 2002. Testing the new animal phylogeny: first use of combined large-subunit and small-subunit rRNA gene sequences to classify the protostomes. Molecular Biology and Evolution 19:289–301.

Mallatt, J. M., J. R. Garey, and J. W. Shultz. 2004. Ecdysozoan phylogeny and Bayesian inference: first use of nearly complete 28S and 18S gene sequences to classify the arthropods and their kin. Molecular Phylogenetics and Evolution 31:178–191.

Manuel, M., C. Borchiellini, E. Alivon, Y. Le Parco, J. Vacelet, and N. Boury-Esnault. 2003. Phylogeny and evolution of calcareous sponges: monophyly of Calcinea and Calcaronea, high level of morphological homoplasy, and the primitive nature of axial symmetry. Systematic Biology 52:311–333.

Martin, M. W., D. V. Grazhdankin, S. A. Bowring, D. A. D. Evans, M. A. Fedonkin, and J. L. Kirschvink. 2000. Age of Neoproterozoic bilaterian body and trace fossils, White Sea, Russia: implications for metazoan evolution. Science 288:841–845.

Martindale, M. Q., J. R. Finnerty, and J. Q. Henry. 2002. The Radiata and the evolutionary origins of the bilaterian body plan. Molecular Phylogenetics and Evolution 24:358–365.

Martindale, M. Q., K. Pang, and J. R. Finnerty. 2004. Investigating the origins of triploblasty: 'mesodermal' gene expression in a diploblastic animal, the sea anemone *Nematostella vectensis* (Phylum Cnidaria; class, Anthozoa). Development 131: 2463–2474.

McHugh, D., and G. W. Rouse. 1998. Life history evolution of marine invertebrates: new views from phylogenetic systematics. Trends in Ecology and Evolution 13:182–186.

Medina, M., A. G. Collins, J. D. Silberman, and M. L. Sogin. 2001. Evaluating hypotheses of basal animal phylogeny using complete sequences of large and small subunit rRNA. Proceedings of the National Academy of Sciences USA 98:9707–9712.

Myrow, P. M., and A. J. Kaufman. 1999. A newly discovered cap carbonate above Varanger-age glacial deposits in Newfoundland, Canada. Journal of Sedimentary Research 69:784–793.

Naora, H., Y. Yang, F. J. Montz, J. D. Seidman, R. J. Kurman, and R. B. S. Roden. 2001. A serologically identified tumor antigen encoded by a homeobox gene promotes growth of ovarian epithelial cells. Proceedings of the National Academy of Sciences USA 98:4060–4065.

Narbonne, G. M. 2004. Modular construction of early Ediacaran complex life forms. Science 305:1141–1144.

Narbonne, G. M., and J. G. Gehling. 2003. Life after snowball: the oldest complex Ediacaran fossils. Geology 31:27–30.

Nei, M., P. Xu, and G. Glazko. 2001. Estimation of divergence times from multiprotein sequences for a few mammalian species and several distantly related organisms. Proceedings of the National Academy of Sciences USA 98:2497–2502.

Nielsen, C. 1998. Origin and evolution of animal life cycles. Biological Reviews of the Cambridge Philosophical Society 73: 125–155.

Nijhout, H. F. 2003. The control of body size in insects. Developmental Biology 261:1–9.

Nützel, A., and J. Fryda. 2003. Paleozoic plankton revolution: evidence from early gastropod ontogeny. Geology 31:829–831.

Olive, P. J. W. 1985. Covariability of reproductive traits in marine invertebrates: implications for the phylogeny of the lower invertebrates. Pp. 42–59 *in* S. Conway Morris, D. George, R. Gibson, and H. M. Platt, eds. The origins and relationships of lower invertebrates.Clarendon, Oxford.

Pasquinelli, A. E., A. McCoy, E. Jiménez, E. Saló, G. Ruvkun, M. Q. Martindale, and J. Baguñà. 2003. Expression of the 22 nucleotide *let-7* heterochronic RNA throughout the Metazoa: a role in life history evolution? Evolution and Development 5: 372–378.

Pawlowski, J., M. Holzmann, C. Berney, J. Fahrni, A. J. Gooday, T. Cedhagen, A. Habura, and S. S. Bowser. 2003. The evolution of early Foraminifera. Proceedings of the National Academy of Sciences USA 100:11494–11498.

Pechenik, J. A. 1979. Role of encapsulation in invertebrate life histories. American Naturalist 114:859–870.

———. 1999. On the advantages and disadvantages of larval stages in benthic marine invertebrate life cycles. Marine Ecology Progress Series 177:269–297.

Peterson, K. J., and D. J. Eernisse. 2001. Animal phylogeny and the ancestry of bilaterians: inferences from morphology and 18S rDNA gene sequences. Evolution and Development 3: 170–205.

Peterson, K. J., R. A. Cameron, and E. H. Davidson. 1997. Set-aside cells in maximal indirect development: evolutionary and developmental significance. Bioessays 19:623–631.

———. 2000. Bilaterian origins: significance of new experimental observations. Developmental Biology 219:1–17.

Peterson, K. J., B. Waggoner, and J. W. Hagadorn. 2003. A fungal analog for Newfoundland Ediacaran fossils. Integrative and Comparative Biology 43:127–136.

Peterson, K. J., J. B. Lyons, K. S. Nowak, C. M. Takacs, M. J. Wargo, and M. A. McPeek. 2004. Estimating metazoan divergence times with a molecular clock. Proceedings of the National Academy of Sciences USA 101:6536–6541.

Pierrehumbert, R. T. 2004. High levels of atmospheric carbon dioxide necessary for the termination of global glaciation. Nature 429:646–649.

Pollard, S. L., and P. W. Holland. 2000. Evidence for 14 homeobox gene clusters in human genome ancestry. Current Biology 10:1059–1062.

Reynolds, C. S. 1984. The ecology of freshwater phytoplankton. Cambridge University Press, New York.

Ridgwell, A. J., M. J. Kennedy, and K. Caldeira. 2003. Carbonate deposition, climate stability, and Neoproterozoic ice ages. Science 302:859–862.

Rigby, S., and C. Milsom. 1996. Benthic origins of zooplankton: an environmentally determined macroevolutionary event. Geology 24:52–54.

———. 2000. Origins, evolution, and diversification of zooplankton. Annual Review of Ecology and Systematics 31:293–313.

Rouse, G. W. 2000. The epitome of hand waving? Larval feeding and hypotheses of metazoan phylogeny. Evolution and Development 2:222–233.

Ruiz-Trillo, I., M. Riutort, D. T. J. Littlewood, E. A. Herniou, and J. Baguñà. 1999. Acoel flatworms: earliest extant bilaterian metazoans, not members of Platyhelminthes. Science 283: 1919–1923.

Ruiz-Trillo, I., J. Paps, M. Loukota, C. Ribera, U. Jondelius, J. Baguñà, and M. Riutort. 2002. A phylogenetic analysis of myosin heavy chain type II sequences corroborates that Acoela and Nemertodermatida are basal bilaterians. Proceedings of the National Academy of Sciences USA 99:11246–11251.

Runnegar, B. 1982. The Cambrian explosion: animals or fossils? Journal of the Geological Society of Australia 29:395–411.

Schaefer, B. F., and J. M. Burgess. 2003. Re-Os isotopic age constraints on deposition in the Neoproterozoic Amadeus Basin: implications for the 'Snowball Earth.' Journal of the Geological Society, London 160:825–828.

Schrag, D. P., and P. F. Hoffman. 2001. Life, geology and snowball earth. Nature 409:306.

Scholtz, C. B., and U. Technau. 2003. The ancestral role of Brachyury: expression of NemBral in the basal Cnidarian *Nematostella Vectensis* (Anthozoa). Development, Genes and Evolution 212:563–570.

Shen, Y., A. H. Knoll, and M. R. Walter. 2003. Evidence for low sulphate and anoxia in a mid-Proterozoic marine basin. Nature 423:632–635.

Shields, G., P. Stille, M. D. Brasier, and N.-V. Atudorei. 1997. Stratified oceans and oxygenation of the late Precambrian environment: a post glacial geochemical record from the Neoproterozoic of W. Mongolia. Terra Nova 9:218–222.

Signor, P. W., and G. J. Vermeij. 1994. The plankton and the benthos: origins and early history of an evolving relationship. Paleobiology 20:297–319.

Sly, B. J., M. S. Snoke, and R. A. Raff. 2003. Who came first—larvae or adults? Origins of bilaterian metazoan larvae. International Journal of Developmental Biology 47:623–632.

Smith, A. B. 1988. Patterns of diversification and extinction in early Palaeozoic echinoderms. Palaeontology 31:799–828.

Smith, A. B., and K. J. Peterson. 2002. Dating the time of origin of major clades: molecular clocks and the fossil record. Annual Review of Earth and Planetary Sciences 30:65–88.

Sohl, L. E., N. Christie-Blick, and D. V. Kent. 1999. Paleomagnetic polarity reversals in Marinoan (ca. 600 Ma) glacial deposits of Australia: implications for the duration of low-latitude glaciation in Neoproterozoic time. Geological Society of America Bulletin 111:1120–1139.

Sommer, U., Z. M. Gliwicz, W. Lampert, and A. Duncan. 1984. The PEG-model of seasonal succession of planktonic events in fresh waters. Archiv für Hydrobiologie 106:433–471.

Stanley, S. M. 1973. An ecological theory for the sudden origin of multicellular life in the late precambrian. Proceedings of the National Academy of Sciences USA 70:1486–1489.

———. 1976a. Fossil data and the precambrian-Cambrian evolutionary transition. American Journal of Science 276:56–76.

———. 1976b. Ideas on the timing of metazoan diversification. Paleobiology 2:209–219.

Steele, R. E., P. Lieu, N. H. Mai, M. A. Shenk, and M. P. J. Sarras. 1996. Response to insulin and the expression pattern of a gene encoding an insulin receptor homologue suggest a role for an insulin-like molecule in regulating growth and patterning in *Hydra*. Development, Genes and Evolution 206:247–259.

Steiner, M., and J. Reitner. 2001. Evidence of organic structures in Ediacara-type fossils and associated microbial mats. Geology 29:1119–1122.

Strathmann, R. R. 1987. Larval feeding. Pp. 465–550 *in* A. C. Giese, J. S. Pearse, and V. B. Pearse, eds. Reproduction of marine invertebrates, Vol. IX. Blackwell Scientific, Palo Alto; Boxwood Press, Pacific Grove, Calif.

———. 1993. Hypothesis on the origins of marine larvae. Annual Review of Ecology and Systematics 24:89–117.

Swofford, D. L. 2002. PAUP* Phylogenetic Analysis Using Parsimony (* and Other Methods) v. 4.0b10 for Macintosh. Sinauer, Sunderland, Mass.

Telford, M. J., A. E. Lockyer, C. Cartwright-Finch, and D. T. J. Littlewood. 2003. Combined large and small subunit ribosomal RNA phylogenies support a basal position of the acoelomorph flatworms. Proceedings of the Royal Society of London B 270:1077–1083.

Thompson, M. D., and S. A. Bowring. 2000. Age of the Squantum "Tillite" Boston Basin, Massachusetts: U-PB zircon constraints on terminal Neoproterozoic glaciation. American Journal of Science 300:630–655.

Tolmacheva, T. J., T. Danelian, and L. E. Popov. 2001. Evidence for 15 m.y. of continuous deep-as biogenic siliceous sedimentation in early Paleozoic oceans. Geology 29:755–758.

Trumpp, A., Y. Rafaeli, T. Oskarsson, S. Gasser, M. Murphy, G. R. Martin, and J. M. Bishop. 2001. c-Myc regulates mammalian body size by controlling cell number not cell size. Nature 414:768–773.

Turner, J. T. 2002. Zooplankton fecal pellets, marine snow and sinking phytoplankton blooms. Aquatic Microbial Ecology 27: 57–102.

Vacelet, J., and N. Boury-Esnault. 1995. Carnivorous sponges. Nature 373:333–335.

Valentine, J. W., and A. G. Collins. 2000. The significance of moulting in ecdysozoan evolution. Evolution and Development 2:152–156.

Valentine, J. W., D. Jablonski, and D. H. Erwin. 1999. Fossils, molecules and embryos: new perspectives on the Cambrian explosion. Development 126:851–859.

Vannier, J., and J.-Y. Chen. 2000. The Early Cambrian colonization of pelagic niches exemplified by *Isoxys* (Arthropoda). Lethaia 33:295–311.

Walker, J. C. G., P. B. Hays, and J. F. Kasting. 1981. A negative feedback mechanism for the long-term stabilization of Earth's surface. Journal of Geophysical Research 86:9776–9782.

Wang, D. Y.-C., S. Kumar, and S. B. Hedges. 1999. Divergence time estimates for the early history of animal phyla and the origin of plants, animals and fungi. Proceedings of the Royal Society of London B 266:163–171.

Williams, G. C. 1966. Adaptation and natural selection: a critique of some current evolutionary thought. Princeton University Press, Princeton, N.J.

Williams, G. E., and M. W. Wallace. 2003. The Acraman asteroid impact, South Australia: magnitude and implications for the late Vendian environment. Journal of the Geological Society, London 160:545–554.

Wolf, Y. I., I. B. Rogozin, and E. V. Koonin. 2004. Coelomata and not Ecdysozoa: evidence from genome-wide phylogenetic analysis. Genome Research 14:29–36.

Wray, G. A., J. S. Levinton, and L. H. Shapiro. 1996. Molecular evidence for deep Precambrian divergences among metazoan phyla. Science 274:568–573.

Xiao, S., Y. Zhang, and A. H. Knoll. 1998. Three-dimensional preservation of algae and animal embryos in a Neoproterozoic phosphorite. Nature 391:553–558.

Xiao, S., X. Yuan, M. Steiner, and A. H. Knoll. 2002. Macroscopic carbonaceous compressions in a terminal Proterozoic shale: a systematic reassessment of the Miaohe biota, south China. Journal of Paleontology 76:347–376.

Xiao, S., H. Bao, H. Wang, A. J. Kaufman, C. Zhou, G. Li, X. Yuan, and H. Ling. 2004. The Neoproterozoic Quruqtagh Group in eastern Chinese Tianshan: evidence for a post-Marinoan glaciation. Precambrian Research 130:1–26.

Young, C. M., and F.-S. Chia. 1987. Abundance and distribution of pelagic larvae as influenced by predation, behavior, and hydrographic factors. Pp. 385–463 *in* A. C. Giese, J. S. Pearse and V. B. Pearse, eds. Reproduction of marine invertebrates, Vol. IX. Blackwell Scientific, Palo Alto; Boxwood Press, Pacific Grove, Calif.

Zhou, C., R. Tucker, S. Xiao, Z. Peng, X. Yuan, and Z. Chen. 2004. New constraints on the ages of the Neoproterozoic glaciations in south China. Geology 32:437–440.

Zrzavy, J., S. Mihulka, P. Kepka, A. Bezdek, and D. Tietz. 1998. Phylogeny of the Metazoa based on morphological and 18S ribosomal DNA evidence. Cladistics 14:249–285.

Paleobiology, 31(2), 2005, pp. 56–76

The competitive Darwin

Hugh Paterson

Abstract.—Although Darwin was not the first to conceive directional selection as a mechanism of phenotypic change, it is his ideas that were received, and that have shaped population biology to this day. A significant change in his theoretical orientation occurred in the mid-1850s. About then he abandoned environmental selection in favor of competitive selection, and adopted relative adaptation with all its consequences as an alternative. These ideas changed his thinking fundamentally and shaped his argument throughout the writing of his great book. It is still these ideas that predominate today.

Here I examine Darwin's ideas in relation to his principle of divergence, sexual selection, and the nature and origin of species. Finally I suggest that had he not misunderstood the function of sexual communication he might well have understood the nature of species and provided a more penetrating resolution to Herschell's "mystery of mysteries," with which he opened his book.

Hugh Paterson. Department of Zoology and Entomology, University of Queensland, St. Lucia, Brisbane, Queensland, Australia. E-mail: h.paterson@mailbox.uq.edu.au

Accepted: 12 September 2004

Introduction

In 1859, *On the Origin of Species* was a revolutionary book that explicated much that had puzzled thinkers since the days of the Greek philosophers. But today the Origin is not particularly widely pored over by most contemporary biologists, which raises the question, To what extent are we dependent on the principal ideas in Darwin's famous book? My aim in this essay is to look again at three ideas that Darwin identified as central. Darwin's concern with the origin of species by natural selection is proclaimed in the title, yet privately Darwin revealed to his publisher, John Murray, that the keystone of his book was his concept of natural selection and his principle of divergence.

Darwin was notoriously secretive with his ideas. His principal confidante and sounding board was Joseph Hooker. Particularly after 1856, Sir Charles Lyell acted as both colleague and mentor. By 1855 Darwin was a well-established and respected leader in British zoology. He was a Fellow of the Royal Society; he had published his "journal of researches" from his voyage around the world aboard H.M.S. *Beagle*, books on the structure and distribution of coral reefs, on volcanic islands, the geology of South America, a pioneering study in four volumes on the barnacles, besides editing the five volumes of the *Zoology of the Voyage of the Beagle*. However, despite having opened his first notebook on the transmutation of species in July 1837, and outlined his ideas on evolution in a pencil "sketch" in 1842, and in a much longer "essay" in 1844 (*in* de Beer 1958), he had not yet published a paper on evolutionary theory.

In September 1855 Wallace, Darwin's junior by 14 years, published an exceptional essay, *On the Law Which Has Regulated the Introduction of New Species,* which Lyell read in November 1855. The impact of Wallace's paper on Lyell was such that he at once opened his very first notebook devoted to "the species problem," some 23 years after writing his influential second volume of his *Principles of Geology.* The first entries in this notebook (Wilson 1970: p. 3) were comments on Wallace's views.

Lyell's first opportunity to discuss Wallace's paper with Darwin came in April 1856 when Sir Charles and Lady Lyell were the guests of the Darwins at Down House. During this visit Darwin let Lyell into the secret of his concept of natural selection. Lyell recorded the following perceptive comment in his species notebook on April 16: "The struggle for existence against other species is more serious than against changes of climate and physical geography."

0094-8373/05/3102-0005/$1.00

More than a century later, Darwin's inclination to regard competition as the principal driving force of evolutionary change was again commented on by Alexander Nicholson (1960), who contrasted Darwin's view with Wallace's. Wallace (1858), he found, favored "environmental selection," which, as a leading advocate of competitive selection, Nicholson regarded as unenlightened. On this point historian of science, Malcolm Kottler (1985), after a detailed study, agreed in essence.

Dov Ospovat (1981: p. 197) noted that Darwin, in his early notebooks, and in his "sketch" of 1842 and his expanded "essay" of 1844, was conventional in advocating environmental selection and in favoring "perfect adaptation," a deist relic. But after 1856, Ospovat pointed out, Darwin advocated competitive selection, which obliged him to recognize that adaptation was generally no more perfect than was needed to resist local competitors (Darwin 1859: pp. 201, 472). Although Darwin maintained this position thereafter, ghostly traces of his past views persisted in the Origin of Species through its six editions.

During his visit to Downe, Lyell had warned Darwin of the danger of Wallace's anticipating his ideas, and had urged him to publish an outline of his views. However, Darwin preferred to start a large work that he entitled *Natural Selection,* which was aborted when Wallace's second paper arrived at Downe from Ternate in the Moluccas in June 1858. An "abstract" of *Natural Selection* appeared on 24 November 1859, as *On the Origin of Species by Means of Natural Selection*.

In this essay I shall consider, rather briefly, Darwin's stand on natural selection, the principle of divergence, and the origin of species, before considering the consequences they have had for evolutionary biology. In particular I shall attempt to determine why, after the 1844 essay, he switched to competitive selection after his early commitment to environmental selection, and what this change led to.

Natural Selection

Preliminary Orientation.—It is perhaps desirable to introduce at this early point a few comments that will help relate Darwin's ideas to our times. Bruce Wallace (1981: p. 216) has provided a neat and informative encapsulation of natural selection that is appropriate for my present purpose: "Survival and reproductive success are the ingredients of natural selection. Selection does *not cause* differential survival and reproduction; selection *is* differential survival and reproduction." No mention is made here of the ways in which survival and reproduction are enhanced, largely because there are many, just as there are many causes of death. Nevertheless, the basic facts are that to be eligible to contribute to a succeeding generation an organism must be alive and must be fertile at maturity. Fertility means that an organism is capable of propagating; fecundity relates to the number of offspring an organism is actually parent to. There are complexities here that we shall need to examine in due course. Natural selection occurs in two principal forms in nature: stabilizing selection, which leads to the stabilization of the phenotype, and directional selection, which results in change in the phenotype. The following comments by well-known population geneticists provide further important insights. Firstly, Crow and Kimura (1970: p. 255) remarked,

> In a natural population only a fraction, and surely only a very small fraction, of the selection is effective in causing the systematic changes in gene frequencies that we think of as evolution. . . . Thus, most genetic selection goes to maintaining the *status quo*, rather than to making progressive evolutionary changes.

And Roger Milkman (1982: p. 105) later said much the same thing in other words: "The main day-to-day effect of natural selection is the maintenance of the *status quo*, the stabilization of the phenotype. To a relatively small directional residue, we attribute the great panorama of evolution."

Of course, these are very broad statements. For example, no reference is made to the environment in which the organisms are found in nature. It is obvious that conditions in an organism's environment generally determine survivorship. But the natural environment of a species varies both daily and seasonally, and superimposed on these are stochastic variations.

When population geneticists write of stabilizing selection they are generally referring to a particular character, such as human birth weight in relation to infant survival as in the classical study made by Karn and Penrose (1951). On the other hand, the stability of phenotype in evolutionary biology refers to the whole organism. It is this phenotypic stability that makes possible the procedures of taxonomy, for example.

The complexities of phenotypic stability are considerable and are receiving growing appreciation (e.g., Raff 1996; Rutherford and Lindquist 1998; Schlichting and Pigliucci 1998). In 1968 Waddington wrote, "The modifications of the character of phenotypes by environmental effects cannot be left out of account, since perhaps the major problem of the whole evolutionary theory is to account for the adaptation of phenotypes to environments" (p. 40).

An organism's restriction to particular environmental conditions characteristic of its species provides us with evidence of adaptation past, and of current stabilizing selection. The way an organism relates to its specific environment involves structural, physiological, and behavioral characteristics.

In the Beginning.—It was stabilizing selection that was familiar to many writers from Lucretius to Lyell and Blyth. Deists, such as Lyell and Blyth, accepted stabilizing selection because it was seen to be a process that preserved the work of the Creator. For example, Blyth (1835: pp. 48–49) wrote as follows:

> It may not be impertinent to remark here, that, as in the brute creation, by a wise provision, the typical characters of a species are, in a state of nature, preserved by those individuals chiefly propagating, whose organization is the most perfect, and which, consequently, by their superior energy and physical powers, are enabled to vanquish and drive away the weak and sickly.

Charles Darwin, we can be sure, encountered such clear statements of stabilizing selection when he read the second volume of Lyell's *Principles of Geology*, Blyth's two principal papers (Blyth 1835, 1837), and de Candolle, as quoted by Lyell (1832: p. 131). It is curious, therefore, that he should have placed relatively little emphasis on stabilizing selection. Perhaps he had not abstracted such statements to the degree we find in the quotation from Bruce Wallace and, hence, did not understand fully that under the normal environmental conditions of a particular species, differential survival preserves the normal adapted individuals, whereas, under changed conditions, new phenotypes are favored, preserved, and finally maintained, by differential survival.

We speak of stabilizing selection when considering the first case and directional selection in the second, but this has the disadvantage of obscuring the fact that the process is the same in both cases: the differential survival of fertile individuals.

The small African antelope *Oreotragus oreotragus*, or klipspringer, is rarely found anywhere but among granite outcrops. Their hooves are specialized for the granite boulders of their habitat, the animals walking on the ends of their hooves. The behavior of klipspringers is also very specialized particularly in relation to their predators, such as the black and martial eagles, baboons, and leopards. Among rocks, aberrant individuals will be at a disadvantage and are likely to fall prey to a predator. Away from rocks, as when males move between rocky outcrops, they are at relatively high risk, for under such conditions their specialized structures are relatively inappropriate. In this way the environment determines the pattern of survival resulting in either stabilization or directional change.

It is not surprising that stabilizing selection is dominant as the population geneticists insist; generally organisms are found in their specific preferred environments, which approximate to the environment in which they evolved, and are thus subject to stabilizing selection. Abnormal individuals are disadvantaged in their usual habitat, but if they are limited to an abnormal environmental, they may, under the new conditions, survive through being somehow more suited to the new circumstances. Thus the environment of an organism controls the pattern of survival.

Ever since Epicurus, and doubtless long before that, humans have noted the correlation that exists between particular species and particular environments. Particular species have long been observed to be associated with particular environmental conditions. Penguins, for example, show many physiological, morphological, and behavioral adaptations to amphibious life in southern lands and the Antarctic. Naturalists of the nineteenth century used the term "station" in referring to, say, the mountain habitat of the chamois. Before Darwin, most naturalists, including Lyell and Blyth, were aware that species disappeared from the fossil record, and that other, new species replaced them. Lyell and Blyth accepted that the loss of species was due to environmental modification for physical reasons, which often could be read from the geological record. However, they believed that such losses were compensated for by the creation of new species appropriate to the newly formed "stations." Such beliefs ensured that "believers" did not consider as possible the transmutation of species, as can be seen in the second volume of Lyell's *Principles of Geology*. Nor did they believe that major adaptive changes to existing species could occur, or were necessary. Minor variation within species was necessarily accepted in the light of abundant evidence, but nothing as great as a station shift was believed possible. Aberrant individuals that did not fit their station did not survive to propagate. The Creator's work was thus preserved in all its original perfection.

Military and other competitive metaphors have often been used. A favorite quotation of Lyell's was a remark by de Candolle (Lyell 1832: p. 131):

> "All the plants of a given country," says Decandolle [sic] in his usual spirited style, "are at war one with another. The first which establish themselves by chance in a particular spot, tend, by the mere occupancy of space, to exclude other species—the greater choke the smaller, the longest livers replace those which last for a shorter period, the more prolific gradually make themselves masters of the ground, which species multiplying more slowly would otherwise fill."

De Candolle, in the space of this brief quotation, outlines the principal idea of natural selection, differential survival. It was to this quotation that Darwin referred (de Beer et al. 1967: pp. 162–163), in his third notebook after reading Malthus's *Essay on the Principle of Population* on 28 September 1838:

> [Sept.] 28th We ought to be far from wondering of changes in numbers of species, from small changes in nature of locality. Even the energetic language of Decandolle [sic] does not convey the warring of the species as inference from Malthus.—increase of brutes must be prevented solely by positive checks, excepting that famine may stop desire.—in nature production does not increase, whilst no check prevail, but the positive check of famine and consequently death. I do not doubt every one till he thinks deeply has assumed that increase of animals exactly proportionate to the number that can live.—Population is increase [sic] at geometrical ratio in FAR SHORTER time than 25 years—yet until the one sentence of Malthus no one clearly perceived the great check amongst men.—there is spring, like food used for other purposes as wheat for making brandy.—Even a *few* years plenty, makes population in man increase & an *ordinary* crop causes a dearth. Take Europe on an average every species must have some number killed year with year by hawks by cold &c.—even one species of hawk decreasing in number must affect instantaneously all the rest.—The final cause of all this wedging, must be to sort out proper structure, and adapt it to changes.—to do that for form, which Malthus shows is the final effect (by means however of volition) of this populousness on the energy of man. One may say there is a force like a hundred thousand wedges trying [to] force every kind of adapted structure into the gaps in the œconomy of nature, or rather forming gaps by thrusting out weaker ones.

This passage provides an indication of what it was in Malthus's book that so influenced Darwin. The particular sentence from Malthus that Darwin refers to was identified by de Beer et al. (1967: p. 162) as, "It may safely be pronounced, therefore, that the population, when unchecked, goes on doubling itself every twenty five years, or increases in a geometrical nature."

Rather than Lyell's and Blyth's references to the "struggle for existence," and de Candolle's bellicose metaphors, it seems that it was this sentence of Malthus's that convinced Darwin of the powers of competition in driving natural selection. Malthus's writings seemed to reify the "struggle for existence" for Darwin. Although he remained faithful to Lyell's term, from then on Darwin saw struggle through Malthus's eyes. Further, it seems quite likely that Malthus's work appealed to both Darwin and Wallace through its quasi-quantitative formulation.

However, years were to pass before Darwin thought all this through and fitted it into place in his theory. As Ospovat (1981) emphasized, between 1837 and 1844, when Darwin wrote out his early views as an extended essay (*in* de Beer 1958), Darwin believed that a species' adaptation to its station is perfect. But after he had completed his work on barnacles on 8 September 1854, competitive selection soon replaced environmental selection as Lyell noticed on 16 April 1856.

Darwin's Change in Direction.—From Lyell's first species notebook it is clear that by mid-April 1856 Darwin had adopted competitive selection in place of environmental selection. He maintained this position consistently from then on. In the *Origin* Darwin outlined the idea behind natural selection as early as page five. Speaking of the struggle for existence, he wrote,

> This is the doctrine of Malthus, applied to the whole of the animal and vegetable kingdoms. As many more individuals of each species are born than can possibly survive; and as, consequently, there is a frequently recurring struggle for existence, it follows that any being, if it vary however slightly in any manner profitable to itself, under the complex and sometimes varying conditions of life, will have a better chance of surviving, and thus be *naturally selected*. From the strong principle of inheritance, any selected variety will tend to propagate its new and modified form.

Ronald Fisher (1958: p. 47) has drawn attention to problems with aspects of Darwin's argument:

> There is something like a relic of creationist philosophy in arguing from the observation, let us say, that a cod spawns a million eggs, that therefore its offspring are subject to Natural Selection; and it has the disadvantage of excluding fecundity from the class of characteristics of which we may attempt to appreciate the aptitude.

One can best gauge the depth of Darwin's commitment to competitive selection from passages like the following from page 68 of the *Origin*:

> Climate plays an important part in determining the average numbers of a species, and periodical seasons of extreme cold or drought, I believe to be the most effective of all checks. I estimated that the winter of 1854–55 destroyed four-fifths of the birds in my own grounds; and this is a tremendous destruction, when we remember that ten per cent. is an extraordinarily severe mortality from epidemics with man. The action of climate seems at first sight to be quite independent of the struggle for existence; but in so far as climate chiefly acts in reducing food, it brings on the most severe struggle between the individuals, whether of the same or of distinct species, which subsist on the same kind of food. Even when climate, for instance extreme cold, acts directly, it will be the least vigorous, or those which have got least food through the advancing winter, which will suffer most.

Thus, Darwin is content to see environmental stress as evidence of competitive selection. Yet, on page 76 of the *Origin* he confesses, "We can dimly see why the competition should be most severe between allied forms, which fill nearly the same place in the economy of nature; but probably in no one case could we precisely say why one species has been victorious over another in the great battle of life." This characteristic feature of competition makes sorting out causality problematic. Tom Park (1962) devoted many years to the controlled study of competition between species of *Tribolium* flour beetles. Summing up Park's classical studies, Simberloff (1980: p. 80) commented, "The most revolutionary part of Park's work . . . was the discovery that under certain environmental conditions a specific outcome could not be predicted; the process was stochastic, and the best prediction one could possibly make . . . was a probability that a particular species would win."

It should be noted that competition is rather frequently invoked as an explanation, but often careful investigation fails to detect it. Rohde (1977) and Gould and Calloway (1980) have provided examples of well-designed studies that failed to support *prima facie* cases of competition. Walter (1988, 2003) and Walter et al. (1984) have provided thoughtful discussions of problems with aspects of competition theory.

Aspects of Natural Selection.—Fisher (1958: p. 149) has drawn attention to another common deficiency in Darwin's argument where Darwin cites the enormous mortality in immature life-history stages as evidence for the intensity of natural selection, without making plain that only characters of those immature stages can be affected—thus heavy mortality of eggs has no influence on characters of the later stages. This is a rather obvious source of error, but one that is frequently overlooked. Fisher wrote,

> It should be observed that if one mature form has an advantage over another, represented by a greater expectation of offspring, this advantage is in no way diminished by the incidence of mortality in the immature stages of development, provided there is no association between mature and immature characters. The immature mortality might be a thousandfold greater, as indeed it is if we take account of the mortality of gametes, without exerting the slightest influence on the efficacy of the selection of the mature form.

Some other aspects of Darwin's evolutionary force, natural selection, should be noted. Darwin was quite consistent in his conclusion that natural selection leads to the differential survival of individual organisms. Almost at the end of the *Origin*, on page 489, he again makes this point clear: "And as natural selection works slowly by and for the good of each being, all corporeal and mental endowments will tend to progress towards perfection." Occasionally Darwin appears to espouse group selection by using phrases like "for the benefit of the species" (e.g., pp. 46, 146, 153), but I believe the general tenor of his argument leads to the conclusion that these are lapses and are not to be taken seriously.

"Natural selection" is a metaphor; consequently the term obscures the underlying process, whereas with "differential survival" the process is explicit. Natural selection is an extraordinarily subtle process, which is difficult enough to understand without additional obfuscation. Darwin's excuse for retaining the term occurs on page 61 of the *Origin*: "I have called this principle, by which each slight variation, if useful, is preserved, by the term Natural Selection, in order to mark its relation to man's power of selection." But "man's power of selection" is fundamentally different from nature's as Darwin (1859: p. 83) well knew: "Man selects only for his own good; Nature only for that of the being which she tends." I believe that instead of drawing attention to a poor analogy, Darwin would have served his readers better by emphasizing the distinctness of natural selection from artificial selection.

Darwin (1859: p. 84) was very insistent on the pervasiveness of natural selection as can be seen from this rather purple patch:

> It may be said that natural selection is daily and hourly scrutinizing, throughout the world, every variation, even the slightest; rejecting that which is bad, preserving and adding up all that is good; silently and insensibly working, whenever and wherever opportunity offers, at the *improvement of each organic being* in relation to its organic and inorganic conditions of life. We see nothing of these slow changes in progress, until the hand of time has marked the long lapse of ages, and then so imperfect is our view into long past geological ages, that we only see that the forms of life are now different from what they formerly were [my emphasis].

This is Darwin's version of Linnaeus's rule, *Natura non facit saltum*. The picture of change conjured up by Darwin epitomizes what has come to be called "phyletic gradualism," or simply "gradualism" (Eldredge and Gould 1972). In carefully considering this quotation, it should be noticed that in it Darwin does not see a role for stabilizing selection. What he anticipates is directional change (= adaptation), as is clear from the words I have emphasized. This is a critical point in evolutionary theory over which contemporary authors differ. Eldredge and Gould (1972) drew attention to the fact that the fossil record seemed generally to illustrate the stability of phenotype over the life of a species, which is in accord with the population geneticists I have cited (e.g., Milkman 1982: p. 105): "To a relatively small directional residue, we attribute the great panorama of evolution." The equating of evolution with change has long been general in evolutionary theory. For example, Fisher (1936: p. 58) remarked, "Evolution is progressive adaptation and consists of nothing else." The term "evolution" was originally derived from the Latin word for "unfolding," as in a bud opening; hence its restriction to "change." However, it can be argued that change results from a form of natural selection, directional selection, and the stabilization of the phenotype results from another, stabilizing selection. To me it is appealing to define evolution in relation to its prime process: evolution is the outcome of natural selection.

Relative Adaptation.—"A cornerstone of the theory of natural selection as it is presented in the *Origin of Species* is the notion of relative adaptation," Ospovat (1981: p. 7) commented. He continued (paraphrasing Darwin 1859: pp. 201–202), "Forms that are successful in the struggle for existence are deemed to be slightly better adapted than those with which they have had to compete for their places in the economy of nature. But since there is always room for improvement, it cannot be said that they are perfectly adapted for those places." Competitive selection together with relative adaptation was seen by Darwin as an improvement on environmental selection. However, he did not appear to appreciate that relative selection entailed problems for the idea of competition as the evolutionary motor.

In a letter to Hooker, dated 11 September 1857, Darwin wrote, "I have been writing an audacious little discussion to show that organic beings are not perfect, only perfect enough to struggle with their competitors." This seems to date this new viewpoint, besides showing that Darwin was rather pleased with his new perspective. The passage referred to by Darwin is to be found on pages 380–382 of his "big book" (Stauffer 1975).

Relative adaptation has particular bearing on the broad concept of adaptation, on sexual selection, and on certain models of the origin of species. However, here we should simply note its general consequence: when every competitive event is distinct, it is hard to see how this can result in a stable species-specific phenotype. I shall return to the consideration of relative adaptation in dealing with situations where it has special bearing.

Phenotypic Variation.—Darwin must have been aware that stability of specific phenotype was a basic prerequisite for the science of taxonomy and, hence, for other sciences such as stratigraphy, biogeography, and phylogeny as well. Stability of phenotype is exactly what is to be expected from stabilizing selection, as Roger Milkman emphasized above. However, this emphasis on stability of species-specific phenotype appears to be in conflict with aspects of Darwin's theory that relate to change. Natural selection requires variable phenotypes for differential survival and differential propagation to be effective in bringing about change. Thus, concern with evolutionary change, and conditions for change, tends to direct attention away from stability of specific phenotypes to the variability of populations. Unsurprisingly, it would seem that one's immediate point of interest is what claims one's exclusive attention. To Lyell stability was central. He had no need for variation in his theorizing in the 1830s. As pointed out by Ospovat (1981), Darwin at first shared this view, but once he accepted the possibility of the transmutation of species, things gradually changed, and the search for variation was on in earnest. Darwin's tendency from then on was virtually to ignore the importance of phenotypic stability.

Sexual propagation can be viewed in a similar way. I have already quoted Darwin's account of natural selection from page five of the *Origin* in which he uses the following statement: "From the strong principle of inheritance, any selected variety will tend to propagate its new and modified form." This assertion is the best that Darwin could do in the light of his ignorance of basic genetics. With hindsight we know that what he wrote is acceptable only if we place due emphasis on his cautious word "tend," because sexual propagation results in variation by recombination. This facilitates the action of natural selection in populations of organisms that are facing changed conditions, but adds to the task of stabilizing selection when conditions are normal. As Blyth remarked above, it is the individuals of normal phenotype, not the variant recombinants that tend to survive and propagate in a species' normal environment. Thus, sexual propagation ensures that the precise genotype of neither parent is ever transmitted to its offspring.

To Darwin, and other biologists of his time, the basis of variation was a mystery. He knew of the phenomenon of "sporting" in plants, and it was clear from the striking achievements of animal breeders that lack of variation was not one of their problems. The problem was to identify and understand the cause of variation: whether it was intrinsic to the organism or induced by the environment.

The first chapter of the *Origin* was entitled "Variation under Domestication," and on the second page Darwin considered the cause of variation. Myths abound vis-à-vis environmental disturbance's imprinting effects on mammalian embryos, and doubtless Darwin, as a countryman, was familiar with most of them. Casting about for a mechanism by which the environment might bear on variation, Darwin (1859: p. 8) wrote,

> Geoffroy St. Hilaire's experiments show that unnatural treatment of the embryo causes monstrosities; and monstrosities cannot be separated by any clear line of distinction from mere variations. But I am strongly inclined to suspect that the most frequent cause of variability may be attributed to the male and female reproductive elements having been affected prior to the act of conception. Several reasons make me believe this; but the chief one is the remarkable effect which confinement or cultivation has on the functions of the reproductive system; this system appearing to be far more susceptible than any other part of the organisation, to the action of any change in the conditions of life.

To this point Darwin was under the misapprehension that variation was environmentally induced; he did not see it as an intrinsic process. Towards the end of the *Origin* Darwin (1859: p. 466) reiterated his views on the variation of animals under domestication (i.e., under "changed circumstances"): "Man does not actually produce variability; he only unintentionally exposes organic beings to new conditions of life, and then nature acts on the organisation, and causes variability."

These mistaken views of Darwin's led to the general opinion expressed in his letter to Asa Gray on 5 September 1857 (Darwin 1858: pp. 51–52 *in* de Beer 1958), "Now take the case of a country undergoing some change. This will tend to cause some of its inhabitants to vary slightly—not but that I believe most beings vary at all times enough for selection to act on them." Contrary to his rather consistent commitment to an extrinsic induction of variation, Darwin has here added a clause that seems to suggest an acceptance of an intrinsic origin for variation. This problem was a chronic one for Darwin; in a letter to Hooker dated 18 March 1862 (Darwin and Seward 1903: p. 198) we read,

> You speak of 'an inherent tendency to vary wholly independent of physical conditions'! This is a very simple way of putting the case (as Dr. Prosper Lucas also puts it); but two great classes of facts make me think that all variability is due to change in the conditions of life—firstly, that there is more variability and more monstrosities (and these graduate into each other) under unnatural domestic conditions than under nature; and, secondly, that changed conditions affect in an especial manner the reproductive organs—those organs which are to produce a new being. But why one seedling out of thousands presents some new character tran-

> scends the wildest powers of conjecture. It was in this sense that I spoke of 'climate,' etc., possibly producing without selection a hooked seed, or any not great variation. I have for years and years been fighting with myself not to attribute too much to Natural Selection—to attribute something to direct action of conditions; and perhaps I have too much conquered my tendency to lay hardly any stress on conditions of life.

Eight days later he wrote again on the subject of variation in response to a reply from Hooker to his earlier letter (my italics):

> Thanks also for your own and Bates' letter now returned. They are both excellent; you have, I think, said all that can be said against direct effects of conditions, and capitally put. But I still stick to my own and Bates' side. Nevertheless I am pleased to attribute little to conditions, *and I wish I had done what you suggest—started on the fundamental principle of variation being an innate principle, and afterwards made a few remarks showing that hereafter, perhaps, this principle would be explicable.*

Nevertheless, Darwin persisted with his views on variation's extrinsic causes through the six editions of the *Origin*. This is important because it is a key point in contemporary theory that variation should be intrinsic and independent of environmental influence (however, see, for example, Jaenisch and Bird 2004). I shall again touch on this matter in considering Darwin's Principle of Divergence.

Sexual Selection.—Another of Darwin's ideas of which he was particularly fond was sexual selection, and here, too, he failed to consider the significance of stabilizing selection and did not notice the significance of relative adaptation.

When Darwin used a phrase such as "varieties of the same kind prefer to pair together," he was summing up a complex signal-response reaction chain that culminates in "like" fertilizing "like." Such systems of coadapted characters are under stabilizing selection and are characterized by stability (Paterson 1993: Chapters 3, 12). Clearly such ideas were not at the front of his mind when he outlined what he meant by sexual selection (1859: p. 87):

> *Sexual Selection* . . . This depends, not on a struggle for existence, but on a struggle between the males for possession of the females; the result is not death to the unsuccessful competitor, but few or no offspring. Sexual selection is, therefore, less rigorous than natural selection. Generally, the most vigorous males, those which are best fitted for their places in nature, will leave most progeny.

This male-male interaction constitutes the first type of sexual selection. The other involves "female choice":

> Amongst birds, the *contest* is often of a more peaceful character. All those who have attended to the subject, believe that there is the severest rivalry between the males of many species to attract by singing the females. The rock-thrush of Guiana, birds of Paradise, and some others, congregate; and successive males display their gorgeous plumage and perform strange antics before the females, which standing by as spectators, at last choose the *most attractive* partner [my italics].

It should be mentioned that here Darwin relied on an inaccurate early account of the propagative behavior of the Cock-of the-rock (*Rupicola rupicola*) (see Snow 1976). Snow's own account fits Darwin's preconceptions less well. In contrast to Darwin's account, when a female is attracted to the vicinity of a display ground, each male descends to his own "court," where he "freezes" in a characteristic stance in which his head is set with one eye looking upward and the other downward—a stance that displays the male's crest. "Choice" was witnessed by Snow on only one occasion despite extended observation. In this case the female, after some time, descended into one of the two courts within her field of view, and commenced "nibbling" at the silken fringes of the male's modified wing feathers. Unfortunately the courtship was then interrupted, so that the act of copulation was not witnessed. From analogy with the manakins and other birds that use leks, which he or others had studied, Snow believes that copulation occurs in the court to which the female descends. In any case, the behavior described by Darwin was never observed and seems peculiarly anthropomorphic to present-day readers.

Darwin saw competition as the key to understanding sexual behavior, which reminds me of his insistence that competition is the key to environmental adaptation. But, when a female chooses to mate with a healthy, rather than an unhealthy, male, or a female fails to respond to an aberrant sexual signal or prefers to pair with a mature male, rather than young and callow one, we are, I should think, confronting *prima facie* examples of stabilizing selection.

To summarize, some implicit assumptions behind Darwin's theory of sexual selection are that:

1. sexual selection is an intraspecific process;
2. males are sufficiently variable to enable females to choose the best male from a group;
3. features of the male phenotype accurately signal the important male qualities;
4. the "choosing" mechanism of the female is invariable; i.e., it is species-specific;
5. the female's choice results in more viable, or at least more, offspring;
6. a male is equally fertile and fecund with any female;
7. in comparing the performance of males, the appropriate basis is life-long reproductive achievement;
8. male attributes that enable the female to select effectively must be largely heritable.

These implications underlying sexual selection theory are not frivolous. For example, with respect to the sixth item, Darwin provided a pertinent anecdote in a letter to Huxley dated 28 December 1862 (Darwin and Seward 1903: p. 226): "I had this morning a letter with the case of a Hereford heifer, which seemed to be, after repeated trials, sterile with one particular and far from impotent bull, but not with another bull."

Because sexual selection is an intraspecific process, it must be preceded by the "recognition" of conspecifics. This primary process was apparently always taken for granted by Darwin; one does not encounter cases where he clearly recognized this process and distinguished it from the processes entailed in male-male encounters or female choice of males. As Darwin noted, clarifying causality in competitive events is difficult, and this certainly applies to sexual selection.

David Snow (1962, 1976) studied in detail the extraordinary reproductive behavior of the Black-and-white Manakin (*Manacus manacus*) in nature. Despite the complex "lek" behavior of the strikingly patterned males, this experienced observer concluded the following (Snow 1976: p. 46):

> Some of the males were much more successful than others in attracting females to their courts. As far as I could see, the position of the court was the key to success; the more central courts in the display ground were the most frequently visited by females. I could find no evidence that the males which occupied the central courts were inherently better in any way than those with peripheral courts; but it would have been impossible to recognize slight differences in the exuberance of their display, since the mere fact of having a central court, with a maximum of stimulation from surrounding males, meant that the owners spent longer at them and displayed more persistently than the more outlying males.

Rivalry among males is constant, and fighting occurs, but established court-holders are rarely ousted. These small birds are remarkably long lived. Six years after Snow's work in Trinidad, the same population was studied by Alan Lill (1974), who found that one male, which had been color-ringed by Snow, was still displaying actively at an age of at least 14 years. Except for a short break of a few weeks during the annual molt, display activity is virtually continuous. A particular individual watched by the Snows throughout the daylight hours of one day spent 90% of his time at his court. Lill's detailed study could find no differences in performance that correlated with mating success. He, too, found that, on the average, birds with more central courts in the display grounds were the most successful.

Being essentially a competitive process, sexual selection as pictured by Darwin seems unlikely to result in a stable, species-specific phenotype as found in the Indian peafowl (*Pavo cristatus*). This species ranges across India and Sri Lanka and yet only the subspecies *cristatus* has been recognized.

I think that I have said enough to support a claim that despite a vast amount of effort devoted to the study of sexual selection, sexual selection is still an insecure theory.

Principle of Divergence

Darwin first revealed his principle of divergence in his letter to Asa Gray, written on 5 September 1857 (de Beer 1958: pp. 264–267):

> Another principle, which may be called the principle of divergence, plays, I believe, an important part in the origin of species. The same spot will support more life if occupied by very diverse forms. We see this in the many generic forms in a square yard of turf, and in the plants or insects on any little uniform islet, belonging almost invariably to as many genera and families as species. We can understand the meaning of this fact amongst the higher animals, whose habits we understand. We know that it has been experimentally shown that a plot of land will yield a greater weight if sown with several species and genera of grasses, than if sown with only two or three species. Now, every organic being, by propagating so rapidly, may be said to be striving its utmost to increase in numbers. So it will be with the offspring of any species after it has become diversified into varieties, or subspecies, or true species. And it follows, I think, from the foregoing facts, that the varying offspring of each species will try (only few will succeed) to seize on as many and as diverse places in the economy of nature as possible. Each new variety or species, when formed, will generally take the place of, and thus exterminate its less well-fitted parent.

Darwin's next sentence provides an indication of the surprising way Darwin arrived at his principle—through his consideration of classification (Ospovat 1981: p. 170):

> This I believe to be the origin of the classification and affinities of organic beings at all times; for organic beings always *seem* to branch and sub-branch like the limbs of a tree from a common trunk, the flourishing and diverging twigs destroying the less vigorous—the dead and lost branches rudely representing extinct genera and families.

The rudiments of Darwin's "tree simile" trace to pages 25 ff. of Darwin's first notebook on the transmutation of species, from well before he had read Malthus's *Essay on the Principle of Population*. Underlying this principle in its final form (see below) is the economic theory of "division of labor," which formed the basis of Adam Smith's first chapter of his book, *The Wealth of Nations*. However, the proximate source of this influence on Darwin was more likely Milne Edwards, whose work had strongly impressed him (Ospovat 1981: p. 210; Darwin 1859: pp. 115–116). However, in discussing Darwin's principle of divergence, it is informative to recall that Adam Smith's famous book opens with the following sentence: "The greatest improvement in the productive power of labour, and the greater part of the skill, dexterity, and judgement with which it is anywhere directed, or applied, seem to have been the effects of the division of labour." An expanded version of the passage quoted from Darwin's letter to Gray is to be found in the *Origin* on pages 112–114. Of plants he wrote the following:

> It has been experimentally proved, that if a plot of ground be sown with one species of grass, and a similar plot be sown with several distinct genera of grasses, a greater number of plants and a greater weight of dry herbage can thus be raised. The same has been found to hold good when first one variety and then several mixed varieties of wheat have been sown on equal spaces of ground. Hence, if any one species of grass were to go on varying, and those varieties were continually selected which differed from each other in at all the same manner as distinct species and genera of grasses differ from each other, a greater number of individual plants of this species of grass, including its modified descendants, would succeed in living on the same piece of

> ground. And we well know that each species and each variety of grass is annually sowing almost countless seeds; and thus, as it may be said, is striving its utmost to increase its numbers. Consequently, I cannot doubt that in the course of many thousands of generations, the most distinct varieties of any one species of grass would always have the best chance of succeeding and of increasing in numbers, and thus of supplanting the less distinct varieties; and varieties, when rendered very distinct from each other, take the rank of species. The truth of the principle, that the greatest amount of life can be supported by great diversification of structure, is seen under many natural circumstances.

The experiments he describes refer to studies made by George Sinclair, head gardener to the Duke of Bedford early in the nineteenth century (Hector and Hooper 2002). This work appealed to Darwin because it appeared to provide a "function" for diversification. Thus, he referred to it as his "doctrine of the good of diversification," in the letter he wrote to Hooker on 26 March 1862. But the proposition requires careful examination and is subject to alternative and contrasting interpretation. To understand the relationship between division of labor and natural selection one should recall that natural selection amounts to the differential survival of fertile *individuals.* The experiments by George Sinclair were designed from the viewpoint of a farmer, and hence the yield of each *plot* is scored in terms of overall biomass.

Darwin conceived the idea of the "principle of divergence" after 1856, following his switch to competitive selection and his acceptance of relative adaptation. Besides viewing the benefit to the farmer as benefit to the individual organism, his principle also espoused sympatric divergence, including sympatric speciation. Impressed by the greater yield of plant biomass in diverse systems, Darwin seemed to discern a benefit for diversification. Thus, Darwin's account seems to suggest that the future benefits of diversification could drive plants in a monoculture to diversify. The scenario appears to require an internal "drive" toward divergence in sympatry. But, of course, as Lewontin (2000) puts it, the process of variation is causally independent of the conditions of selection. It is of interest that Hooker should have argued so firmly for the innateness of variation in his 1862 correspondence with Darwin. I shall return to the problem of sympatric divergence in the following section.

Divergence of Populations (Subspecies and Species)

By divergence I mean change at the level of the population. This is a consequence of adaptation, and one that embraces the origin of subspecies and species. I have already noted Fisher's remark, that "Evolution is progressive adaptation and consists of nothing else," with which this opinion concurs.

The subtlety of Darwin's discussion of the nature of species has been revealed in an essay by John Beatty (1985). In his conclusion, Beatty wrote,

> In order to communicate any more than a verbal disagreement with members of one's scientific community, it is necessary to respect their language rules, at least in part. But when the community's theory-laden definitions undermine the rival position being proposed, then those particular language rules cannot be respected—some other language rules of the community must be adopted instead as common ground for discourse. Those other rules may include actual examples of language use within the community.
>
> For instance, Darwin's theory of the evolution of species was undermined by the non-mutationist and non-transmutationist definitions of "species" to which his fellow naturalists adhered. He clearly could not defend the evolution of species, in either of these senses of "species." However, he could and did defend the evolution of what his fellow naturalists actually *called* "species"—on the supposition that what they called "species" did not satisfy their non-evolutionary definitions of "species."

Beatty provided two illustrative quotations from Darwin's aborted *Natural Selection,* from pages 98 and 97 respectively: "In the following pages I mean by species, those collections of individuals, which have been so designated by naturalists"; and, "We have to discuss in this work whether forms called by all naturalists distinct species are not lineal descendents of other forms." Here Darwin had in mind examples such as the "missel thrush," song thrush, and blackbird, all of which were generally accepted as species by naturalists.

In reading *The Origin of Species* we need to realize that, like most biologists, Darwin used the term "species" in different ways. At times he spoke of species in taxonomy—species categories in the hierarchical Linnean system at which level species taxa are assigned. But at other times he was thinking in evolutionary terms, viewing species and subspecies as populations. In the second chapter of the *Origin*, entitled "Variation under Nature," Darwin wrote,

> Certainly no clear line of demarcation has as yet been drawn between species and sub-species—that is, the forms which in the opinion of some naturalists come very near to, but do not quite arrive at the rank of species; or, again, between sub-species and well-marked varieties, or between lesser varieties and individual differences.
>
> These differences blend into each other in an insensible series; and a series impresses the mind with the idea of an actual passage. Hence I look at individual differences, though of small interest to the systematist, as of high importance for us, as being the first step towards such slight varieties as are barely thought worth recording in works on natural history. And I look at varieties which are in any degree more distinct and permanent, as steps leading to more strongly marked and more permanent varieties; and at these latter, as leading to sub-species, and to species.

"Individual variants" are "mutants" to us. As Darwin's term suggests, these arise sporadically, first appearing as aberrant individual organisms. In 1859 Darwin illustrated varieties by citing the primrose (*Primula vulgaris*) and the cowslip (*Primula veris*). These structurally distinct organisms occur in nature as populations with somewhat different distributions. As Darwin points out, these plants possess distinct scents, they differ a little in flowering period, though they overlap in this, and in his words, "they ascend mountains to different heights." He pointed out that according to Gärtner, a most careful observer, they can be crossed only with much difficulty. Darwin commented on these plants:

> We could hardly wish for better evidence of the two forms being specifically distinct. On the other hand, they are united by many intermediate links, and it is very doubtful that these links are hybrids; and there is, as it seems to me an overwhelming amount of experimental evidence, showing that they descend from common parents, and consequently must be ranked as varieties.

Darwin (1859: p. 312) rejected what he called the "common view of the immutability of species" in favor of "their slow and gradual modification, through descent and natural selection." That is, through the action of "competitive adaptation."

Another point to consider on entering into a discussion of Darwin's views on species is that he regarded species as well-marked varieties (1859: p. 481). In 1859 he considered the cowslip and the primrose as varieties, and the oxlip as a hybrid between these two varieties. Although in 1859 he believed that no clear line of demarcation had yet been drawn between species and varieties, it is interesting to note that by 1877 he had come to accept the present-day view that all three represent distinct species.

Darwin was also insecure in recognizing subspecies, though in the *Origin*, in relation to the origin of pigeon breeds, he did at times sound quite contemporary in his use of the term: I am fully convinced that the common opinion of naturalists is correct, namely that all have descended from the rock-pigeon (*Columba livia*), including under this term several geographical races or sub-species." This comment brings out the key point that subspecies are geographically discrete, which is evidence that they arise in isolation, and that they cannot coexist because they "intercross."

Darwin's views on subspecies incorporated the rudiments of a concept of species. He realized that both subspecies and species occur in nature as populations. However, unlike subspecies, which are geographically segregated, populations of species are able to coexist, as did the three *Turdus* species in Darwin's own garden. He recognized that for some unapparent reason, and in contrast to subspecies, species are generally able to coexist in nature without "intercrossing."

Early in his first transmutation notebook of July 1837, Darwin considered the nature of species in genetic terms: "A species as soon as once formed by separation or change in part of country, repugnance to intermarriage—settles it." Defining species in terms of "repugnance" was common at the time and did not originate with Darwin. Young Darwin's confident adoption of species delineated in terms of "repugnance" had only a brief life; he wrote the following passage in his 1844 essay (de Beer 1958: p. 125):

> It has often been stated that different species of animals have a sexual repugnance towards each other; I can find no evidence of this; it appears as if they merely did not excite each other's passions. I do not believe that in this respect there is any essential distinction between animals and plants; and in the latter there cannot be a feeling of repugnance.

"Repugnance" does not appear at all in the *Origin*. This early dalliance with "repugnance" is important for the light it throws on his thinking about species. He was struggling to grasp the biological nature of species, and here he clearly rejected what was later to be called "reproductive isolation" as the basis for recognizing species. Virtually alone in his time he also saw that sterility was no basis for defining species, and he realized that sterility could not be the product of natural selection.

After critically reviewing the extensive pollination studies of Gärtner and Kölreuter he concluded (Darwin 1859: p. 248):

> It is certain, on the one hand, that the sterility of various species when crossed is so different in degree and graduates away so insensibly, and, on the other hand, that the fertility of pure species is so easily affected by various circumstances, that for all practical purposes it is most difficult to say where perfect fertility ends and sterility begins . . . It can thus be shown that neither sterility nor fertility affords any clear distinction between species and varieties; but that the evidence from this source graduates away, and is doubtful in the same degree as is the evidence derived from other constitutional and structural differences.

Furthermore, Darwin was firmly of the belief that sterility could not possibly be the product of natural selection. Consider this passage on page 245 of the *Origin*:

> On the theory of natural selection the case is especially important, inasmuch as the sterility of hybrids could not possibly be of any advantage to them, and therefore could not have been acquired by the continued preservation of successive profitable degrees of sterility. I hope however, to be able to show that sterility is not a specially acquired or endowed quality, but is incidental on other acquired differences.

This was a notable conclusion drawn from his basic theory of natural selection, and one that demonstrated his remarkable trust in his theory. In his day this was a very counterintuitive conclusion, yet it is as solid today as in 1859. It was, furthermore, shared by neither Wallace nor Huxley, who exasperated Darwin with their blindness to his logic. Later (1877: p. 345) Darwin extended this argument to "illegitimate" crosses within heterostylous species such as *Primula vulgaris*:

> There is reason to believe that the sterility of these unions has not been especially acquired, but follows as an incidental result from the sexual elements of the two or three forms having been adapted to act on one another in a particular manner, so that any other kind of union is inefficient, like that between distinct species.

It can be seen that Darwin, after reviewing the evidence, saw no reason to rely on "reproductive isolation" to account for the ability of related species to coexist. However, he did not arrive at any alternative, although a solution was tantalizingly close. Thus, Darwin was fully aware that the "missel thrush," the song thrush, and the blackbird have distinct biological characteristics. The male songs are distinct, the three species have comparable breeding ranges (Sims 1978), and they pair positive-assortatively. He knew of many such cases. He would certainly have known of Gilbert White's famous trio of cryptic species in the warbler genus *Phylloscopus*, which White distinguished by their songs and preferred habitats (see Darwin's 1863 acquiescence *in* Barrett 1977: Vol. II, p. 92). In fact, in 1859 Darwin had all the pertinent information to understand the nature of species, yet he failed to see it in perspective and therefore was blocked from understanding the process that lies behind the origin of species. On page 103 of the *Origin* we find the following comments on how closely related but distinct "varieties" can persist together: "I can bring a considerable catalogue of facts, showing that within the same area, varieties of the same animal can long remain distinct, from haunting different stations, from breeding at slightly different seasons, or from varieties of the same kind preferring to pair together."

The three "facts" cited here by Darwin each contribute to positive-assortative mating, thereby quite incidentally eliminating intercrossing. Each is a characteristic, either acting together or separately, that delimits species by inducing like to breed with like. In fact, they answer Darwin's question, "Why are not all organic beings blended together in an inextricable chaos?" Darwin provided no illustrations in support of his claim; it would have been valuable to have at least a single actual example cited. He might well have treated White's three cryptic *Phylloscopus* species—the chiffchaff, the wood warbler, and the willow warbler—as varieties of the same animal that remain distinct within the same area by haunting different stations, by breeding at slightly different seasons, or by preferring to pair together with their own kind.

Each of the three factors specified by Darwin is clearly genetically determined and, hence, subject to stabilizing selection. With the benefit of decades of subsequent research we can see from this example how close Darwin was to justifying the title of his book.

But, of course, Darwin did not achieve a clear understanding of the nature of species, or of how they arise. However, he did achieve an understanding of the essential basis of the process under which new species arise, namely, adaptation through differential survival of fertile organisms, which I shall discuss below.

Discussion

The importance Darwin attributed to the genesis of species is evident from the title of his book and from his early citing of John Herschell's expression, "the mystery of mysteries," referring to the replacement of extinct species with others, in the first few lines of the book. In letters he spoke of natural selection and his Principle of Divergence as the keystone of his book. After a century and a half, natural selection and sexual selection are still attracting attention, but interest in Darwin's principle of diversity has declined.

Natural Selection.—Natural selection is usually interpreted as *differential survival and differential propagation.* Survival is read as *survival to sexual maturity,* and propagation is scored in terms of offspring achieving sexual maturity. Thus, we see that sterile organisms resemble organisms that die before achieving sexual maturity, in making no contribution to the following generation. Fertile individuals that fail to propagate are also non-contributors. Above, I provided an example, mentioned by Darwin, of a heifer that was infertile with a particular fertile bull but quite fertile with another. In such a case, two physically normal individuals may fail to contribute to the next generation. For such reasons it is those successful in leaving sexually mature offspring, despite all existing stresses, that are evolutionary contributors.

In the normal habitat to which a species is adapted, modal individuals survive and propagate, thereby maintaining the species' phenotype. This is stabilizing selection, which was well known to biologists of Darwin's day and much earlier ("Nature's broom"). Population geneticists have emphasized that the main day-to-day effect of natural selection is the maintenance of the status quo—the stabilization of the phenotype. A little thought will show that maintaining the normal phenotype has the consequence that biologists can classify organisms at the species level. Thus, the classification of organisms is dependent on stabilizing selection.

Evolutionary change *is less frequent* than stabilizing selection in nature, but on it depends adaptation. According to Fisher, and many others, the word evolution is restricted to adaptation: "Evolution is progressive adaptation and consists of nothing else." Darwin and Wallace rather played down stabilizing selection and emphasized directional selection and change. But, as I have already explained, stabilizing selection, occurring in the normal habitat of a species, constitutes the main day-to-day effect of natural selection. I have suggested that when a small population of a species is restricted to new environmental conditions, either it will become extinct or directional selection will adapt the population to the new conditions. When adapted to the new conditions the organisms will once again come under stabilizing selection.

The phenotype of a species is quite stable as I have argued. But this stability masks underlying genetic variation. It seems reasonable to suggest that the extent of change under directional selection depends on how different the two habitats are—the original and the new one. Not all aspects of the phenotype are modified under directional selection. Those features that are relatively inadequate under the changed conditions are most subject to change. For example, a human population in Africa that becomes restricted to an area where the malaria parasite *Plasmodium falciparum* is common might evolve a phenotype tolerant to this dangerous species without any superficial change in phenotype. In contrast, animal or plant species restricted to an area that is subject to progressive drying over extensive periods may well change considerably and eventually stabilize under the new desert conditions. Of course, under such conditions of change many, perhaps most, species are likely to become extinct.

Competitive selection is a process that Darwin perceived but did not fully explore. After considering the pros and cons of small and large populations in adaptive change, Darwin decided in favor of the divergence of large populations because they are more variable, and because this greater variability meant greater stress from competition. However, because every competitive bout is different it is hard to see how a stable population phenotype can result from such a scenario of competitive directional selection. As Darwin realized, small populations occupy a small area and have relatively low numbers. Although less variable, small populations are more propitious for the fixation of a phenotype because the environment is more likely to be uniform. Of course, small populations are also more apt to become extinct. After a small population has attained the fixation of a phenotype under directional selection, it returns to stabilizing selection. Under the less stringent conditions of stabilizing selection, the population may grow. It is likely first to occupy locally available stretches of suitable environment, but later it could be expected to extend its range to more distant areas of appropriate habitat (Hengeveld 1992). Small numbers of organisms blown off a mainland and out to sea, and ending up on distant island that is ecologically quite distinct from the organisms' mainland habitat, are naturally much more constrained by the tightly limited borders. Migration from the island is difficult. Darwin's insistence on perpetual selection is worrying. From the context it seems he is discussing directional selection, not stabilizing selection. This specification underlies his commitment to gradualism.

The final aspect of natural selection to be discussed is sexual selection. Because sexual selection has become so abstracted it is difficult for us to view it through the eyes of Darwin's contemporaries. Thus I shall restrict myself to what Darwin wrote in the *Origin*, and little more. In other words, I shall attempt to look at Darwin's ideas as they were proposed.

The first point about sexual selection is that it does not happily accommodate the principal function of sexual behavior, which is surely the effective achievement of fertilization. Instead, it is concerned centrally with competition, and it follows from Darwin's commitment to competitive selection.

The two forms of sexual selection are intermale competition that often involves actual conflict, whether ritualized or real, and intermale competition based on female "choice," whether real or merely so conceived by human observers. The sexually mature males of the stickleback *Gasterosteus aculeatus* have dark backs and red ventral surfaces. They occupy and defend a territory centered on a nest, which the male builds on the substrate. By means of a characteristic display above the territory, a gravid female is attracted and led to the nest where she deposits a number of eggs. The male then enters the nest himself, to fertilize the eggs. The female stickleback, which lacks the coloration of the male, recognizes a potential mating partner by his red belly well displayed with a particular ritual. A territorial male is able to recognize readily trespassing males before launching an attack on them. Theoretically a female is able to compare and choose from a number of males holding more or less contiguous territories. Milinski and Bakker (1990) have provided evidence that males infected with a ciliate parasite are less colorful and are discriminated against by females on the basis of color. Whether this should be regarded as sexual selection or stabilizing selection is open to question.

Darwin (1874: pp. 323–324) reviewed a number of cases that were difficult to classify as either "natural selection" or "sexual selection," concluding finally, "But in most cases of this kind it is impossible to distinguish between the effects of natural and sexual selection." This suggests a lack of clarity in Darwin's distinctions between his two forms of selection. The process in both natural selection and sexual selection involves differential survival and differential propagation. Darwin (1859: p. 88), spoke of the outcome of sexual selection: "the result is not death to the unsuccessful competitor, but few or no offspring." Of course, leaving no offspring has the same evolutionary effect as dying prematurely: no contribution to the next generation.

He was also very impressed by the disadvantage of bright colors and extravagant plumes (1874: p. 758), "It is evident that the brilliant colours, top-knots, fine plumes, etc., of many male birds cannot have been acquired as a protection; indeed they sometimes lead to danger." After dwelling on the refined beauty of the Argus pheasant, he wrote (1874: p. 609) with an anticipated critic in view, "but he will then be compelled to admit that the extraordinary attitudes assumed by the male during the act of courtship, by which the wonderful beauty of his plumage is fully displayed, are purposeless; and this is a conclusion which I for one will never admit." And nor should he. But, what Darwin does not say is that competitive charming of females can be looked at quite differently. Other interpretations are possible. In a detailed study of the Argus pheasant in the wild, Davison (1981) showed that male courting sites were separated by some miles in the jungle where he worked. It seems unlikely that females could effectively compare the finery and activity of males under such circumstances.

All sexual species are equipped to find or attract, and recognize, conspecific mating partners, and then to cooperate adequately to achieve insemination. A rough outline of the well-analyzed courtship of sticklebacks was provided above (Tinbergen 1951; Bastock 1967). This statement is as true for cryptic species as for birds-of-paradise and peacocks. The extravagance of the courtship of the Argus pheasant is perhaps excessive, but leaving this judgment aside, it clearly is effective in achieving fertilization under jungle conditions. And achieving fertilization is a necessity in all biparental organisms.

Darwin made it quite clear that he regarded sexual selection as an intraspecific process. This means that before intermale competition or competition involving females can occur the organisms must have recognized each other as conspecifics and have recognized them by sex. Darwin failed to see this point and makes no provision for it: to him all reproductive behavior was interpreted in terms of competition. In fact, the huge edifice of modern sexual selection rests on the acceptance of competition and its consequences. Among these consequences are relative adaptation and what it in turn leads to. For example it intersects the problem of how to achieve a uniform phenotype when each act of choice must be seen as an independent one. This would follow from the uniqueness of individuals, with regard to both fine details of the display in the male and the mechanism of choice in the female.

Brief though this analysis is, I think that we can say that sexual selection was not one of Darwin's better ideas. At the root of the problem is understanding how signaling systems come to be established—not just how a signal arises, but how it becomes a signal. This entails coadapting a new signal with a new receiver in the opposite sex. Some progress is being made now on the mode of establishment of signal and receiver in chemical recognition in the ova and sperm of marine organisms, insects, mammals, and angiosperms (Dodds et al. 1996; Grosberg and Hart, 2000; Neuberger and Novartis, 2002; Roelofs and Rooney 2003; Swanson et al., 2003). A review of signals among biparental organisms will reveal their opportunistic nature. Among birds, for example, visual signals can be as simple as the black facial streak of the male flicker, which dramatically identifies its sex (Noble 1936), to the extravagances of the birds-of-paradise or the pheasants and peafowl. Similarly, the extremes of vocal signals are just as great and, doubtless, so are the chemical signals (Dulac and Axell 1995; Holy et al. 2000) if only we were better equipped to assess them. The time is ripe for a neurophysiological attack on coadapted signal systems in the Eukarya and a switch in focus to the fundamental matter of the role of behavior in achieving effective fertilization.

Principle of Divergence.—Darwin conceived his principle of divergence in connection with branching in his tree simile, which occurs in his first notebook on the transmutation of species. However, it was soon co-opted for wider purposes. Ultimately he saw in it a reason for diversification. Ecological ideas to do with vacant niches and the filling of vacant niches are derivatives of Darwin's idea. Obviously, the filling of vacant niches to avoid competition can only occur sympatrically. Divergence usually entails the origin of new species, and thus the new species must diverge from its parent population without physical isolation. However, unequivocal cases of sympatric speciation are not known. It is difficult to see how a species could ever be recognized as having arisen sympatrically. When Darwin spoke to Hooker of his "doctrine of the good of diversification," he had the experiments of Sinclair in mind, which showed that a plot of land produces a greater biomass of herbage if planted with several species rather than just one. With the economic principle of the division of labor in mind, Darwin concluded that diversifying was beneficial and saw in these experiments, and others like them, a biological principle that diversification was beneficial (or "good"). But it was not the individual plant that was shown to have benefited by diversification in Sinclair's experiments, but the Duke of Bedford—the farmer. In nature, where there is no plan, natural selection prevails and the more efficient exploitation of the natural world is not of concern.

The Origin of Species.—Although Darwin was tantalizingly close to understanding the nature of species and subspecies, the insight finally eluded him. This failure was not unconnected with his enthusiasm for sexual selection and, ultimately, with his switch to competitive selection. It is interesting that he was fully aware of the inadequacy of sterility as a basis for defining species, and that in the *Origin* he rejected the popular Victorian concept of isolation through "repugnance." In 1862 Darwin published *The Various Contrivances by Which Orchids Are Fertilised by Insects,* which he opened with a significant first sentence: "The object of the following work is to show that the contrivances by which Orchids are fertilised, are as varied and almost as perfect as any of the most beautiful adaptations in the animal kingdom; and, secondly to show that these contrivances have for their main object the fertilization of the flowers with pollen brought by insects from a distinct plant."

Unfortunately Darwin never conceived the idea of writing a book on the contrivances by which animals achieve fertilization. Carrying out such a task with his great experience would have focused his mind on the critical facts, and would perhaps have led him to viewing reproductive behavior in a more biological way than in terms of competition. This in turn may well have solved his problem of the nature of species.

A final point concerns Darwin's insight that directional selection changes varieties (individual variants) to species, via subspecies. This view is quite in keeping with Fisher's dictum that "evolution is progressive adaptation and consists of nothing else" (Fisher 1936: p. 56). The contemporary habit of calling the process of deriving new species from older ones "speciation" is unfortunate, because it implies that there is a special process whose function is to produce species. In fact, as Fisher's dictum implies, species arise through the action of directional selection (adaptive change). Whether a small population adapting to new environmental conditions ends up as a new species or a new subspecies depends on which characters are disadvantageous and therefore subject to adaptation. As I emphasized above, successful fertilization in biparental organisms of all sorts depends on behavioral adaptations and recognition of like by like. Successful fertilization of the eggs of a female three-spined stickleback initially involves a gravid female recognizing a territorial stickleback male, and vice versa. This is followed by an exchange of signals that elicit appropriate responses from the recipient, leading ultimately to the fertilization of the ova by the male. Such reciprocal signaling systems are stable because they are coadapted; a mutant signal or a mutant receiver results in a diminished response or no response and, consequently, a failure to fertilize the ova. Now, such signal-response chains relate to the environment, in this case clear water unobstructed by weed. Forest birds like the Australian Lyre Bird or the Malaysian Argus Pheasant possess loud calls that attract females from a distance. In both species, when a female is attracted to the male's display site, her appearance stimulates the famous display by the male. The male's call and his signals are specific, and only conspecific females respond. Ultimately fertilization follows a successful negotiation of all the stages. Typically, these fertilization systems, which lead to positive assortative fertilization, are not only distinct but also appropriate to the environment of the species. Should a small population of a species become restricted to a new and different environment, part of its adaptive adjustment under directional selection will involve the signaling system if it is no longer appropriate to the new circumstances. In this way the adaptation of the fertilization system to new conditions can lead to the signaling system becoming too distinct for it to function with individuals of the ancestral species. In this case we would say that a new species had arisen. The members of the new population now mate positive-assortatively just as the different *Turdus* species did, and no doubt still do, in Darwin's garden at Downe.

Consider a small population of birds, isolated under conditions that are different from their normal habitat, and to which they adapt in the usual way. The stresses posed by the new habitat affect the food available, or the predators they must face, but in these new conditions the existing signal-response fertilization system is fully operational and therefore does not change. At the end of the process such a population would be regarded as a subspecies, which could not coexist with the parental population because the individuals would not distinguish their own members from the parental ones, even though they might have different feeding habits and be adapted to avoiding different predators.

If these scenarios are realistic, it will be seen that progression from individual variant to subspecies to species is not essential or usual. It is true that a subpopulation of a subspecies could become isolated in a different environment which would lead to the formation of a new species, but in this it is no different from any other population. The point is, subspecies should not automatically be regarded as incomplete species. They are not *necessarily* preliminary stages to the formation of species.

By focusing on competition Darwin missed the significance of the courtship of the Argus Pheasant, for example. It has long been quite clear that the aesthetics that Darwin discussed, the appreciation of beauty, and so on, are most improbable. Seeing the peacock's finery in terms of signals to conspecific males or females changes the whole picture that Darwin conjured up and one wonders to what extent Darwin's hypothesis of sexual selection was shaped by the false description of the display of the Cock-of-the-rock on which he relied. I believe that we will not understand the extravagance of some signals until we come to understand how new signals are entrained and become coadapted with an appropriate receiver.

An appealing aspect of the proposals made above on the nature of species and how, in principle, they may have evolved is that the process is free of goals for species. Although speciation arises as a consequence of directional selection, it cannot be said that "speciation" is an adaptive process. Rather, new species arise as a fortuitous consequence of adaptive change. Species are by-products, not products, one might say. If these ideas are not wrong, it can be seen that subspecies is not an intermediate stage in the process of a new species arising; subspecies are generally by-products of adaptive change under directional selection just as species are.

Conclusion

Darwin's switch to competitive selection has had a lasting and significant impact on evolutionary theory, a significance that I suspect has not been fully understood. Emphasizing competition led Darwin away from an understanding of species, laid the foundations of competition in ecology, and determined the course of evolution theory thereafter. Another influence has been to initiate the unmatched effort devoted to sexual selection in its various guises. An aspect of this is the tendency to regard anything to do with propagation as under the purview of sexual selection and hence connected to competition. However, if we read carefully Darwin's account of sexual selection, we see that this is not a deficiency that can be laid at Darwin's door; it derives from later generations.

All said and done, Darwin's book has had an amazing impact on the world of thought. What is disappointing to me is that Wallace never carried out his early intention to write his version of the evolutionary story based on environmental selection. However, as it is, by 1858, he had written two remarkable papers that provide a generous sample of what he might have said.

Dedication

It is a privilege for me to dedicate this essay to Steve Gould, who did so much to stir my imagination through his insights and his masterly use of language. With Niles Eldredge he opened all our eyes to the limitations of gradualism and the prevalence of punctuated equilibrium, both made vivid by his choice of terms. His essays led me to paleontology, to understanding de Vries, Buckland, Lyell and Hutton, and to the cathedral of San Marco, and in doing so provided me with a vivid reification of "effects." He inspired my students and illuminated my lectures with the insights that I gleaned from his writings. Scientists find it hard, it seems, to encourage their fellow scientists. But Steve Gould did this. His stand against the obfuscation of wonder by fundamentalists was unique. Steve always was pleased to encounter new wonders he had not yet met with in his omnivorous reading and explorations. With Jack Calloway, Steve looked into the assumed competition of brachiopods and bivalves during the Paleozoic, and he elegantly revealed not competition, but "ships that pass in the night." I miss his original and restless mind.

Literature Cited

Barrett, P. H. 1977. The collected papers of Charles Darwin. University of Chicago Press, Chicago.

Bastock, M. 1967. Courtship: a zoological study. Heinemann, London.

Beatty, J. 1985. Speaking of species. Pp. 265–282 *in* D. Kohn, ed. The Darwinian heritage. Princeton University Press, Princeton, N.J.

Blyth, E. 1835. An attempt to classify the 'varieties' of animals, with observations on the marked seasonal and other changes which naturally take place in various British species and which do not constitute varieties. Magazine of Natural History 8:40–53.

———. 1837. On the psychological distinctions between man and all other animals; and the consequent diversity of human influence over the inferior ranks of creation, from any mutual or reciprocal influence exercised among the latter. Magazine of Natural History, new series, 1:1–9, 77–85, 131–141.

Crow, J., and M. Kimura. 1970. An introduction to population genetics theory. Harper and Row, New York.

Darwin, C. 1858. On the variation of organic beings in a state of nature. Journal of the Proceedings of the Linnean Society of London (Zoology) 3:45–52.

———. 1859. On the origin of species. John Murray, London.

———. 1862. The various contrivances by which orchids are fertilised by insects. John Murray, London.

———. 1874. The descent of man and selection in relation to sex. John Murray, London.

———. 1877. The different forms of flowers on plants of the same species. John Murray, London.

Darwin, F., and A. C. Seward. 1903. More letters of Charles Darwin. Appleton, New York.
Davison, G. W. H. 1981. Sexual selection and the mating system of *Argusianus argus* (Aves: Phasianidae). Biological Journal of the Linnean Society 15:91–104.
de Beer, G. 1958. Evolution by natural selection. Cambridge University Press, Cambridge.
de Beer, G., M. J. Rowlands, and B. M. Skramovsky. 1967. Darwin's notebooks on transmutation of species, Part VI. Pages excised by Darwin. Bulletin of the British Museum (Natural History) 3(5):131–176.
Dodds, P. N., A. E. Clarke, and E. Newbigin. 1996. A molecular perspective on pollination in flowering plants. Cell 85:141–144.
Dulac, C., and R. Axell. 1995. A novel family of genes encoding putative pheromone receptors in mammals. Cell 83:196–206.
Eldredge, N., and S. J. Gould. 1972. Punctuated equilibrium: an alternative to phyletic gradualism. Pp. 82–115 *in* T. J. M. Schopf, ed. Models in paleontology. Freeman, Cooper, San Francisco.
Fisher, R. 1936. The measurement of selective intensity. Proceedings of the Royal Society of London B 121:58–62.
———. 1958. The genetical theory of natural selection. Dover, New York.
Gould, S. J., and C. B. Calloway. 1980. Clams and brachiopods —ships that pass in the night. Paleobiology 6:383–396.
Grosberg, R. K., and H. W. Hart. 2000. Mate selection and the evolution of highly polymorphic self/nonself recognition genes. Science 289:2111–2113.
Hector, A., and R. Hooper. 2002. Darwin and the first ecological experiment. Science 295:639–640.
Hengeveld, R. 1992. Dynamic biogeography. Cambridge University Press, Cambridge.
Holy, T. E., C. Dulac, and M. Meister. 2000. Responses of vomeronasal neurons to natural stimuli. Science 289:1562–1572.
Jaenisch, R., and A. Bird. 2004. Epigenetic regulation of gene expression: how the genome integrates intrinsic and environmental signals. Nature Genetics 33(Suppl.):245–254.
Karn, M. N., and L. S. Penrose. 1951. Birth weight and gestation time in relation to maternal age, parity and infant survival. Annals of Eugenics 16:147–164.
Kottler, M. 1985. Charles Darwin and Alfred Russel Wallace: two decades of debate over natural selection. Pp. 367–432 *in* D. Kohn, ed. The Darwinian heritage. Princeton University Press, Princeton, N.J.
Lewontin, R. C. 2000. The triple helix. Harvard University Press, Cambridge.
Lill, A. 1974. Sexual behaviour of the lek-forming White-bearded Manakin (*Manacus manacus trinitatis* Hartert). Zeitschrift für Tierpsychologie 36:1–36.
Lyell, C. 1832. Principles of geology, II. John Murray, London.
Milinski, M., and T. C. M. Bakker. 1990. Female sticklebacks use male colouration in mate choice and hence avoid parasitized males. Nature 332:330–333.
Milkman, R. 1982. Towards a uniform selection theory. Pp. 105–118 *in* R. Milkman, ed. Perspectives on evolution. Sinauer, Sunderland, Mass.
Neuberger, M. S., and T. Novartis. 2002. Antibodies: a paradigm for the evolution of molecular recognition. Biochemical Society Transactions 30:341–350.
Nicholson, A. J. 1960. The role of population dynamics in natural selection. Pp. 477–521 *in* S. Tax, ed. Evolution after Darwin. I. The evolution of life. University of Chicago Press, Chicago.
Noble, G. K. 1936. Courtship and sexual selection of the flicker (*Colaptes auratus luteus*). Auk 53:269–282.
Ospovat, D. 1981. The development of Darwin's theory. Cambridge University Press, Cambridge.
Park, T. 1962. Beetles, competition, and populations. Science 138:1369–1375.
Paterson, H. E. H. 1993. Evolution and the recognition concept of species. Johns Hopkins University Press, Baltimore.
Raff, R. A. 1996. The shape of life. Chicago University Press, Chicago.
Roelofs, W. L., and A. P. Rooney. 2003. Molecular genetics and evolution of pheromone biosynthesis in Lepidoptera. Proceedings of the National Academy of Sciences USA 100:9179–9184.
Rohde, K. 1977. A non-competitive mechanism responsible for restricting niches. Zoologischer Anzeiger 199:164–172.
Rutherford, S. L., and S. Lindquist. 1998. Hsp90 as a capacitor for morphological evolution. Nature 396:336–342.
Schlichting, C. D., and M. Pigliucci. 1998. Phenotypic evolution: a reaction norm perspective. Sinauer, Sunderland, Mass.
Simberloff, D. 1980. A succession of paradigms in ecology: essentialism to materialism and probabalism. Pp. 63–99 *in* E. Saarinen, ed. Conceptual issues in ecology. D. Reidel, Dordrecht.
Sims, E. 1978. British thrushes. Collins, London.
Snow, D. 1962. A field study of the Black-and-white Manakin, *Manacus manacus*, in Trinidad. Zoologica 47:65–104.
———. 1976. The web of adaptation: bird studies in the American tropics. Collins, London.
Stauffer, R. C. 1975. Charles Darwin's natural selection. Cambridge University Press, Cambridge.
Swanson, W. J., R. Nielsen, and Q. Yang. 2003. Pervasive adaptive evolution in mammalian fertilization proteins. Molecular Biology and Evolution 20:18–20.
Tinbergen, N. 1951. The study of instinct. Oxford University Press, Oxford.
Waddington, C. H. 1968. The paradigm for the evolutionary process. Pp. 37–45 *in* R. C. Lewontin, ed. Population biology and evolution. Syracuse University Press, Syracuse.
Wallace, A. R. 1855. On the law which has regulated the introduction of new species. Annals and Magazine of Natural History 16(2):184–196.
———. 1858. On the tendency of varieties to depart indefinitely from the original type. Proceedings of the Linnean Society of London (Zoology) 3:53–62.
Wallace, B. 1981. Basic population genetics. Columbia University Press, New York.
Walter, G. H. 1988. Competitive exclusion, coexistence and community structure. Acta Biotheoretica 37:281–313.
———. 2003. Insect pest management and ecological research. Cambridge University Press, Cambridge.
Walter, G. H., P. H. Hulley, and A. J. F. K. Craig. 1984. Speciation, adaptation and interspecific competition. Oikos 43:146–148.
Wilson, L. G., ed. 1970. Sir Charles Lyell's scientific journals on the species question. Yale University Press, New Haven, Conn.

Paleobiology, 31(2), 2005, pp. 77–93

Key innovations, convergence, and success: macroevolutionary lessons from plant phylogeny

Michael J. Donoghue

Abstract.—Improvements in our understanding of green plant phylogeny are casting new light on the connection between character evolution and diversification. The repeated discovery of paraphyly has helped disentangle what once appeared to be phylogenetically coincident character changes, but this has also highlighted the existence of sequences of character change, no one element of which can cleanly be identified as *the* "key innovation" responsible for shifting diversification rate. In effect, the cause becomes distributed across a nested series of nodes in the tree. Many of the most conspicuous plant "innovations" (such as macrophyllous leaves) are underlain by earlier, more subtle shifts in development (such as overtopping growth), which appear to have enabled the exploration of a greater range of morphological designs. Often it appears that these underlying changes have been brought about at the level of cell interactions within meristems, highlighting the need for developmental models and experiments focused at this level. The standard practice of attempting to identify correlations between recurrent character change (such as the tree growth habit) and clade diversity is complicated by the observation that the "same" trait may be constructed quite differently in different lineages (e.g., different forms of cambial activity), with some solutions imposing more architectural limitations than others. These thoughts highlight the need for a more nuanced view, which has implications for comparative methods. They also bear on issues central to Stephen Jay Gould's vision of macroevolution, including exaptation and evolutionary recurrence in relation to constraint and the repeatability of evolution.

Michael J. Donoghue. Department of Ecology and Evolutionary Biology and Peabody Museum of Natural History, Yale University, New Haven, Connecticut 06520. E-mail: michael.donoghue@yale.edu

Accepted: 7 September 2004

Introduction

Much of Stephen Jay Gould's work was concerned, directly or indirectly, with patterns of character evolution, patterns of clade diversification, and the causal link between these two. Although Gould did not take an explicitly phylogenetic approach to these problems, others have in recent years. In any case, our knowledge of the Tree of Life has expanded enormously (Cracraft and Donoghue 2004) and it is worth considering how phylogenetic insights may be influencing our views on macroevolution and especially the link between character evolution and diversification. In this essay I provide the perspective of someone working on plant evolution, together with a few concrete plant examples. Gould was not, of course, especially interested in plants, but his ideas were clearly intended to apply to organisms of all sorts.

Specifically, I begin by briefly characterizing what we have learned recently about the fundamental structure of green plant phylogeny, drawing a few generalizations about the nature of that progress. Then I consider how this progress has been, or at least should be, affecting our understanding of the connection between character evolution and diversification. My basic argument is that recent phylogenetic findings are making it increasingly difficult to sustain the traditional view of key innovations and also to maintain standard comparative approaches to detecting the effects of character change on diversity. These realizations suggest several new methodological needs and research strategies. In closing I briefly consider how these ideas relate to some of Gould's views on macroevolution.

Progress in Understanding Plant Phylogeny

Figure 1 provides an overview of our present knowledge of phylogenetic relationships among the major lineages of green plants. This is simplified, of course, and consciously rendered pectinate to serve my purposes (see O'Hara 1992, on the representation of trees). Readers are referred to other recent reviews (Bateman et al. 1998; Chapman et al. 1998;

0094-8373/05/3102-0006/$1.00

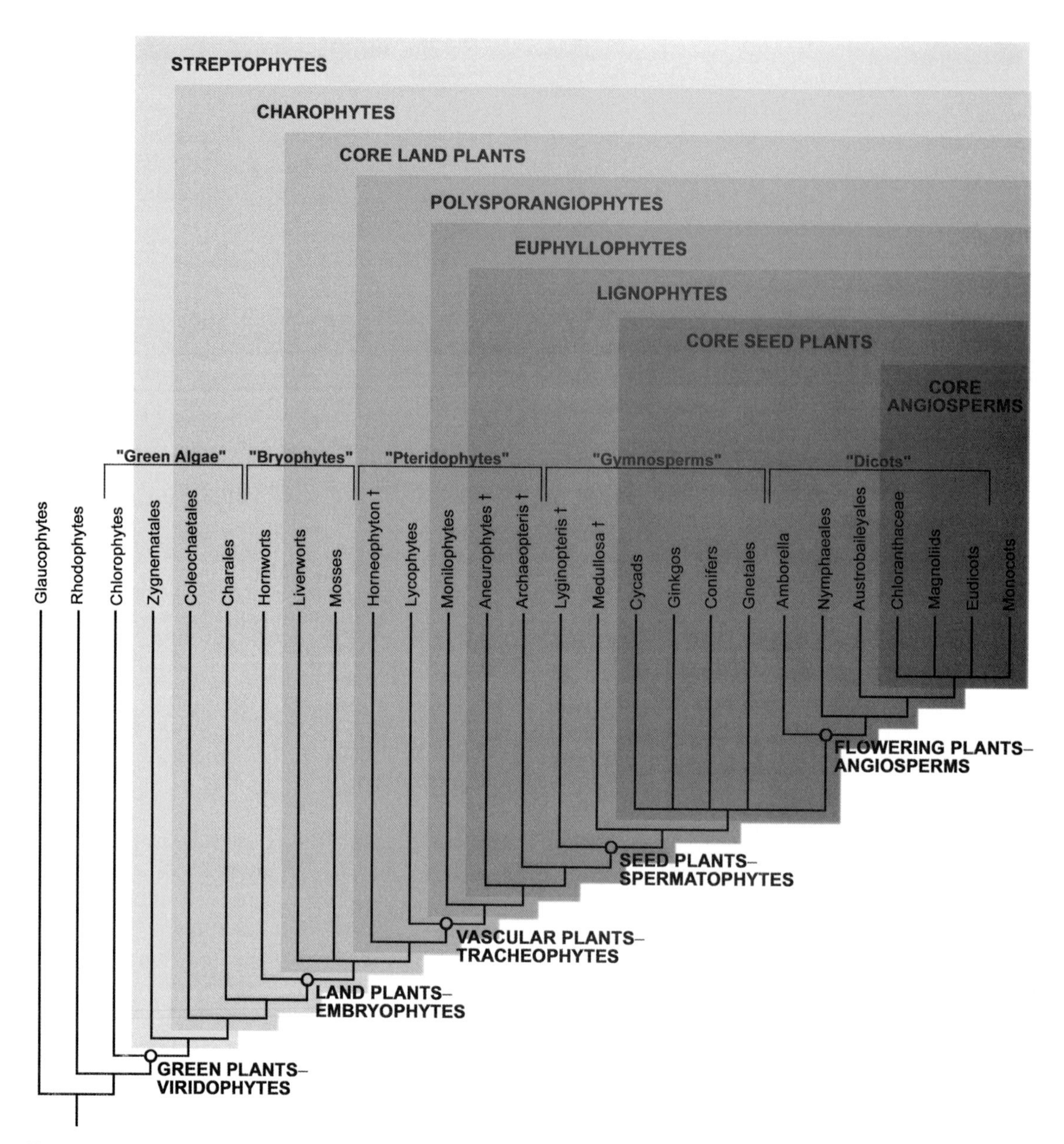

FIGURE 1. An overview of green plant phylogeny, illustrating the recent discovery of major clades (shaded groups); the monophyly of some traditionally recognized groups (shown at nodes with open circles) has been upheld, whereas others are now seen to be paraphyletic (names in quotation marks). † marks denote extinct groups. See text for references and discussion.

Doyle 1998; Kenrick 2000; Donoghue 2002, 2004; Judd et al. 2002; Delwiche et al. 2004; Pryer et al. 2004; Soltis et al. 2004) for references to the primary literature underpinning Figure 1, and for levels of support, commentary on remaining controversies, and a wide variety of evolutionary implications not touched upon here.

Several familiar and long-recognized taxa are strongly supported as monophyletic. These include the entire green plant clade (the viridophytes), land plants (embryophytes), vascular plants (tracheophytes), seed plants (spermatophytes), flowering plants (angiosperms), and monocotyledons (monocots). Conveniently, these clades are marked by characters that relate to their names: green plants by chlorophyll b, land plants by a resting embryo stage in the life cycle (hence embryophytes), vascular plants by vascular tissue with specialized cells for the transport of water (tracheids), seed plants by seeds (integumented megasporangia), flowering plants by one or more carpels in the shortened reproductive axes that we call flowers, and monocots by embryos with just a single seed leaf (cotyledon).

Phylogenetic analyses conducted over the last two decades have also shown that several other traditionally recognized major groups are not monophyletic, but instead represent grades of organization. Specifically, traditional "green algae," "bryophytes," "pteridophytes" (seedless vascular plants), "gymnosperms" (naked-seed plants), and "dicotyledons" appear to be paraphyletic. These had each been diagnosed on the basis of what we now recognize to be ancestral traits. For example, green algae are green plants that lack the specialized characteristics of the land plant clade (they live in the water, lack a resting embryo, etc.). In bryophytes the sporophyte phase is unbranched and lacks vascular tissues of the sort found in tracheophytes. As the names implies, "seedless vascular plants" are vascular plants that lack seeds, "gymnosperms" are seed plants that lack carpels, and so forth.

Recognition that these traditional groups are paraphyletic has, of course, resulted from the discovery of new major clades that unite one or more of the lineages traditionally assigned to the grade group directly with an included clade. For instance, the dismantling of the traditional green algae came about through the recognition that some groups formerly treated as green algae are actually more closely related to land plants than they are to other green algal lineages. Specifically, it was discovered (initially on the basis of ultrastructural features, but now with much molecular support; [e.g., Karol et al. 2001]) that the Charales and several other lineages (e.g., Klebsormidiales, Zygnematales, Coleochaetales) are more closely related to lands plants than they are to Chlorophyceae, Trebuxiophyceae, and Ulvophyceae (the latter three making up the Chlorophyte clade in the strict sense). The name "streptophytes" has now been widely applied to this newly discovered clade (Delwiche et al. [2004], prefer the name "charophytes;" see Donoghue 2004).

Similarly, in the first phylogenetic analyses of land plants, hornworts and mosses were found to be more closely related to vascular plants than to liverworts, the other major lineage of "bryophytes" (Mishler and Churchill 1985). The term "stomatophytes" was coined for this clade, reflecting the presence of stomates in hornworts, mosses, and vascular plants. In recent years, however, several alternative hypotheses have surfaced, especially the idea that the first split was between hornworts and a clade containing the other three clades (e.g., Nickrent et al. 2000; Renzaglia et al. 2000). In any case, phylogenetic analyses that have sampled a sufficient number of representatives of these groups have supported the view that bryophytes do not form a clade but rather represent a grade of organization within land plants.

The name euphyllophytes has recently been applied to the clade including horsetails, whisk-ferns (psilophytes), various fern lineages, and seed plants (e.g., Kenrick and Crane 1997). These are more closely related to one another than to the other extant lineage of seedless vascular plants, the lycophytes. The name "anthophytes" was applied to the hypothesized clade including the "gymnosperm" group Gnetales along with the flowering plants, to the exclusion of cycads, ginkgos, and conifers (Doyle and Donoghue 1986). As in the bryophyte case, many recent analyses (reviewed by Donoghue and Doyle 2000) do not support this anthophyte clade (Gnetales instead being allied with conifers). In any case, however, "gymnosperms" remain paraphyletic relative to angiosperms when fossil groups (e.g., Paleozoic and Mesozoic "seed ferns") are considered (Donoghue and Doyle 2000; Pryer et al. 2004); that is, the first seed plants clearly lacked carpels. Finally, within flowering plants, the recently discovered eudicot clade (containing more than 160,000 species) and a re-circumscribed magnoliid clade (containing magnolias, black peppers, avocados, etc.) are found to be more closely related to monocots than they are to some other lineages of "dicotyledons," such

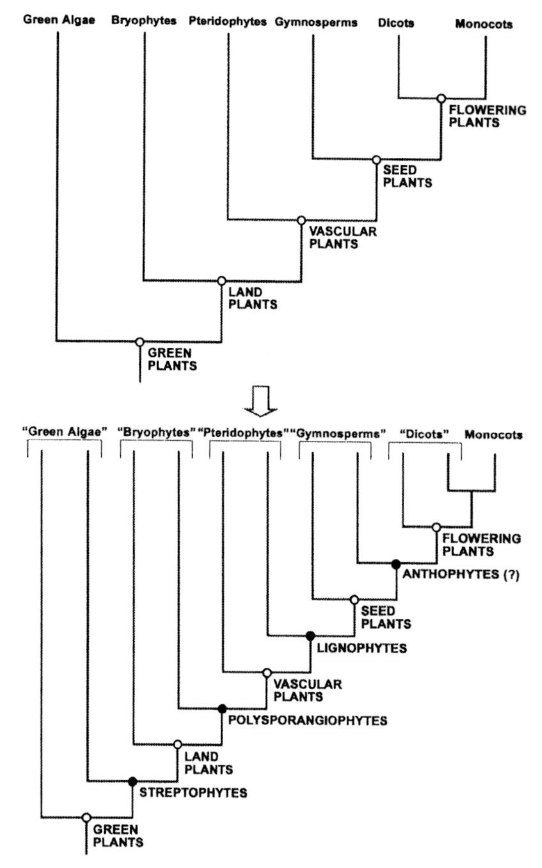

FIGURE 2. The nature of progress in resolving plant phylogeny. The upper tree shows the standard view as of the 1970s; the lower tree depicts current understanding. Major clades supported as monophyletic are marked by open circles at the nodes; newly discovered clades are marked by black circles.

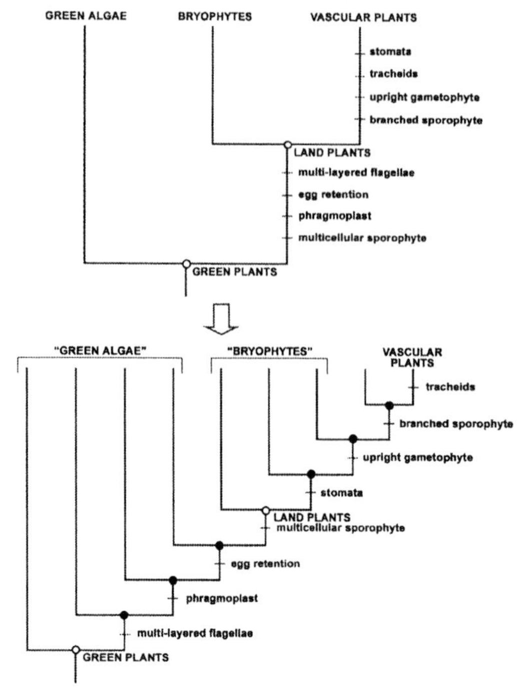

FIGURE 3. An example of the impact of new phylogenetic knowledge (discovery of the paraphyly of "green algae" and "bryophytes") on our understanding of character evolution. What once appeared to be clumped changes at key nodes (upper tree) can now be sorted into a sequence of character changes (lower tree) that clarify the transition to land and the origin of vascular plants.

as *Amborella* and the water-lilies (Zanis et al. 2002; Soltis et al. 2004).

Figure 2 presents a cartoon summary of these results to highlight the nature of the progress that has been made since the 1980s. In general, our advances have entailed confirmation of the monophyly of some long-recognized major clades, along with the recognition of a number of paraphyletic taxa through the discovery of new major clades. Names such as "green algae," "bryophytes," and "dicots" are now either being dropped altogether or being used only to refer to particular life styles or grades of organization. Meanwhile, names such as euphyllophytes and eudicots are finding their way into introductory textbooks (e.g., Judd et al. 2002) and are beginning to orient the way we think about plant diversity and conduct research. The discovery and abandonment of paraphyletic groups is, in general, what progress is all about in phylogenetic systematics (Donoghue 2004).

Character Sequences and Developmental Enablers

How have these advances changed our understanding of plant evolution? The most obvious impact has been on our ability to dissect the evolutionary sequence of events surrounding the greatest transformations in green plant history. For example, consider the transition from life in the water to life on land (see Graham 1993). When green algae and bryophytes were both viewed as clades, this transition appeared to entail a very large number of steps that could not be placed in any particular temporal order (Fig. 3, top). This implied either a large number of extinctions of intermediary taxa and, consequently, major gaps in our knowledge, or a wholesale correlated transformation from one life form to another. Under these circumstances several alternative theories remained viable to explain the evolution of features such as the land plant life cycle, entailing the alternation of multicellular haploid (gametophyte) and diploid (sporophyte) phases. Was a multicellular haploid phase or a multicellular diploid phase added to an ancestral non-alternating life cycle? Or, perhaps the ancestor of land plants belonged to a lineage within which alternation of generations had already evolved. Did the precursors of land plants live in salt water, fresh water, or even on land (several "green algal" lineages independently made the transition to land)? What was the basic body plan from which land plants evolved? After all, "green algae" present an impressive number of alternatives, from unicells, to colonies, to filaments, to pseudo-parenchymatous forms, with or without cell walls separating the nuclei. With no way to sort out the sequence of events, the transition to land largely remained a mystery.

Knowing now that both the traditional green algae and bryophytes are paraphyletic, and having succeeded in identifying the closest living relatives of land plants (Charales and Coleochaetales [Karol et al. 2001]), we can start to establish the sequence of events from the origin of the first green plants through their movement onto land (Fig. 3, bottom). On this basis, we can be quite certain that land plants arose within a lineage of "green algae" living in fresh water, probably quite near the shore. Their ancestors probably had rather complex parenchymatous construction, with gametes (and then zygotes) borne on the parent plant in specialized containers. Perhaps most importantly, we can infer that the land plant life cycle originated through the intercalation of a multicellular diploid phase (by delaying the onset of meiosis) into a life cycle resembling that retained in Coleochaetales and Charales (wherein the diploid zygote undergoes meiosis directly to form haploid spores). Likewise, we can infer that the first land plants had a bryophyte-like life cycle in which the gametophyte was the dominant phase and the sporophyte was smaller and parasitic on the gametophyte.

Moving within land plants, the discovery of the polysporangiophyte clade (Kenrick and Crane 1997; see Pryer et al. 2004) implies that enlargement and branching of the sporophyte preceded the acquisition of tracheids (Fig. 3). Moreover, fossil reconstructions of the gametophytes of the first polysporangiophytes (Remy et al. 1993) suggest that the transition to sporophyte dominance moved through a stage in which gametophyte and sporophyte phases were more or less similar in structure (so-called isomorphic alternation of generations [Kenrick and Crane 1997]).

I provide this level of detail to draw attention to the great significance of recent phylogenetic advances, which have basically settled many major questions about plant evolution. But the main point I want to make here is that recent phylogenetic discoveries don't just help us to choose among existing hypotheses, but also shed genuinely new light on such problems. Many of the newly discovered green plant clades serve to focus our attention on seemingly minor—but in retrospect apparently quite profound—shifts in the nature of plant development. Prime examples concern meristem structure and function in relation to branching. The polysporangiophyte clade is marked by the ability of the sporophyte plant to branch dichotomously, as compared to the ancestral unbranched condition retained today in the bryophytic lineages (Fig. 4). Dichotomous branching made it possible for a given sporophyte to produce more sporangia and more spores per fertilization event, and perhaps generally to become larger (Mishler and Churchill 1985; Knoll et al. 1986). This ability of the apical meristem to branch apparently set the stage for a series of changes that now mark the tracheophyte clade, notably the evolution of differentiated vascular tissues for the flow of water and nutrients through a larger upright plant body. In retrospect, dichotomous branching may have established the conditions for—or enabled—the evolution of increased size, of vascular tissue, and of many other downstream character changes.

The same line of reasoning applies to the evolution in euphyllophytes of the differentiation between a main axis, or trunk portion of stem, and lateral branches (Fig. 4)—so-called

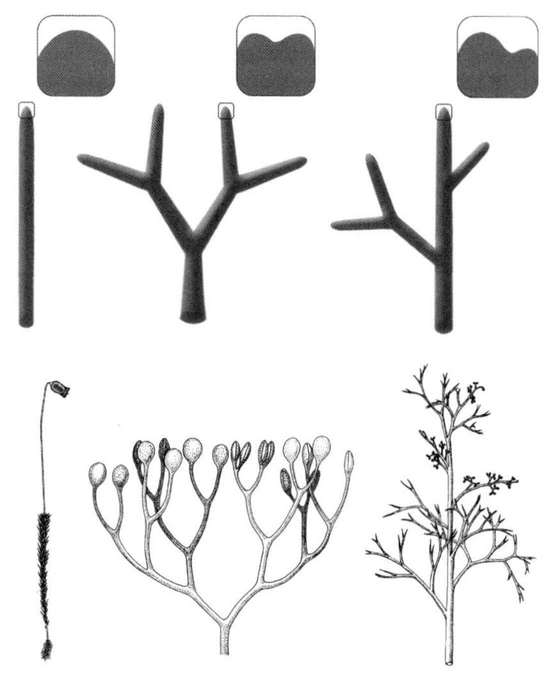

FIGURE 4. A comparison of sporophyte branching among early-branching lineages of land plants. In the bryophytic lineages (left) the sporophyte is unbranched; dichotomous branching evolved at the base of the polysporangiophytes (center); overtopping (or pseudomonopodial growth) evolved at the base of the euphyllophytes (right). Insets at the top represent these differences in branching at the level of the apical meristem. (Drawings at the bottom are from Stewart and Rothwell 1993.)

overtopping or pseudomonopodial growth (Zimmermann 1965). This seemingly minor shift at the level of the shoot apical meristem appears to have enabled the evolution (most likely independently in several lineages [e.g., Boyce and Knoll 2002; but see Schneider et al. 2002; Pryer et al. 2004]) of the determinate lateral organs that we call leaves (or, more specifically, megaphyllous leaves, as distinct from the so-called microphyllous leaves of lycophytes), and, in turn, the evolution of seeds and flowers. These derived traits (e.g., leaves, seeds) are often viewed as the key innovations responsible for the evolutionary success (usually measured in terms of the number of species) of their respective lineages. Recent phylogenetic discoveries have the effect of highlighting subtle, but crucial, underlying developmental shifts at the level of the apical meristem that made possible the evolution of the more obvious characters.

These observations have an important bearing on the identification of "key innovations." In two obvious ways the identification of key innovations becomes easier. First, as already noted, recent progress has distributed inferred character changes across a series of branches as opposed to having them piled up at particular nodes (Fig. 3). The problem with having character changes concentrated at a node is that it is unclear which one (or which combination) of the changes might have triggered a shift in diversification rate. Decomposing such a set of characters can help single out the character(s) associated most directly with shifts in diversification. Second, decomposing paraphyletic groups reduces the number of species in the sister group of the focal clade, thereby increasing the magnitude of the diversity contrast. For example, Charales and Coleochaetales contain many fewer species than did the traditionally circumscribed "green algae" (with probably more than 35,000 species). The discovery that Charales (with approximately 500 species) are sister to land plants (with over 300,000 species), and in turn that the Coleochaetales (with about 30 species) are sister to the clade containing these two, greatly accentuates the contrast in diversity between land plants and the several lineages to which they are most closely related. In general, this sort of change makes it easier to locate a significant shift in diversification rate (Moore et al. 2004) and therefore increases the inclination to explain it with reference to a key character change.

However, a third impact of phylogenetic discoveries challenges the very notion of key innovations. The existence in our classifications of major groups such as tracheophytes, spermatophytes, and angiosperms has drawn our attention to the obvious traits of these clades—vascular tissue, seeds, and flowers—as potential drivers of diversification. The discovery of a set of new major clades, including polysporangiophytes, euphyllophytes, lignophytes, etc., likewise focuses our attention on their somewhat more subtle features—dichotomous branching, pseudomonopodial growth, bifacial cambium, etc. The intercalation of these new clades between the traditional groups, I predict, will bring about a subtle but fundamental shift in how we view the link between character evolution and success. Despite the increased ease (just noted) with which we may be able to associate particular character changes with shifts in diversification, I suspect that we will become increasingly less comfortable about phylogenetically localizing "key" innovations. Instead, because there are often causal links between characters that evolved earlier and later in a sequence, it will seem increasingly natural to think from the outset about a series of changes culminating in a combination of traits that together served to increase diversification. Appreciating the interdependencies and the combined effects of character changes doesn't just relocate the cause to another node in the tree, but instead distributes the causation across a series of nodes. As we become increasingly aware of the ways in which apparently minor developmental changes early in a chain rendered new morphological designs accessible, we might even be tempted to view early steps as actually necessitating later ones. But the causal links will generally be much more subtle. Overtopping growth did not, we presume, necessitate the evolution of macrophyllous leaves. Instead, it enabled the exploration of a new set of morphological designs, which eventually set the stage for the evolution of leaves.

This refinement in outlook will, I suspect, reveal some important new evolutionary generalities. For example, in the several cases we have been considering (dichotomous branching, overtopping) the enabling changes appear to have been developmental shifts at the level of apical meristems, which presumably involved shifts in gene expression and the localization of signals at the level of cells and cell layers within the meristem. These underlying changes appear now to be highly conserved, in the sense of showing little homoplasy, which perhaps implies that the derived state was somewhat difficult to achieve in the first place and/or that the derived condition rather quickly became burdened by the evolution of dependent traits. Paradoxically, despite the current entrenchment of such traits, they may initially have conferred greater flexibility, opening up new design possibilities and consequently the exploitation (or "creation") of new environments.

So far, the basic apical meristem features highlighted here (Fig. 4) have attracted rather little attention from molecular developmental biologists. These characters are, after all, deeply embedded within the phylogeny of green plants, a very great distance from the popular model organisms, and relevant mutations have rarely been recorded. At this stage, even the formulation of credible developmental models, and perhaps the identification of candidate genes and appropriate study organisms, would be quite useful. Along these lines, Geeta (2003) has recently sketched such a model for the origin of dichotomous branching. This entails a duplication in the location of the normal activity of the shoot apical meristem (regulated in part by the KNOX gene pathway), possibly brought about by the periodic expression of so-called MYB genes in the center of the meristem (specifically the ARP genes AS1/rs2/phan). My hope is that speculation of this type will encourage more careful comparisons and experimental work in the relevant organisms (e.g., the apex of the moss sporophyte, branching in lycophytes).

Convergence and Equivalence

New phylogenetic results will also, I believe, bring about a shift in how we interpret the significance of the recurrence of similar character states in different lineages. We have rightly viewed such cases as providing opportunities to test the effect of the evolution of a trait of interest on the evolution of other traits or on diversification rate. A repeated association between the evolution of a trait and elevated diversification rates suggests a causal connection. This seems reasonable, so long as we also appreciate that the effect might be somewhat indirect, or a function of the accumulation of characters, as discussed above. However, rather little attention has been paid to negative results—for example, where a trait is associated with increased diversification in one or a few lineages, but not in other lineages, and the correlation ends up looking weak with respect to a predicted consistent effect. One has the sense that such mixed results are the norm, although this is difficult to assess because such "insignificant" results tend not to be published.

What are we to make of such cases? One interpretation has recently been discussed by de Queiroz (2002), namely that the influence of a particular sort of character change is contingent on other factors. That is, for example, the origin of a particular state in a particular environment (say, the herbaceous habit in a terrestrial setting) may have a positive effect on diversification, whereas the evolution of the "same" trait in a different environment (say, herbaceousness in an aquatic habitat) might have little influence on diversification, or maybe even a negative effect. Feild et al. (2004) emphasized the critical role of the environmental context in understanding the function and the effect on diversification of such "key" angiosperm characters as vessels and closed carpels.

This is an excellent point, but another interpretation also comes to mind. Maybe some ways of making a trait are really somehow "better" than others. After all, traits that evolved independently in separate, distantly related lineages are apt to be truly convergent (as opposed to parallel) in the sense of having been constructed from different starting points, and possibly in quite different ways. Those differences might ultimately be of great significance in terms of both the subsequent evolutionary changes that they enable and lineage "success." Some ways of "solving" a problem might ultimately be better than others in the sense of allowing greater evolutionary flexibility.

Several cases from plant evolution come to mind, especially related to changes in organism size and longevity. The tree growth habit (tall plants, with a thickened single trunk, branching well above ground level) evolved many times independently—in lycophytes, equisetophytes, and lignophytes (Fig. 1), to name a few prominent cases (others cases are discussed briefly below). These cases all involved the same basic mechanism, namely the production in the stem of a cylinder of cambium—a secondary meristematic tissue that produces new cells to the inside and/or the outside of the stem, thereby increasing the girth of the stem. This, in concert with the evolution of a variety of mechanical support systems, allowed the evolution of large trees (Niklas 1997). Importantly, however, painstaking paleobotanical studies have shown that the cambium functioned differently in these different clades. In lignophytes (including the "progymnosperms," such as *Archaeopteris*, and seed plants) we find the familiar situation, in which the cambium is "bifacial"—producing secondary xylem tissue toward the center of the stem and secondary phloem tissue toward the outside (Fig. 5, top). By contrast, extinct tree lycophytes and equiseto-

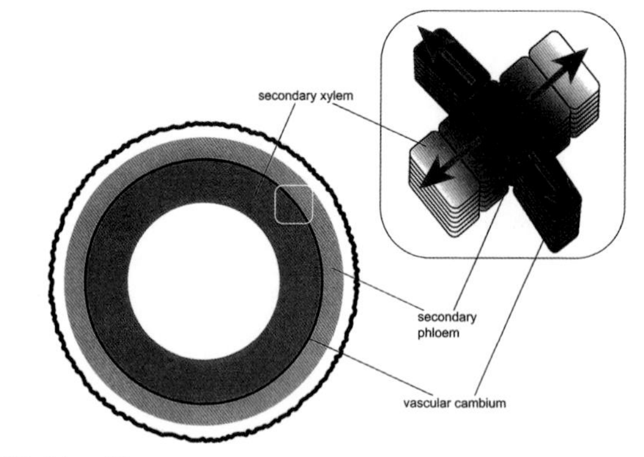

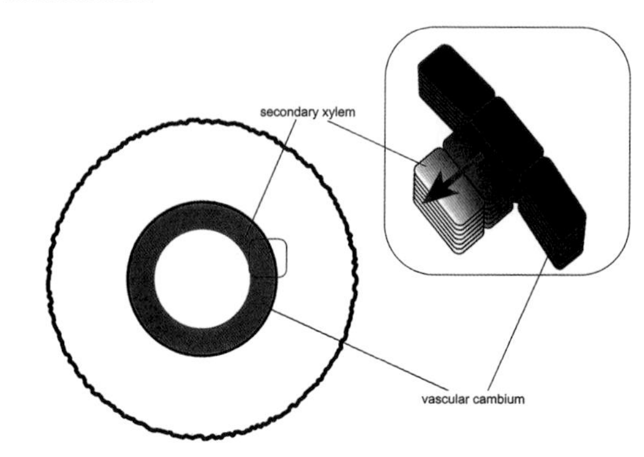

FIGURE 5. Differences between the bifacial cambium in the lignophyte lineage (including seed plants) and the unifacial cambium found in extinct tree lycophytes (e.g., *Lepidodendron*). The bifacial cambium produces both secondary xylem and secondary phloem, and the cambial initials are able to divide both periclinally (producing cells that differentiate in secondary tissues) and anticlinally (producing new cambial initials). The unifacial cambium produced only secondary xylem and the cambial initials divided only periclinally, limiting expansion of the cambial cylinder and the production of wood. These seemingly minor differences translated into major differences in evolutionary flexibility and "success" (see text).

phytes are reconstructed as having had a unifacial cambium (Eggert 1961, 1962; Cichan and Taylor 1990). They produced cells only to the inside, which differentiated as secondary xylem tissue, but not toward the outside to be differentiated as secondary phloem (Fig. 5, bottom). Secondary phloem appears to have been entirely lacking in these plants (Eggert 1972; Eggert and Kanemoto 1977).

Both types of construction allowed the evolution of large trees, but the differences in detail appear to have had profound consequences. The familiar seed plant cambium originated in the Devonian at the base of the lignophyte clade. Rather shortly thereafter, by the end of the Devonian, the major lineages of seed plants (aside from the angiosperms) had come into existence, including a variety of "seed fern" groups (looking rather like modern tree ferns), cycad-like plants, conifer-like plants, etc. This radiation spawned highly successful lineages of woody plants from the standpoint of their longevity, structural diversity, and species numbers.

By contrast, today there are only perhaps 1200 species of lycophytes, the vast majority of which (e.g., *Selaginella*, with approximately 700 species) represent lineages that retained the ancestral herbaceous habit and never included trees. Large lycophyte trees evolved (perhaps several times) within a clade characterized by heterospory (i.e., the production of two kinds of spores) and a flaplike "ligule" associated with each leaf, and they diversified and became widespread especially during the Carboniferous. *Isoetes* (so-called quillworts), containing perhaps 150 species of small rosette plants, is the only living descendant of the lycophyte line in which secondary growth evolved—these plants have retained a cambium and "rootlets" that resemble those of the extinct trees (Gifford and Foster 1989). However, *Isoetes* probably originated within the so-called cormos line of lycophytes (including *Chaloneria* and *Pleuromeia*), which never attained the great size of *Lepidodendron* and the other very large lycophyte trees. There are no living representatives of this "rhizomorphic" lineage. Similarly, the equisetophyte lineage (horsetails and relatives), which was diverse and produced large trees in the Carboniferous, is represented today by just 15 species of *Equisetum* (Des Marais et al. 2003), all of them relatively small plants lacking wood.

Although the down-sizing of the lycophyte and equisetophyte clades (in both plant stature and species number) may not be directly or entirely attributable to the structure of their wood, the unifacial cambium does seem to have placed significant functional constraints on the evolution of these plants—constraints that are reflected in a variety of other characteristics. First, relative to lignophytes with the bifacial cambium, the unifacial plants produced rather little wood. This was not a function of the unifacial cambium per se, but rather of the apparent inability of these plants to expand greatly the circumference of the cambial cylinder (Cichan and Taylor 1990). Cambial cells in lignophytes can undergo both periclinal and anticlinal cell divisions, the periclinal ones adding xylem and phloem and the anticlinal ones adding extra cells to the ring of cambium (Fig. 5). By contrast, cambial cells in unifacial plants apparently did not divide anticlinally. Consequently, any increases in the cambial ring were brought about by the growth of cambial initials in length, spreading apart the cambial initials situated just above and below them in the cambial cylinder. This mechanism can produce only very limited circumferential increases, and the girth of these plants may have resulted largely from something analogous to the primary thickening meristem found today in palm trees (see below; Bateman et al. 1992; Bateman 1994). The paucity of wood formed by these plants apparently had several other consequences. For one thing it meant that the wood that was produced had to be especially efficient, and achieving this entailed structural changes in the vascular tissue and the tracheary elements themselves (Cichan 1986). Also, because the wood of these plants could provide only minimal mechanical support (as compared with lignophyte trees), in the lycophyte line a peculiar barklike "periderm" tissue (situated in the outer cortex, beneath the persistent leaf bases) was "invented" to stiffen the trunk.

Other peculiar attributes of these plants reflect the lack of secondary phloem. In lignophytes, secondary phloem makes possible the

FIGURE 6. A sample of growth forms in extinct lycophytes. Two drawings on the left (from Phillips and DiMichele 1992) show early stages in the life cycle—establishment of the stigmarian "root" system with possibly photosynthetic "rootlets" prior to rapid stem elongation. Three drawings on the right (from Stewart and Rothwell 1993) show reconstructed forms of the determinate stems (not drawn to the same scale); from left to right: *Sigillaria*, *Pleuromeia*, and *Lepidodendron*.

transport of carbohydrates from sites of photosynthesis (typically leaves) to distant parts of the plant, such as the roots. In the absence of secondary phloem, such long-distance transport would be severely limited, which would necessitate the maintenance of photosynthesizing structures in the vicinity of tissues that needed to stay alive in order to function. Consequently, in the unifacial lineages we find several highly unusual strategies. Focusing now on lycophyte trees, we see the maintenance of photosynthesizing leaf bases all over the stems (and, consequently, the absence of normal bark as in seed plant trees). These plants probably also provisioned their massive so-called stigmarian "root" systems by producing photosynthesizing "rootlets" (probably leaf homologs), some of which appear to have been deployed above ground or into shallow water in the swamps that most of these plants occupied (Fig. 6) (Phillips and DiMichele 1992).

The truly weird life cycles inferred for these plants (Andrews and Murdy 1958; Eggert 1961; DiMichele and Phillips 1985; Phillips and DiMichele 1992) are also probably related to the lack of secondary phloem. It appears that the large lycophyte trees grew very little in height for many years, instead remaining stumplike while the stigmarian system became well established underground (Fig. 6). Then they bolted up to great heights, quickly produced their spores (repeatedly, or only once), and then died. In effect, the enormous above-ground stems behaved like the inflorescences of other plants. This highly unusual determinate growth mode (also found in equisetophyte trees) may have been, in part, a means of avoiding the long-term maintenance of dispersed functioning tissues without secondary phloem and the production of costly mechanical support tissues.

As noted above, the tree habit evolved in other lineages as well (Fig. 7). For example, arborescent forms are found among extinct marattialean (*Psaronius*) and filicalean (*Tempskya*)

FIGURE 7. Diversity of form among extinct treelike plants from the Devonian and Carboniferous (not drawn to the same scale). From left to right: *Archaeopteris* (an early lignophyte); *Calamites* (an equisetophyte); *Psaronius* (a marattialean "fern"), in which the trunk was formed by a mantle of adventitious roots; *Tempskya* (a filicalean "fern"), in which the trunk was formed by numerous smaller stems embedded in a tangle of adventitious roots.

ferns, and in modern tree ferns (Cyatheaceae). Trees were also re-evolved several times within the ancestrally herbaceous monocotyledonous flowering plants, with palm trees providing a prime example. In each of these cases the tree habit was achieved in a distinctly different way, and again in each case with obvious downstream consequences (Niklas 1997). In the upper Paleozoic *Psaronius* (Fig. 7) and in extant tree ferns, a cambium is lacking, and increased girth and mechanical support are provided by a mantle of adventitious intertwining roots. Cross-sections of the trunk of *Tempskya* reveal yet another way to make a tree—its "false" stems were made of many smaller ramifying stems (each lacking secondary growth) packed in among a dense thicket of adventitious roots (Andrews 1948). Similar construction is found today in the osmundaceous fern *Todea*. Palms also lack a vascular cambium, and increases in diameter are largely due to what is called a primary thickening meristem, situated in a zone where the young leaves attach to the stem (Rudall 1991; Tomlinson 1995). Their mechanical strength is provided by a combination of a greater density of vascular bundles in the outer cortical tissue and thickening of the cell walls in that region (Niklas 1997). Some other monocots, within a variety of separate lineages living mostly in arid regions (e.g., Agavaceae, Convalariaceae, Iridaceae, Xanthorrhoeaceae [Chase et al. 2000]), have also become trees (Tomlinson and Zimmermann 1969; Tomlinson 1995). As in the palms, the presence of vascular bundles that appear "scattered" in the stem (an atactostele) and of individual bundles that are "closed" to further growth (both conditions associated with the origin of the monocots) effectively precluded the re-evolution of a "normal" ring of cambium. Instead, these plants invented a novel form of unifacial cambium (the "etagen" cambium), situated near the periphery of the stem, which yields derivatives that differentiate as additional ground tissue and into whole new vascular bundles containing both xylem and phloem (Rudall 1991; Tomlinson 1995).

The point of giving these details about tree construction is to illustrate that different ways of attaining a given condition, evolved convergently in different lineages, can be constrained by prior circumstances (e.g., the reinvention of a cambium in monocots with scattered bundles) and, most importantly for present purposes, can sometimes have significant consequences for subsequent evolution in those lineages (e.g., the bizarre structures and life cycles of lycophytes). Such differences among lineages might translate into "negative" or only weakly positive results in standard phylogenetic comparative tests for character correlations or for correlations with diversity. As explained in the next section, I do not intend this as an argument against attempting to identify common evolutionary responses to convergent characters across lineages. Instead, I hope to highlight the potential power of negative results in such tests in helping to pinpoint consequential differences in "the same" structure, thereby refining the initial causal hypothesis.

Some Methodological Implications

Phylogenetic discoveries have been affecting macroevolutionary studies in a variety of completely obvious ways. In general, in trying to make sense of the tempo and mode of macroevolution it helps to know how species are related to one another. The point of my paper is that some much less obvious, but ultimately more fundamental, effects are on the horizon. Presently, we tend to want to pin the cause of the "success" of a clade on a "key innovation"—used here to refer to a trait responsible for increasing the rate of diversification (see Givnish 1997; Sanderson 1998; and Hunter 1998; for alternative views on "key innovation" and "adaptive radiation"). Tests of evolutionary character relationships and key innovation hypotheses hinge on phylogenetic correlations. Does the character of interest really correlate with a shift in diversification? Do we see repeated instances, in different clades, of such a correlation?

I made the case above that key innovations may not happen at a point in a tree, but over a region. Likewise, shifts in diversification may ratchet upward (or downward) not at a single spot in a tree, but over a series of nodes. New comparative methods need to be designed with this image in mind. We need tests that attempt to identify particular sequences of change that may have impacted diversification, as well as clusters of positive, but perhaps individually less than significant, shifts in diversification rate (see Moore et al. 2004 for some methodological developments along these lines).

Likewise, in testing for repeated evolutionary correlations, more attention needs to be paid to potentially significant character differences in different clades. In the case of truly convergent characters, as illustrated by the evolution of the tree habit, differences in constructional details can have profound effects on subsequent evolution and, ultimately, on what we judge to be clade success. Failure to identify a significant correlation in a phylogenetic test could reflect such underlying differences and might help to refine the comparison. Ultimately, of course, it is critical to specify a particular mechanistic connection between the evolution of a trait and the evolution of other traits and/or diversification rate. Formulating the causal hypothesis as precisely as possible will more clearly circumscribe which instances of "the same" character are relevant in performing a test (see Coddington 1994).

In many cases I imagine that an initial phylogenetic test will narrow the set of comparisons to characters with more specific similarities, perhaps often to cases of the parallel evolution of states in the strict sense (involving the same structural modifications and presumably the same genes, and therefore perhaps in more closely related organisms; see discussion in the next section). But, this is not to say that phylogenetic correlation tests are properly applied only to parallel changes. Instead, because the outcome, whether one uses wildly convergent or only strictly parallel changes, is potentially of interest, I am suggesting a nested data exploration strategy, beginning perhaps with obviously convergent traits and narrowing down the comparison depending on the results. For example, it seems well worth testing whether the tree habit, regardless of how it was actually attained, had a significant effect on the evolution of other traits or on patterns of diversification. Likewise, to mention another popular case in the plant literature (e.g., Donoghue 1989; Heilbuth 2000; Vamosi et al. 2003), it is worthwhile testing whether dioecy and fleshy, bird-dispersed propagules are correlated, or whether either one has influenced diversification, regardless of major structural differences (e.g., whether the actual fleshy structure is the wall of the seed, the wall of the fruit, or some accessory structure). But, where very different structures are involved, we should not be surprised or disappointed by negative or ambiguous results. Instead, we should learn from such experiences that structural details might make a difference with respect to the presumed mechanistic hypothesis, and then design more refined comparisons. Such refinements should take account of different organismal and environmental contexts (de Queiroz 2002), but they also should take more seriously the distinction between convergence and parallelism, which is often glossed over in such work.

Some Connections to Gould

These observations connect to Steve Gould's thoughts in a variety of ways. Gould presumably would have appreciated the idea of developmental enablers—changes early in a sequence that opened up new design options. But, exactly how such traits relate to Gould's concepts and terminology is a bit complicated. In my examples the underlying changes that set the stage for later, more obvious changes are themselves, I presume, adaptations. They are what once would have been labeled "preadaptations," a term that Gould rejected on the grounds of its being " 'prepackaged' for inevitable trouble and misunderstanding" (Gould 2002: p. 1232). Gould and Vrba (1982) introduced the term "exaptation" to cover any instance of co-optation, whether from a previous adaptation or from a nonaptation, but they emphasized that "exaptations that began as nonaptations represent the missing concept" (Gould and Vrba 1982: p. 12). Unfortunately, they left this more specific concept unnamed. Gould (2002: p. 1278), therefore, recently distinguished between what he called "franklins" ("alternative potential functions of objects now being used in another way") and "miltons" ("currently unused material organs and attributes") as the basic elements of the "exaptive pool." As he pointed out, "franklin" captures the concept behind the term preadaptation and "milton" captures the notion of nonaptations available for co-opting. Where do my plant examples fall in this expanded terminology? If I'm forced to use Gould's terms (which I must admit I have a hard time taking seriously), then my examples are very likely "franklins." That is, the underlying traits that I have described as developmental enablers (e.g., dichotomous branching, overtopping) were probably adaptations in their own right, but they also clearly provided inherent potential for future exaptive changes (e.g., to pseudomonopodial growth, leaves).

Having claimed that these cases are franklins, I hasten to note that I think there are also important miltons in plants, which have also been brought to light in phylogenetic analyses. For example, in recent studies of the angiosperm clade Dipsacales (a group of around 1100 species of Asteridae), we have discussed the evolution of a specialized structure called an "epicalyx" (Donoghue et al. 2003). It appears that the epicalyx evolved (possibly twice, in Dipsacaceae and in Morinaceae) through modification of several sets of subtending "supernumerary" bracts, which we interpret as having been "left over" from the earlier loss of flowers in the inflorescence (Donoghue et al. 2003). If so, the supernumerary bracts are miltons that were co-opted to form the epicalyx. It is difficult to quantify at this stage, but in view of the nature of plant morphology, and especially the evolutionary use and reuse of "leaves" for a very wide variety of purposes, I suspect that the co-optation of miltons has been quite common in plant evolution. As for Gould's distinctions between "spandrels," "manumissions," and "insinuations" (Gould 2002: p. 1278), I won't attempt to further categorize the epicalyx. In this case, and in the other real examples that come to mind, these categories do not seem mutually exclusive enough to warrant the formality.

My discussion of convergence and success intersects another area of relevance to Gould's thought, namely the distinction between parallelism and convergence, which he portrayed as critical for properly understanding the notion and the extent of "historical constraint." Parallelisms, he argued, reveal historical constraints in the evolving system—the same condition originates again and again within a lineage owing to something about the structure and development of the shared ancestor. Convergences, on the other hand, demonstrate the power of natural selection to fashion similar forms from very different starting points. Here again there are terminological issues. Gould (2002) provided a fine analysis of the convergence-parallelism distinction but settled on a terminology that I think may not be ideal. As he stressed, E. Ray Lankester, who coined the term "homoplasy" in 1870, viewed it (ironically) as a form of homology (equivalent to Owen's "general homology"). Specifically, Lankester meant to apply it to "independently evolved, but historically constrained, similarities—what we would now call parallelisms" (Gould 2002: p. 1073). Nevertheless, Gould chose to follow standard practice in applying "homoplasy" very broadly to all sorts of non-homology, including both parallelism and convergence.

My own preference is to use "analogy" for all non-homologous similarities (e.g., as Osborn did in 1905), and to use "homoplasy" in the more restricted (and original) sense to refer to parallelisms. "Homoplasy" would then refer precisely to the sorts of recurrent similarities detected in phylogenetic analyses. That is, it would refer to recurrences in the states of characters that are actually included in phylogenetic analyses on the working assumption that they are truly homologous because they pass Remane's positional, structural, and developmental tests of homology (Patterson 1982; Donoghue 1992). By contrast, convergences fail such tests and are excluded at the outset (as individual characters) from phylogenetic analyses. Applying the terms in this way would serve to connect these abstract discussions directly to work on levels of homoplasy in the phylogenetic literature (e.g., Sanderson and Donoghue 1989, 1996). Metrics of the extent of parallelism (e.g., the consistency index) could then help to quantify the importance of historical constraint.

But, leaving aside these terminological issues, I quite agree with Gould that the parallelism-convergence distinction is important from the standpoint of what it implies about the mechanisms underlying character change. However, the point of my examples is different, namely that the distinction is also important because parallelisms and convergences may have rather different long-term evolutionary consequences. The tree habit as manifested by lycophytes had very different consequences (in terms of the evolution of other characters, and long-term success) than did the tree habit as it evolved in lignophytes. Making (or failing to make) the convergence-parallelism distinction can have important consequences for comparative tests, and I have suggested a strategy of nested tests beginning with clear instances of convergence and working toward parallelisms.

This last statement implies the existence of a continuum between parallelism and convergence, which Gould also clearly appreciated and used to his advantage. His basic argument was that (1) parallelisms are important because they reveal constraints due to deeper homology, and (2) recent developmental studies have revealed that many instances of supposed convergence are actually at least in part cases of parallelism. Therefore, (3) constraint has been even more pervasive than we might have supposed. The critical link in his argument is the contention that many real cases show signs of both convergence and parallelism, a point he illustrated with examples such as the role of the *Pax-6* gene in the evolution of eyes in different animal lineages. So, what begins as a plea for paying more attention to the convergence-parallelism distinction ends up stressing that the distinction is a blurry one at best. My guess is that this blurriness is even more pervasive than Gould imagined. In plants, at least, with their modular, open developmental systems, David Baum and I (Baum and Donoghue 2002) have argued that cases of mixed or partial homology (see Sattler 1984, 1991) may be common owing to "transference of function," especially between adjacent organs, brought about by shifts in the location where genes are expressed (what we termed "homeoheterotopy"). The epicalyx in Dipsacales, mentioned above, may provide a concrete example. That is, the calyx-like appearance and function of the epicalyx might reflect the activation of calyx identity genes in a newly formed structure adjacent to the calyx (Donoghue et al. 2003). In the end, it may be difficult to sustain the notion of pure convergence, a thought that I suppose Gould would have enjoyed.

Finally, these thoughts about recurrence also bear on the issue of the role of convergence vis-à-vis the repeatability of evolution (Gould 1989; Conway Morris 1998, 2003; Conway Morris and Gould 1998). Convergences, parallelisms, and mixtures of the two surely will occur in any evolving systems, and at least for parallelisms we can make concrete predictions about the frequency of occurrence (depending on the number of branching events, the number of character states, and rates of character evolution [Donoghue and Ree 2000]). But the mere fact of recurrence, I would argue, does little to guarantee convincing repeat performances in running the tape of Life over again. The idea of convergence is that structures are put together in different ways from different starting points in different lineages. If my argument is correct that differences in construction (even seemingly minor ones) can have major effects on downstream evolutionary changes and patterns of diversification, then convergence on the same basic form in different iterations might yield wildly different outcomes. Large size, for example, may be selected again and again, but depending on the details of how large size is actually attained, we might end up with very different sorts of organisms. In one iteration we might get familiar-looking lignophyte-like trees (e.g., imagine a pine tree, or a maple), but in the next iteration we might see giant clubmosses or horsetails, and in a third go-around the world might fill up with palm trees. Although there may be commonalities in what is selected for, different mechanisms underlying the response could translate into enormous differences in structure, life cycles, patterns of "success," ecological communities, and so on. So, approaching the problem from a different angle, I end up squarely on Gould's side of this particular argument.

Acknowledgments

I am very grateful to Elizabeth Vrba and Niles Eldredge for inviting me to contribute to this volume, to J. Cracraft, S. Smith, R. Geeta, D. Stevenson, C. Delwiche, J. Doyle, L. Hickey, and W. DiMichele for helpful discussions, and to T. Feild and an anonymous reviewer for their comments on the manuscript. B. Moore deserves special thanks for critically reading the manuscript, and for his insights based partly on having reached similar conclusions about trait interdependence in relation to key innovations in his master's work at the University of Toronto. He also kindly produced the figures. Although I would argue that Steve Gould never fully embraced phylogenetic thinking, phylogenetic biologists (even botanical ones) are deeply indebted to him for developing macroevolutionary ideas that will keep us occupied for many years to come.

Literature Cited

Andrews, H. N. 1948. Fossil tree ferns of Idaho. Archaeology 1: 190–195.

Andrews, H. N., and W. H. Murdy. 1958. *Lepidophloios* and ontogeny in arborescent lycopods. American Journal of Botany 45:552–560.

Bateman, R. M. 1994. Evolutionary-developmental change in the growth architecture of fossil rhizomorphic lycopsids: scenarios constructed on cladistic foundations. Biological Reviews of the Cambridge Philosophical Society 69:527–597.

Bateman, R. M., W. A. DiMichele, and D. A. Willard. 1992. Experimental cladistic analyses of anatomically preserved arborescent lycopsids from the Carboniferous of Euramerica: an essay in paleobotanical phylogenetics. Annals of the Missouri Botanical Garden 79:500–559.

Bateman, R. M., P. R. Crane, W. A. DiMichele, P. R. Kenrick, N. P. Rowe, T. Speck, and W. E. Stein. 1998. Early evolution of land plants: phylogeny, physiology and ecology of the primary terrestrial radiation. Annual Review of Ecology and Systematics 29:263–292.

Baum, D. A., and M. J. Donoghue. 2002. Transference of function, heterotopy, and the evolution of plant development. Pp. 52–69 *in* Q. Cronk, R. Bateman, and J. Hawkins, eds. Developmental genetics and plant evolution. Taylor and Francis, London.

Boyce, C. K., and A. H. Knoll. 2002. Evolution of developmental potential and the multiple independent origins of leaves in Paleozoic vascular plants. Paleobiology 28:70–100.

Chapman, R. L., M. L. Buchheim, C. F. Delwiche, T. Friedl, V. A. R. Huss, K. G. Karol, L. A. Lewis, J. Manhart, R. M. McCourt, J. L. Olsen, and D. A. Waters. 1998. Molecular systematics of green algae. Pp. 508–540 *in* D. Soltis, P. Soltis, and J. Doyle, eds. Systematics of plants II. Kluwer Academic, New York.

Chase, M. W., D. E. Soltis, P. S. Soltis, P. J. Rudall, M. F. Fay, W. J. Hahn, S. Sullivan, J. Joseph, M. Molvray, P. J. Kores, T. J. Givnish, K. J. Sytsma, and J. C. Pires. 2000. Higher-level systematics of the monocotyledons: an assessment of current knowledge and a new classification. Pp. 3–16 *in* K. Wilson and D. Morrison, eds. Monocots: systematics and evolution. CSIRO Publishing, Collingwood, Victoria, Australia.

Cichan, M. A. 1986. Conductance of the wood of selected Carboniferous plants. Paleobiology 12:302–310.

Cichan, M. A., and T. N. Taylor. 1990. Evolution of cambium in geological time—a reappraisal. Pp. 213–228 *in* M. Iqbal, ed. The vascular cambium. Wiley, Somerset, U.K.

Coddington, J. A. 1994. The role of homology and convergence in studies of adaptation. Pp. 53–78 *in* P. Eggleton and R. Vane-Wright, eds. Phylogenetics and ecology. Academic Press, London.

Conway Morris, S. 1998. The crucible of creation. Oxford University Press, Oxford.

———. 2003. Life's solution: inevitable humans in a lonely universe. Cambridge University Press, Cambridge.

Conway Morris, S., and S. J. Gould. 1998. Showdown on the Burgess shale. [The Challenge by Conway Morris and the Reply by S. J. Gould.] Natural History 107:48–55.

Cracraft, J., and M. J. Donoghue, eds. 2004. Assembling the tree of life. Oxford University Press, New York.

Delwiche, C. F., R. A. Anderson, D. Bhattacharya, B. D. Mishler, and R. M. McCourt. 2004. Algal evolution and the early radiation of green plants. Pp. 121–137 *in* Cracraft and Donoghue, 2004.

de Queiroz, A. 2002. Contingent predictability in evolution: key traits and diversification. Systematic Biology 51:917–929.

Des Marais, D. L., A. R. Smith, D. M. Britton, and K. M. Pryer. 2003. Phylogenetic relationships and evolution of extant horsetails, *Equisetum*, based on chloroplast DNA sequence data (*rbc*L and *trn*L-F). International Journal of Plant Sciences 164:737–751.

DiMichele, W. A., and T. L. Phillips. 1985. Arborescent lycopod reproduction and paleoecology in a coal-swamp environment of Late Middle Pennsylvanian age (Herrin Coal, Illinois, USA). Review of Paleobotany and Palynology 44:1–26.

Donoghue, M. J. 1989. Phylogenies and the analysis of evolutionary sequences, with examples from seed plants. Evolution 43:1137–1156.

———. 1992. Homology. Pp. 170–179 *in* E. Keller and E. Lloyd, eds. Keywords in evolutionary biology. Harvard University Press, Cambridge.

———. 2002. Plants. Pp. 911–918 *in* M. Pagel, ed. Encyclopedia of evolution, Vol. 2. Oxford University Press, Oxford.

———. 2004. Immeasurable progress on the Tree of Life. Pp. 548–552 *in* Cracraft and Donoghue, 2004.

Donoghue, M. J., and J. A. Doyle. 2000. Demise of the anthophyte hypothesis? Current Biology 10:R106–R109.

Donoghue, M. J., and R. H. Ree. 2000. Homoplasy and developmental constraint: a model and an example from plants. American Zoologist 40:759–769.

Donoghue, M. J., C. D. Bell, and R. C. Winkworth. 2003. The evolution of reproductive characters in Dipsacales. International Journal of Plant Sciences 164:S453–S464.

Doyle, J. A. 1998. Phylogeny of the vascular plants. Annual Review of Ecology and Systematics 29:567–599.

Doyle, J. A., and M. J. Donoghue. 1986. Seed plant phylogeny and the origin of angiosperms: an experimental cladistic approach. Botanical Review 52:321–431.

Eggert, D. A. 1961. The ontogeny of Carboniferous arborescent Lycopsida. Palaeontographica Abteilung B 108B:43–92.

———. 1962. The ontogeny of Carboniferous arborescent Sphenopsida. Palaeontographica 110B:99–127.

———. 1972. Petrified *Stigmaria* of sigillarian origin from North America. Review of Paleobotany and Palynology 14:85–99.

Eggert, D. A., and N. Y. Kanemoto. 1977. Stem phloem of Middle Pennsylvanian *Lepidodendron*. Botanical Gazette 138:102–111.

Feild, T. S., N. C. Arens, J. A. Doyle, T. E. Dawson, and M. J. Donoghue. 2004. Dark and disturbed: a new image of early angiosperm ecology. Paleobiology 30:82–107.

Geeta, R. 2003. Variation and diversification in plant evo-devo [book review]. American Journal of Botany 90:1257–1261.

Gifford, E. M., and A. S. Foster. 1989. Morphology and evolution of vascular plants, 3d ed. W. H. Freeman, New York.

Givnish, T. J. 1997. Adaptive radiation and molecular systematics: issues and approaches. Pp. 1–54 *in* T. Givnish and K. Sytsma, eds. 1997. Molecular evolution and adaptive radiation. Cambridge University Press, Cambridge.

Gould, S. J. 1989. Wonderful life: the Burgess shale and the nature of history. Norton, New York.

———. 2002. The structure of evolutionary theory. Harvard University Press, Cambridge.

Gould, S. J., and E. S. Vrba. 1982. Exaptation—a missing term in the science of form. Paleobiology 8:4–15.

Graham, L. E. 1993. Origin of the land plants. Wiley, New York.

Heilbuth, J. C. 2000. Lower species richness in dioecious clades. American Naturalist 156:221–241.

Hunter, J. P. 1998. Key innovations and the ecology of macroevolution. Trends in Ecology and Evolution 13:31–36.

Judd, W. S., C. S. Campbell, E. A. Kellogg, P. F. Stevens, and M. J. Donoghue. 2002. Plant systematics: a phylogenetic approach, 2d ed. Sinauer, Sunderland, Mass.

Karol, K. G., R. M. McCourt, M. T. Cimino, and C. F. Delwiche. 2001. The closest living relatives of land plants. Science 294: 2351–2353.

Kenrick, P. 2000. The relationships of vascular plants. Philosophical Transactions of the Royal Society of London B 355: 847–855.

Kenrick, P, and P. R. Crane. 1997. The origin and early diversification of land plants: a cladistic study. Smithsonian Institution Press, Washington, D.C.

Knoll, A. H., S. W. F. Grant, and J. W. Tsao. 1986. The early evolution of land plants. *In* T. Broadhead, ed. Land plants: notes for a short course. Studies in Geology 15:45–63. Department of Geology,University of Tennessee, Knoxville.

Lankester, E. R. 1870. On the use of the term homology in modern zoology, and the distinction between homogenetic and homoplastic agreements. Annals of the Magazine of Natural History, 4th series, 6:34–43.

Mishler, B. D., and S. P. Churchill. 1985. Transition to a land flora: phylogenetic relationships of the green algae and bryophytes. Cladistics 1:305–328.

Moore, B. R., K. M. A. Chan, and M. J. Donoghue. 2004. Detecting diversification rate variation in supertrees. Pp. 487–533 *in* O. Bininda-Emonds, ed. Phylogenetic supertrees: combining information to reveal the tree of life. Kluwer Academic, New York.

Nickrent, D., C. L. Parkinson, J. D. Palmer, and R. J. Duff. 2000. Multigene phylogeny of land plants with special reference to bryophytes and the earliest land plants. Molecular Biology and Evolution 17:1885–1895.

Niklas, K. J. 1997. The evolutionary biology of plants. University of Chicago Press, Chicago.

O'Hara, R. J. 1992. Telling the tree: narrative representation and the study of evolutionary history. Biology and Philosophy 7: 135–160.

Osborn, H. F. 1905. The ideas and terms of modern philosophical anatomy. Science 21:959–961.

Patterson, C. 1982. Morphological characters and homology. Pp. 21–74 *in* K. Joysey and A. Friday, eds. Problems of phylogenetic reconstruction. Academic Press, London.

Phillips, T. L., and W. A. DiMichele. 1992. Comparative ecology and life-history biology of arborescent lycopsids in Late Car-

boniferous swamps of Euramerica. Annals of the Missouri Botanical Garden 79:560–588.
Pryer, K. M., H. Schneider, and S. Magallón. 2004. The radiation of vascular plants. Pp. 138–153 *in* Cracraft and Donoghue, 2004.
Remy, W., P. G. Gensel, and H. Hass. 1993. The gametophyte generation of some early Devonian land plants. International Journal of Plant Sciences 154:35–58.
Renzaglia, K. S., R. J. Duff, D. L. Nickrent, and D. J. Garbary. 2000. Vegetative and reproductive innovations of early land plants: implications for a unified phylogeny. Philosophical Transactions of the Royal Society of London B 355:768–793.
Rudall, P. J. 1991. Lateral meristems and stem thickening growth in monocotyledons. Botanical Review 57:150–161.
Sanderson, M. J. 1998. Reappraising adaptive radiation. American Journal of Botany 85:1650–1655.
Sanderson, M. J., and M. J. Donoghue. 1989. Patterns of variation in levels of homoplasy. Evolution 43:1781–1795.
———. 1996. The relationship between homoplasy and confidence in a phylogenetic tree. Pp. 67–89 *in* M. Sanderson and L. Hufford, eds. Homoplasy and the evolutionary process. Academic Press, San Diego.
Sattler, R. 1984. Homology—a continuing challenge. Systematic Botany 9:382–394.
———. 1991. Process homology: structural dynamics in development and evolution. Canadian Journal of Botany 70:708–714.
Schneider, H., K. M. Pryer, R. Cranfill, A. R. Smith, and P. G. Wolf. 2002. Evolution of vascular plant body plans: a phylogenetic perspective. Pp. 330–363 *in* Q. Cronk, R. Bateman, and J. Hawkins, eds. Developmental genetics and plant evolution. Taylor and Francis, London.
Soltis, P. S., D. E. Soltis, M. W. Chase, P. K. Endress, and P. R. Crane. 2004. The diversification of flowering plants. Pp. 154–167 *in* Cracraft and Donoghue, 2004.
Stewart, W. N., and G. W. Rothwell. 1993. Paleobotany and the evolution of plants, 2d ed. Cambridge University Press, New York.
Tomlinson, P. B. 1995. Non-homology of vascular organization in monocotyledons and dicotyledons. Pp. 589–622 *in* P. Rudall, P. Cribb, D. Cutler, and C. Humphries, eds. Monocotyledons: systematics and evolution, Vol. II. Royal Botanic Gardens, Kew, U.K.
Tomlinson, P. B., and M. H. Zimmermann. 1969. Vascular anatomy of monocotyledons with secondary growth—an introduction. Journal of the Arnold Arboretum 50:159–179.
Vamosi, J. C., S. P. Otto, and S. C. H. Barrett. 2003. Phylogenetic analysis of the ecological correlates of dioecy in angiosperms. Journal of Evolutionary Biology 16:1006–1018.
Zanis, M. J., D. E. Soltis, P. S. Solits, S. Mathews, and M. J. Donoghue. 2002. The root of the angiosperms revisited. Proceedings of the National Academy of Sciences USA 99:6848–6853.
Zimmermann, W. 1965. Die Telomtheorie. Fischer, Stuttgart.

Paleobiology, 31(2), 2005, pp. 94–112

Wonderful strife: systematics, stem groups, and the phylogenetic signal of the Cambrian radiation

Derek E. G. Briggs and Richard A. Fortey

Abstract.—Gould's *Wonderful Life* (1989) was a landmark in the investigation of the Cambrian radiation. Gould argued that a number of experimental body plans ("problematica") had evolved only to become extinct, and that the Cambrian was a time of special fecundity in animal design. He focused attention on the meaning and significance of morphological disparity versus diversity, and provoked attempts to quantify disparity as an evolutionary metric. He used the Burgess Shale as a springboard to emphasize the important role of contingency in evolution, an idea that he reiterated for the next 13 years. These ideas set the agenda for much subsequent research. Since 1989 cladistic analyses have accommodated most of the problematic Cambrian taxa as stem groups of living taxa. Morphological disparity has been shown to be similar in Cambrian times as now. Konservat-Lagerstätten other than the Burgess Shale have yielded important new discoveries, particularly of arthropods and chordates, which have extended the range of recognized major clades still further back in time. The objective definition of a phylum remains controversial and may be impossible: it can be defined in terms of crown or total group, but the former reveals little about the Cambrian radiation. Divergence times of the major groups remain to be resolved, although molecular and fossil dates are coming closer. Although "superphyla" may have diverged deep in the Proterozoic, "explosive" evolution of these clades near the base of the Cambrian remains a possibility. The fossil record remains a critical source of data on the early evolution of multicellular organisms.

Derek E. G. Briggs. Department of Geology and Geophysics, Yale University, Post Office Box 208109, New Haven, Connecticut 06520-8109. E-mail: derek.briggs@yale.edu
Richard A. Fortey. Department of Palaeontology, The Natural History Museum, Cromwell Road, London SW7 5BD, United Kingdom. E-mail: raf@nhm.ac.uk

Accepted: 24 April 2004

Introduction

Of the popular books written by S. J. Gould, none enjoyed more global success than *Wonderful Life* (1989). The book, at heart, was a detailed account of the investigation of the Middle Cambrian "soft-bodied" fossils from the Burgess Shale of British Columbia discovered by Charles Doolittle Walcott 80 years earlier. It celebrated the kind of patient morphological work carried out by Harry Whittington and his colleagues over several decades, and for many paleontologists provided a high-profile vindication of what might be termed "classical" research. Thanks to *Wonderful Life,* such esoteric animals as *Hallucigenia* and *Anomalocaris* became almost as familiar to the scientifically literate as *Brontosaurus*. Gould also proposed several major new ideas, some derived from the work of the students of the Cambrian fossils, others entirely his own. These ideas stimulated criticism and research programs in equal measure. Fifteen years later, it is timely to examine how Gould's ideas on the Cambrian have fared in the light of subsequent thought and discoveries. Whatever the outcome, few would question that *Wonderful Life* served as an inspiration at best, a goad at worst. It was a catalyst for the examination of issues that are still under discussion today.

The main focus of *Wonderful Life* was the meaning of the Cambrian evolutionary "explosion." Gould apparently had no doubt that the appearance of major groups of animals at the base of the Cambrian was a product of rapid evolution of novel clades, rather than, for example, a change in their fossilization potential due to the acquisition of skeletons and an increase in body size. The notion of an unrivaled period of morphological experimentation in the Cambrian led to a controversial corollary: that the range of designs—what Gould termed "disparity"—at some fundamental level was actually *greater* in the Cambrian than it was subsequently.

Although early claims about the numbers of

0094-8373/05/3102-0007/$1.00

Cambrian body plans are not taken seriously today, the phylogenetic "explosiveness" or otherwise at the base of the Cambrian has remained an issue ever since, with new fossil discoveries and new kinds of evidence, notably molecular, being brought to bear on the topic. What range of form evolved in the Cambrian, and how quickly? How should we classify fossils that appear to fit on the stem-group leading to a "modern" clade? What, after all, *is* a phylum, and how do we recognize one? How long was what Cooper and Fortey (1998) termed the "phylogenetic fuse" prior to the "explosion"? If the range of designs was so wide in the Cambrian then the chance extinction of one animal rather than another might have served to reset the subsequent direction of evolution. To what extent is the shape of life contingent upon the chance extirpation of what might have proved to be another great group *in statu nascendi*? We might expect answers to some of these questions by now, not least because many new fossil faunas have been discovered since *Wonderful Life* was written. In this paper we examine the extent to which Gould's views have stood the test of new evidence, and discuss how subsequent research has raised new questions since his assessment of the meaning of the Burgess Shale.

Classification of the Burgess Shale Animals

Gould championed the significance of fossils in understanding the early history of animal groups and their interrelationships, a view that contrasted with the stance taken by more extreme "pattern cladists" who argued that fossils can provide little or no useful information about the phylogeny of modern organisms (Patterson 1981). In *Wonderful Life* (1989) Gould emphasized the importance of fossils, and particularly Cambrian fossils, in providing evidence of morphologies that have been lost as a result of extinction. The initial emphasis of interpretations of the Burgess Shale fossils was on their peculiarities. There were indeed aspects of the design of some of the animals that were unique. These fossils did not—still do not—fit *readily* into categories based solely on the living fauna. At the time *Wonderful Life* was published, a current perception was that the peculiarities that such animals exhibited was an indication that they must have originated separately from different putative soft-bodied Precambrian ancestors. This evolved into a view that at least some of the "weird wonders," such as *Hallucigenia* and *Opabinia*, were preserved fragments of "new phyla"—experiments in design that failed to prosper, and were weeded out by subsequent extinctions. This view of evolution came under scrutiny almost from the outset (Fortey 1989; Briggs and Fortey 1989), and much progress has been made subsequently in fitting even some of the most bizarre animals into a reasonable tree of descent, without invoking polyphyly. Progress toward understanding the phylogenetic relationships and classification of Cambrian soft-bodied animals has come from various sources.

Critical restudy of some of the animals showed that they were not as peculiar as first claimed. The *cause célèbre* in this regard was *Hallucigenia*, a definitively peculiar animal "with an anatomy to match its name" (Gould 1989: p. 14) which, when further prepared (Ramsköld 1992), showed paired series of legs that demonstrated its nature as some kind of lobopod. In the process, the original reconstruction was turned upside down, and its "unknown phylum" status relinquished. Similar attempts were made to rehabilitate the enigmatic *Opabinia* (Budd 1996) as a stem-group arthropod, by reinterpretation of its appendages and the identification of leglike structures on its ventral surface, although the evidence for these remains equivocal.

New discoveries served to link formerly problematic taxa more closely with recognized clades, revealing the possibility of their rational classification. This was particularly the case when new taxa from the Chengjiang fauna of China and the Sirius Passet fauna of Greenland added to the picture of Cambrian diversification. It had been appreciated since Conway Morris's (1977b) study of priapulids from the Burgess Shale that some groups of minor significance today were of considerable importance and diversity in the Cambrian. Following his 1993 description of *Kerygmachela* from Greenland (Budd 1993), Budd argued that the lobopods were also comparatively diverse in these early faunas and made a case for their seminal role in the evolution of arthropods (Budd 2002). Whether or not his ideas prove to be correct in detail, they serve to link a variety of "weird" fossils with more familiar morphologies. Some other enigmatic Cambrian fossils also turned out to be lobopods—for example, the plates known as *Microdictyon* are the "body armor" of one of these creatures (Chen et al. 1989; Ramsköld and Chen 1998). Other fossils that first appeared to be unique to the Burgess Shale proved to have close relatives from other Cambrian Konservat-Lagerstätten, thus turning an oddity into a clade. The giant animal *Anomalocaris* excited much speculation as to whether it was, or was not, an arthropod. It had an arthropod-like pair of cephalic grasping limbs, but its trunk appendages appeared unique. Gould, following Whittington and Briggs (1985), considered that it represents a separate phylum. The discovery of anomalocaridids from China (Hou et al. 1995), and additional forms from British Columbia that remain to be described (Collins 1996) increased the disparity of the group, and for all their peculiarities, brought the anomalocaridids within the compass of the discussion of arthropod origins (definitive evidence of segmented trunk limbs in anomalocaridids remains to be published—*Parapeytoia* may be a "great-appendage" arthropod). At the same time, these new animals suggested that the anomalocaridids may have constituted a clade, Dinocarida (Collins 1996). Whether *Halkieria* (Conway Morris and Peel 1995) from Greenland is a fossil that links molluscs and brachiopods is still under debate, but it can be added to the inventory of Cambrian animals that provide evidence of character combinations that link high-level taxa rather than suggesting "separate origins" from some unknown ancestor.

Gould did not embrace the cladistic method (Levinton 2001), relying on a more traditional approach to phylogeny, which tended to emphasize differences between taxa rather than the similarities that indicate relationship. Many of the Burgess Shale animals were sufficiently well preserved to be incorporated into cladistic analyses of relationship. With their complex characters, arthropods were the most suitable candidate for combining data from fossils and Recent animals together in a single analysis. Wills et al. (1994) demonstrated that the great majority of Cambrian taxa can be accommodated with their modern counterparts in a plausible hypothesis of relationships—by building a phylogeny from the bottom up rather than trying to place fossils into higher taxa defined exclusively on the basis of living organisms. This approach shifted the emphasis from autapomorphy—the much-bruited "weirdness"—to synapomorphy, the characters shared between one animal and another. The logic in ranking of taxa demands that a classification is based on an analysis of this kind, in contrast to Hou and Bergström's (1997) erection of 16 new higher taxa (as well as six new families) when describing the Chengjiang arthropods, without rigorous character analysis. The first cladistic analysis that attempted to order the Burgess Shale arthropods by using parsimony (Briggs and Fortey 1989) has been followed by many analyses using progressively more sophisticated character sets (e.g., Wills et al. 1998) that combine fossil and living taxa. At the same time, phylogeny of the arthropods as a whole—including sister taxa such as Onychophora and Tardigrada—has been being tackled from the molecular systematics database (Wheeler 1998). Increased levels of resolution have been obtained by analyzing molecular and morphological evidence in a single database (Giribet et al. 2001). Fossil-based cladistics, and the "top down" approach of using molecular sequences from living animals, constitute a double assault on the arthropod tree. Even so, it would be incorrect to claim that even the major features of the tree are uncontroversial; for example, different versions of the deep branches in arthropod phylogeny were published in adjacent papers in *Nature* (Hwang et al. 2001; Giribet et al. 2001), and paleontologists continue to disagree about the homologies to be recognized in coding morphological characters.

Phylogenetic analyses of the Burgess Shale arthropods based on parsimony (Briggs and Fortey 1989; Briggs et al. 1992a; Wills et al. 1994, 1998) were based on all available morphological characters. Whether the characters were unweighted, or weighted on the basis of successive approximations, the gross topology was retained. However, consistency indices were relatively low and the positions of some individual taxa fluctuated, perhaps as a natural consequence of the character combinations found in the early representatives of groups. We do not understand enough yet about the homology and polarity of characters to determine the phylogeny of the Cambrian forms in detail. Developmental studies of modern arthropods offer the best hope for determining the significance of characters and how to use them.

Regardless of ongoing contention, a number of generalizations revealed by the cladistic approaches are relevant to the scenario outlined in *Wonderful Life.* Comparatively few fossils have a morphology that prevents their being incorporated into a tree with living forms. Most of the arthropod genera can be considered as stem crustaceans or stem arachnomorphs, and there is good consensus on the placing of most of the important taxa. The Trilobita, for example, are securely placed within a larger arachnomorph clade. Dinocarida form a sister group of Euarthropoda. Onychophora lie outboard of the euarthropods. Since the work of Averof and Akam (1995) Insecta and Crustacea are increasingly regarded as allied, and the former have neither any Cambrian record nor plausible sister taxon of that age. It is not necessary to invoke phylum-level taxa to accommodate the Cambrian fossil material. However, the autapomorphies exhibited by some of the early arthropod stem groups are open to different interpretations: some workers consider them "worth" a high-level taxon, whereas cladistic analyses show that many of them constitute plesions on the tree leading to the crown group.

The Question of Disparity

As a response to what might be termed cladistic reductionism, Gould (1991) emphasized the quantification of morphospace as a measure of "disparity"; i.e., he considered the morphological reach of autapomorphies to be as significant as the collective compass of synapomorphies. In a universe of possible designs—without prime regard to phylogenetic origin—was he correct in asserting that the Cambrian forms occupied a larger volume of morphospace than living animals? Does the evidence of the Burgess Shale animals invert the conventional cone of increasing diversity? This turned out to be a hard question to address, not least because of the difficulties of quantifying design (Briggs et al. 1992a). Of the Cambrian faunas, the arthropods are perhaps the most obvious group to use in tackling this question, because they exhibit striking differences in appendages and tagmatization. Wills et al. (1994) used principal component analysis to determine the distribution of a range of living and Cambrian arthropods in morphospace, supplemented by a variety of measures of disparity, and concluded that there was no difference in morphospace compass between Cambrian and Recent faunas. The cone of increasing diversity was not inverted (sensu Gould 1989: p. 46) but might be represented rather as a "tube" with a diameter that remained roughly constant following its establishment in the early Cambrian (see Wills and Fortey 2000). There was certainly no evidence for greater disparity of design in the Cambrian, one of the main conclusions of *Wonderful Life.* The analysis of Wills et al. (1994) was vulnerable to the criticism that the Recent animals had been selected for their taxonomic "spread" rather than randomly sampled—such as was supposed to have happened in the fossilization process (Briggs et al. 1992b; Foote and Gould 1992; Lee 1992). Lofgren et al. (2003) showed that disparity among Carboniferous arthropods was approximately 90% of that in both the Cambrian and Recent, but they emphasized that the important difference is not the amount of morphospace but the particular regions occupied through time. A second study by Wills (1998) on the priapulid worms was able to sample more comprehensively the Recent fauna of this species-poor group, and found that the Cambrian taxa occupied *less* morphospace than the Recent. In this case, there was also a shift in the morphospace regions occupied between the Cambrian and Recent. There is no study that reports to the contrary, and enthusiasm for "greater disparity" in the Cambrian appears to be on the wane. Discoveries of younger, unusual morphologies in the Silurian Konservat-

Lagerstätte of Herefordshire, England (e.g., Sutton et al. 2001) and the Devonian Hunsrück Slate of Germany (Briggs and Bartels 2001), which had already yielded *Mimetaster*, part of a clade with *Marrella*, also serve to diminish the apparent distinctiveness of the Cambrian faunas. The lack of what Gould would have called weird wonders in the Ordovician and Silurian may be at least partly a matter of the scarcity of exceptional preservation (Allison and Briggs 1993; Briggs et al. 1996). This more prosaic view of the Cambrian radiation is now widely accepted but, even if disparity were not greater during the Cambrian, by that time the radiation had filled as much morphospace as is occupied by all the arthropods today. The rapid evolution of form still remains to be explained, and it is only the amount of evolution implied by Gould's view that has diminished.

What, If Anything, Is a Phylum? The Crown Group–Stem Group Debate

Gould's (1989) thesis on the Cambrian radiation was based around the concept of phyla or body plans. The focus on phyla has shifted from considering them as a measure of morphological separation to attempting to understand their relationships, and much of this hinges on the use of molecular data. However, the problem of conceptualizing phyla continues to permeate debates about the nature of the Cambrian explosion. Budd and Jensen (2000) reassessed the fossil record of the bilaterian phyla in an attempt to clear up "confusion in the definition of a phylum" (p. 253). In simple terms their view was that the earliest members of a lineage leading to the modern members of any phylum cannot, by definition, have acquired the diagnostic body plan, and therefore cannot be identified, on the basis of morphology, as belonging to that particular phylum. So at what point does a lineage merit the term phylum?

Budd and Jensen (2000; see also Budd 2003) resorted to a crown-versus stem-group definition (Fig. 1A), a method provided much earlier (Jefferies 1979) to accommodate both fossil and Recent taxa in cladograms. The crown group within a clade consists of the last common ancestor of all living forms in the phylum and all of its descendants. The stem group is the remainder of the clade, i.e., a series of extinct organisms lying "below" the crown group on the cladogram. Budd and Jensen argued that the origin of the crown group is the only point in the cladogram that can be objectively defined, so they equated the appearance of the phylum with the origin of the crown group.

Defining the crown group as the phylum is convenient mechanistically, as it is straightforward to apply and avoids the difficulty of how to separate members of sister phyla near their divergence from a common ancestor. It focuses on that part of the tree of life that includes extant animals and for which, therefore, molecular evidence is available. Restricting the concept of a phylum to the crown group as advocated by Budd and Jensen (2000) also has a number of drawbacks, however, that reduce its utility as a measure of the timing and evolution of character complexes (Wills and Fortey 2000). There are, by definition, no extinct phyla—only clades with living representatives merit phylum status. Curiously, then, some fossil taxa do not belong to a phylum. Membership of the crown group varies with different hypotheses of relationship (although it is not a phylum, the crown clade Mammalia provides a good example [Benton 2000]). Phyla are defined on the basis of an arbitrarily selected time line (the Recent), the autapomorphies present in the living forms determine how many and which characters define the phylum, and fossils that do not share these autapomorphies are not members. The inclusion of fossil taxa within the crown group usually depends on the chance survival of a few primitive organisms today, and this in turn inevitably determines how early the phylum originates. The survival of modern horseshoe crabs, for example, pulls the extinct eurypterids into the crown clade of chelicerates (Selden and Dunlop 1998). On this basis a phylum is defined by the possession of an arbitrary number of autapomorphies, and an organism with one fewer ($n - 1$ rather than n), lying immediately below the crown group, is disqualified (Fig. 1A). Were the horseshoe crabs to go extinct, the chelicerate crown group would be defined by scorpions + sister taxa; eurypter-

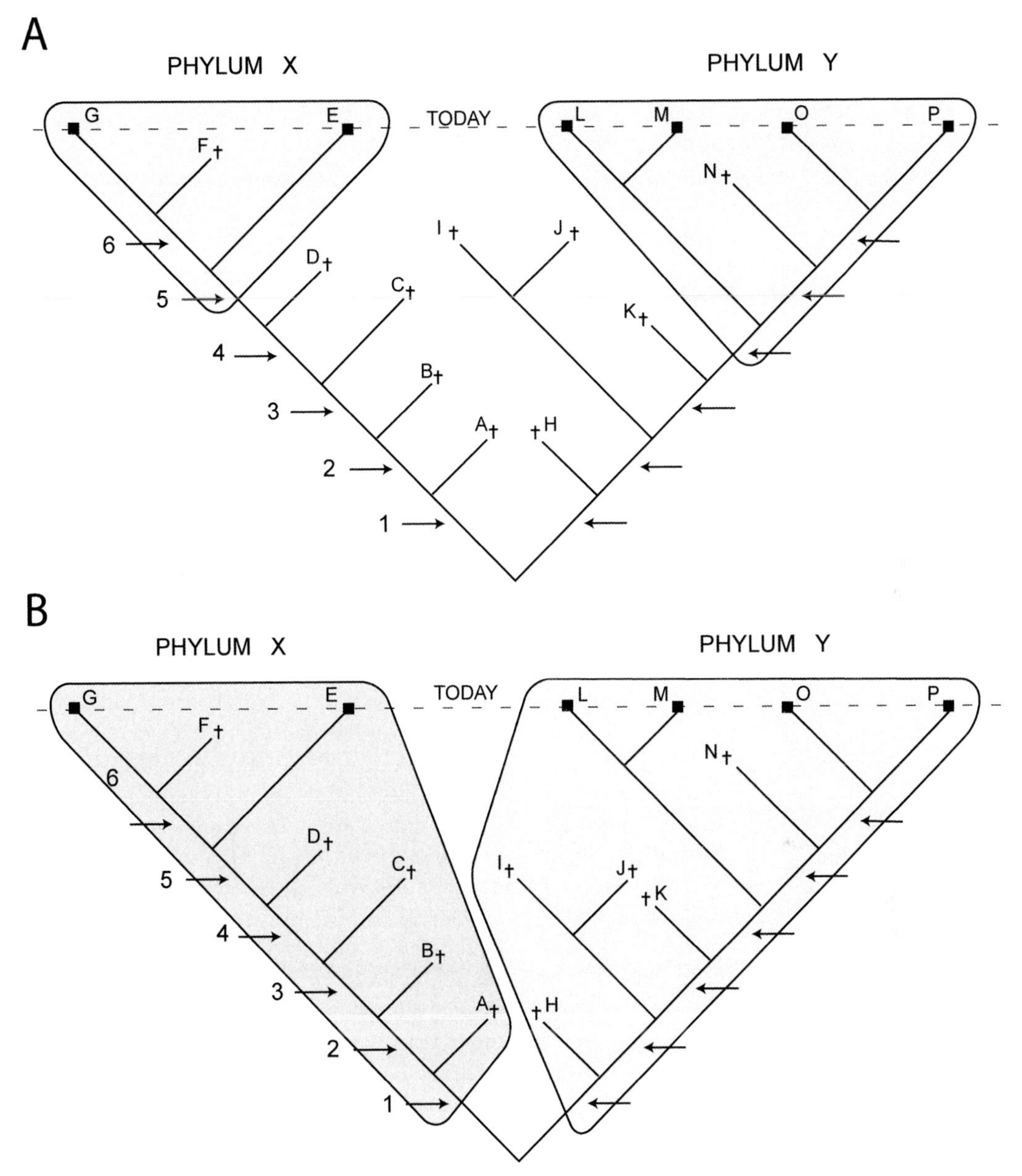

FIGURE 1. Diagrammatic illustration of two definitions of phyla. A, The crown-clade definition advocated by Budd and Jensen (2000). Phylum X contains three taxa, E to G, of which one is extinct. Extinct taxa A to D belong to the stem group, but are not part of the phylum. The phylum is defined by five synapomorphies; extinct taxon D, which has just one fewer, is disqualified from membership. Similar considerations apply to Phylum Y. B, The total-clade definition, preferred here. Phylum X contains seven taxa, living and extinct, and is defined by one synapomorphy. Similar considerations apply to Phylum Y. Taxa A and H may be difficult to assign, as their proximity to the origin is reflected in similarities of primitive morphology, but this difficulty does not bear on the reality of their position to one side of the node or the other. Note that the taxa in the diagram could represent any level below phylum, and "synapomorphy" implies a suite of characters that define a node. (Based on Jefferies 1979; Budd and Jensen 2000; Wills and Fortey 2000.)

ids and horseshoe crabs would be excluded, despite the many characters they share with the scorpions. There seems something very arbitrary about defining a major group on a whim of history. Budd and Jensen's definition implies that neither the Phylum Arthropoda nor the Mollusca are represented until the Late Cambrian, the Annelida until the Ordovician, or the Priapulida until the Carboniferous, although Chordata and Brachiopoda still originate in the Early Cambrian. But few will agree that this helps us to track the evolution of major groups; most still adhere to Gould's (2002: p. 1155) view that "all major bilaterian phyla with conspicuously fossilizable hard parts make their first appearance in the fossil record within a remarkably short interval . . . of the so-called Cambrian explosion."

Budd and Jensen (2000) reduced the significance of the Cambrian "explosion" not by making more time available for evolution in the Precambrian (the cryptic fossil record implied by some molecular clock estimates), nor by diminishing the amount of evolution required (by quantifying and comparing morphological separation among sample taxa), but by arguing that much of the morphological evolution required to give rise to the modern phyla actually took place later than is normally acknowledged, during the Phanerozoic. On the other hand, the more new Cambrian arthropods that are discovered, the more (morphologic) evolution seems to have happened already by the early Cambrian. Although we endorse the need to interpret the history of body plans in the light of well constrained phylogenies, Budd and Jensen made no attempt to quantify post-Cambrian evolution. They suggested "that a great amount of body-plan reorganization must have taken place post Cambrian, to generate such distinctive taxa as for example the spiders and flying insects" (p. 259). However, the insects considered in the analysis of Cambrian and Recent arthropod disparity performed by Wills et al. (1994) plotted closer to the global centroid than did 16 of the Cambrian taxa (spiders were omitted from the study). Likewise the living *Acanthopriapulus* lies closer to the overall centroid for fossil and living Priapulida than six of the ten fossil taxa included in Wills's (1998) analysis of this phylum, and further from it than the Carboniferous *Priapulites*, which is included in the crown group by Budd and Jensen. Budd and Jensen's redefinition of phylum in terms of stem and crown groups, though helpfully objective, does not improve our understanding of the Cambrian radiation. Phylogenies tell us about the probable sequence of events, but quantification of disparity (Briggs et al. 1992a; Wills et al. 1994), the approach inspired by Gould's *Wonderful Life*, remains a better way of measuring amounts of evolution.

It has been suggested that species might be identified on the basis of a DNA "bar code," the sequence of single gene, but there is some doubt that any single gene could resolve all animal species (Pennisi 2003). A higher taxon might be defined by a control gene, if one were unique to a single clade. Hox genes control the expression of structures such as limbs, so that their mutations may have a profound effect on the phenotype of the adult organism. It is tempting to correlate homeotic changes with the diversification of segmentation patterns in arthropods, for example, the group that provides the most diverse and disparate Cambrian sample and has also received substantial attention from developmental biologists today. The main problem with evoking such a model, however, is that abrupt ("saltational") changes might result in a new organism that is not integrated functionally (Budd 1999). It is possible, therefore, that changes in the features controlled by Hox genes accumulated by increments, and the control hierarchy was imposed subsequently to ensure an increase in efficiency in building the body plan (but see Ronshaugen et al. 2002). This is the interpretation favored, for example, for the origin of insect wings (Carroll et al. 1995). Thus the most primitive members of a clade may not have evolved the diagnostic control genes just as they also lack the morphological characters that identify the living members of a phylum.

How then to define a phylum? The total group (Fig. 1B), i.e., all the members of a clade, both stem and crown, has been used by those generating molecular phylogenies. The difficulty of separating the earliest representatives of the stem groups of two diverging phyla does not negate the validity of the branching point as the origin of the clade. Nor does it diminish the usefulness of analyses of early fossil taxa that specify the suite of characters displayed and where the taxa lie on the stem group. Trilobites may be extinct but even cladistics can accommodate them in the Phylum Arthropoda. What we know of the fossils present in the Lower Cambrian indicates that they had already diverged a considerable morphological distance from the ancestral condition when phyla might have been confusingly similar and impossible to classify on morphology alone. It remains a feasible research program to decide on which side of a fundamental dichotomy a given fossil lies.

The Importance of Other Cambrian Konservat-Lagerstätten

The interval since the publication of *Wonderful Life* (Gould 1989) has witnessed significant new research on the Cambrian, and in particular many important discoveries from the Lower Cambrian of Chengjiang in southwest China, the Sirius Passet fauna of North Greenland, and the Upper Cambrian phosphatized "Orsten" of southern Sweden and elsewhere. The number of genera described from the Chengjiang fauna is now only slightly less than that from the Burgess Shale, and discoveries are still being made. Discoveries of new soft-bodied fossil occurrences naturally extend the ranges of taxa without hard parts, owing to the taphonomic bias in the fossil record against non-biomineralized taxa.

Our understanding of metazoan phylogeny has advanced considerably since 1989, largely through molecular data. The cladogram (Fig. 2) looked very different prior to the recognition of the Lophotrochozoa and Ecdysozoa. Regardless of phylogenetic hypotheses, however, new fossil discoveries have extended or provided more convincing evidence for the stratigraphic range of taxa, and have reduced the length of ghost ranges.

The first attempt to use cladistic methods to unravel the phylogeny of the Burgess Shale arthropods was made by Briggs and Whittington in 1981, but this was before the widespread use of computer algorithms and was based on judgments about the significance of character attributes. At that time the few taxa that showed a suite of characters diagnostic of one of the major living groups (particularly the arrangement of head appendages) were assigned accordingly, but the majority were considered problematica. On this basis *Canadaspis* was interpreted as a crustacean (Briggs 1978) and *Sanctacaris* as a primitive sister group of the chelicerates (Briggs and Collins 1988). The remainder were the basis for the "twenty unique designs of arthropod" recognized by Gould (1989: p. 209).

The pivotal role claimed for the Burgess Shale bivalved arthropod *Canadaspis* as the earliest unequivocal crustacean (Briggs 1978, 1992a) was undermined by cladistic analysis using a complete range of characters (Briggs et al. 1992a; Wills et al. 1994, 1998). More-likely candidates were discovered among the tiny phosphatized arthropods from the Orsten of the Alum Shale of Sweden (Walossek 1999), but these are Upper Cambrian in age. It appears that *Canadaspis* originated lower in the arthropod stem group than the Orsten stem Crustacea. However, a fossil from Shropshire, England, reported by Siveter et al. (2001), extended the range of the phosphatocopid crustaceans to the Lower Cambrian, confirming that the Eucrustacea had originated by this time (the phosphatocopids are a sister taxon to the Eucrustacea sensu Walossek 1999). Cladistic analyses (Briggs et al. 1992a; Wills et al. 1994, 1998) placed *Sanctacaris* within the Arachnomorpha (Wills et al. 1998), but not in proximity to the living chelicerates. However, the recent description of a larval pycnogonid from the Swedish Orsten provides evidence for a Cambrian origin of chelicerates (Waloszek and Dunlop 2002). (Waloszek and Dunlop [2002] considered pycnogonids to form a sister group to the euchelicerates, but Giribet et al. [2001] placed them as a sister group to all other arthropods.)

Gould (1989: p. 171) considered that Whittington was incorrect to regard *Aysheaia* as a representative of a separate group, considering that it "should be retained among the Onychophora." In this respect Gould appears to have been correct—the discovery of several Cambrian lobopods, particularly in the Chengjiang biotas, has revealed a diverse group that includes the modern onychophorans (Ramsköld and Hou 1991; Ramsköld and Chen 1998). Ironically, of course, *Hallucigenia* also belongs here (Ramsköld and Hou 1991), although the incomplete information available to Conway Morris (1977a), and consequently Gould, prevented this realization.

Ramsköld and Chen (1998) included living

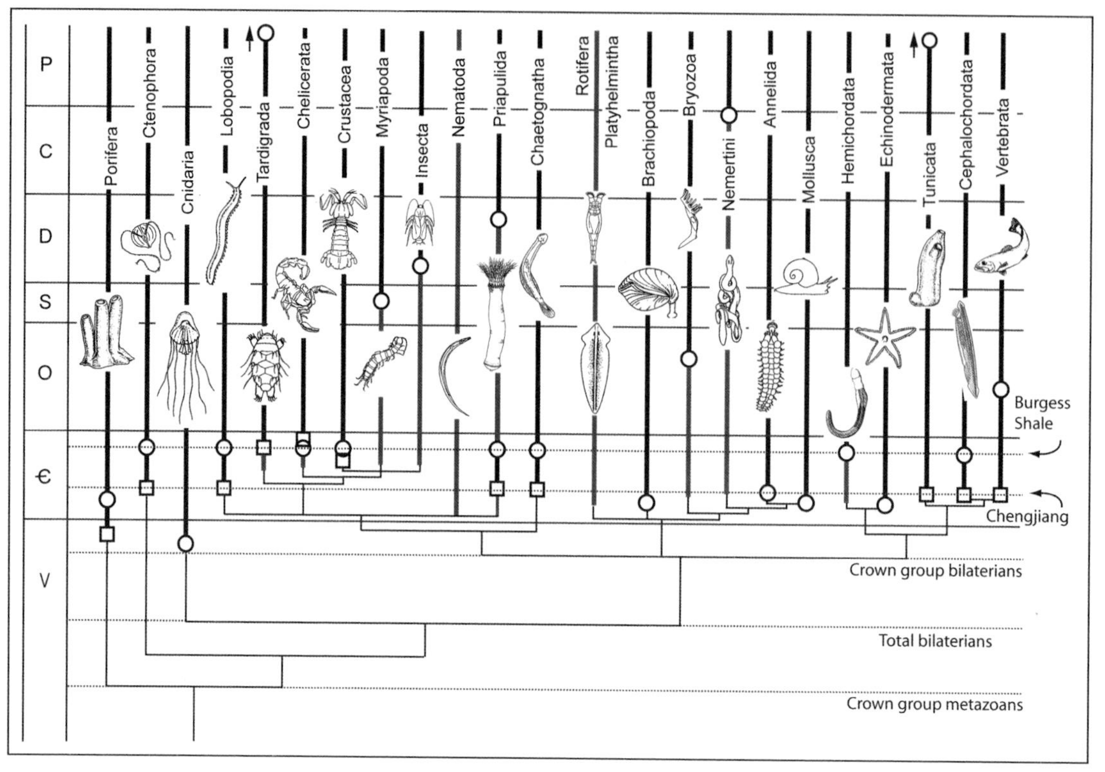

FIGURE 2. The fossil record and metazoan phylogeny. Relationships are illustrated as in Pennisi 2003. Dark lines represent the stratigraphic range of major metazoan taxa from their first appearance in the fossil record, indicated by a square, to the present. Light lines represent ghost ranges implied by a previous occurrence of a related taxon. The data on Lobopodia, Brachiopoda, Mollusca, Hemichordata, and Echinodermata, as well as phyla that first appear after the Cambrian, are from Benton 1993. Other first appearances are discussed in the text. Since *Wonderful Life* (1989), when first appearances were as indicated here by circles, discoveries in the Chengjiang fauna have pushed various Burgess Shale occurrences back 20 million years—Ctenophora, Lobopodia, Priapulida, Chaetognatha, Cephalochordata—and the vertebrates back from the Ordovician. The range of the tardigrades has extended from the Tertiary to the Cambrian, and the tunicates have acquired a reliable Cambrian record. Sponges have been discovered in the Vendian.

Onychophora (a single datum) and the Carboniferous taxon *Ilyodes* in an analysis with the Cambrian lobopods. Their strict consensus revealed the living Onychophora and *Ilyodes* as a clade in a polychotomy with two clades of Cambrian lobopodians (one containing *Hallucigenia*). In three of their nine solutions, however, the clade that includes the Recent forms is a sister group to one including all the Cambrian lobopodians. These solutions place all the Cambrian lobopodians in the crown group onychophorans and support Gould's view that *Aysheaia* does indeed belong here. In any event Ramsköld and Chen's (1998) analysis shows that at least one of the two clades of fossil lobopods must be a sister group of modern onychophorans and lie within the crown group. Budd and Jensen (2000: p. 259) regarded *Aysheaia* as "probably best considered not to lie in even the onychophoran stem lineage," but this assertion was not supported by a parsimony analysis (Budd 1996). The place of the tardigrades in arthropod evolution remains controversial. Dewel and Dewel (1998) placed them in the arthropod stem group between lobopods and anomalocaridids but conceded that this placement was conjectural. Importantly, however, Müller et al. (1995) described a stem-group tardigrade from the Middle Cambrian of Siberia extending the origins of the group back to the Cambrian.

Gould illustrated his *Wonderful Life* thesis with *Pikaia*, a laterally compressed, ribbon-shaped animal that displays the chordate features of myomeres and a notochord, but bears an unusual pair of tentacles on the "head." Gould argued that humans exist "because *Pikaia* survived the Burgess decimation" (1989: p. 323). Although he emphasized (p. 322) that he would not "be foolish enough to state that all opportunity for a chordate future resided with *Pikaia* in the Middle Cambrian," the fact that *Pikaia* was then the *only* chordate known from the Burgess Shale added a certain piquancy to the story and made the argument more easily accessible. *Pikaia* is now regarded as representing "a grade comparable to cephalochordates" (Shu et al. 1999: p. 46) and not directly ancestral to the vertebrates (Shu 2003: p. 733).

Gould doubted (1989: p. 149) that conodonts should be assigned to the chordates: *Wonderful Life* was written before evidence for their chordate affinities (Aldridge et al. 1986) was widely accepted by paleontologists (see Briggs 1992b; Ahlberg 2001). Gould (p. 322) believed that "fossils of true vertebrates, initially represented by agnathan, or jawless, fishes, first appear in the Middle Ordovician . . . ." He predicted (p. 322), however, that "other chordates, as yet undiscovered, must have inhabited Cambrian seas," a prophesy fulfilled by the subsequent discovery of two new poorly known forms from the Burgess Shale (Simonetta and Insom 1993; Smith et al. 2001) and, more importantly, by a remarkable diversity in the Chengjiang biota. Arthropods are diverse in both Burgess and Chengjiang biotas, but chordates are much more abundant in the Chinese material.

Haikouichthys (Fig. 3A) (Shu et al. 1999, 2003a) is fishlike, with well-documented generalized vertebrate characters, and it has been accepted rapidly as an early craniate. The head, with large eyes and paired nostrils, and the apparent vertebral elements (Shu et al. 2003a) demonstrate its craniate affinities—and indicate that vertebrate evolution was well advanced by the Early Cambrian. Shu et al. (1999) originally placed *Haikouichthys* directly crownward of hagfishes. However, recent molecular-phylogenetic studies indicate that hagfishes are not stem craniates, as previously thought, but are highly derived vertebrates that group with lampreys as cyclostomes (Takezaki et al. 2003; Mallatt and Sullivan 1998). This leaves *Haikouichthys* as a stem-group craniate (Shu et al. 2003a). *Myllokunmingia* (Shu et al. 1999) is similar to *Haikouichthys* and has been interpreted as a senior synonym (Hou et al. 2002; but see Conway Morris 2003b). An additional Chengjiang agnathan, *Zhongjianichthys* (Shu 2003), is also similar, suggesting that there was not much morphological diversity among these Early Cambrian vertebrates. Hard parts first appear in the Middle Cambrian Euconodonta.

Vetulicola (Fig. 3B), a large Chengjiang animal with a carapace-like anterior part and segmented tail-like posterior part, was first described as a large bivalved arthropod (Hou 1987). Chen and Zhou (1997) commented on its enigmatic nature and assigned it to a new class of stem-group arthropods, Vetulicolida. They argued that *Banffia* from the Burgess Shale belongs to the same group. A new taxon *Didazoon* was described by Shu, Conway Morris, and others (Shu et al. 2001b) and it, together with additional data on *Xidazoon* and *Vetulicola*, prompted them to erect a new phylum Vetulicolia. This should have delighted Gould—after all *Wonderful Life* was based on discoveries of new Cambrian phyla and here, over ten years later, examples were still turning up. The interpretation of perforations in the anterior body region as precursors of gill slits led Shu et al. (2001b, 2003b) to regard vetulicolians as primitive deuterostomes. Gee (2001), however, remarked on the similarity of the vetulicolian body plan to that of a tunicate tadpole larva and suggested that they may occupy a more crownward position and be the sister group of chordates. Lacalli (2002) and Jefferies (personal communication 2003) considered the possibility that vetulicolians are stem tunicates. In contrast, Butterfield (2003) and Mallatt et al. (2003) argued that *Vetulicola* has a cuticle and is therefore more likely to have been an arthropod. A tunicate was described from the Chengjiang Lagerstätte by Shu et al. (2001a). Doubts were cast on the tunicate nature of this fossil by Chen et al. (2003), who recognized additional specimens with large lophophore-like tentacles. They described another animal from the Lower Cambrian of China as a tunicate, on the basis of a larger number of specimens. New phyla of

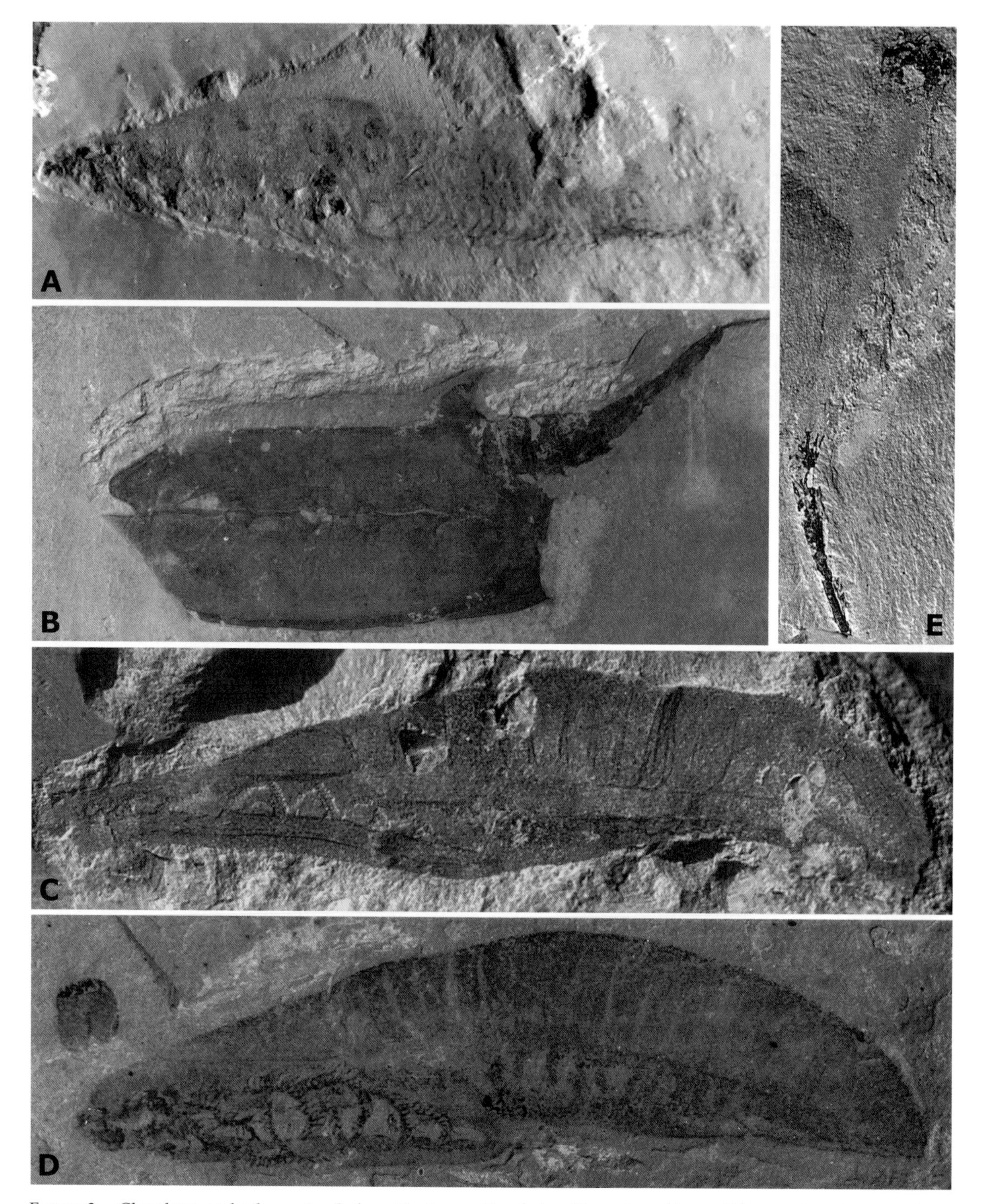

FIGURE 3. Chordates and other animals from the Lower Cambrian Chengjiang biota of Yunnan Province, southern China. A, *Haikouichthys ercaicunensis*, HZ-f-12-127 (Yunnan Institute of Geological Sciences, Kunming), ×3.85. B, *Vetulicola cuneata*, ELRC 19397a (Early Life Research Center, Chengjiang), ×1.2. C, *Yunnanozoon lividum*, ELRC 52004, ×5.49. D, *Haikouella lanceolatum*, ELRC 00258, ×5.23. D, Possible chaetognath *Eognathacantha ercainella*, ELRC 02001a, ×3.73. Photographs kindly provided by J.-Y. Chen (Fig. 3A with the permission of L.-J. Chen, Yunnan Institute of Geology).

modern metazoans are described rarely, and when they are, such a designation usually reflects a very simple morphology with a lack of diagnostic features. Their affinities can be tested by using molecular and embryological evidence (Winnepenninckx et al. 1998 on Cycliophora; Bourlat et al. 2003 on *Xenoturbella;* Ruiz-Trillo et al. 2002 on Acoela). New phylum status is not proposed often for fossils either. Given the need to rely on morphology, allied to the vagaries of taphonomy and the difficulty of interpreting unfamiliar form, the erection of new extinct phyla is fraught with difficulty. At the very least it demands explicit hypotheses of character distribution and an analysis of relationships within a cladistic context.

Yunnanozoon (Fig. 3C), a blade-shaped, soft-bodied, segmented animal from the Chengjiang fossil fauna, was first described by Hou et al. (1991) as a wormlike problematicum. Chen et al. (1995) described additional material, including evidence of a notochord and myomeres, and interpreted *Yunnanozoon* as a cephalochordate. They noted that this predicted that "other chordate clades (tunicates and craniates) had evolved by the Late Atdabanian in the main burst of the Cambrian Explosion" (p. 720). Dzik (1995) considered that *Yunnanozoon* "belongs to a completely extinct group of the earliest chordates" (p. 352). He erected a Class Yunnanozoa, Order Yunnanozoida, Family Yunnanozoidae. A problem with the chordate interpretation of the yunnanozoans is the unusual ventral position of the notochord relative to the myomeres. Shu et al. (1996b) collected additional material and reinterpreted *Yunnanozoon* as the earliest known hemichordate, on the basis of their interpretation of features as proboscis, collar and trunk. Chen and Li (1997) redescribed *Yunnanozoon* and reinterpreted it as a primitive chordate, arguing that the hemichordate features are artifacts of decay. They also considered that the single specimen described by Shu et al. (1996a) as *Cathaymyrus*, a *Pikaia*-like chordate, is, in fact, an example of *Yunnanozoon*. Chen et al. (1999) discovered abundant well-preserved specimens of a new yunnanozoan, which they described as *Haikouella* (Fig. 3D), that appeared more craniate-like than cephalochordate-like. Holland and Chen (2001) reviewed the evidence for the origin and early evolution of the vertebrates and demonstrated, by means of a cladistic analysis, that yunnanozoans are the sister group of the vertebrates, lying crownward of the cephalochordates. Shu et al. (2003b) described a new species of *Haikouella*, adding new information on the morphology of the yunnanozoans, including external gills. They argued that there is no evidence for a chordate affinity, but that these animals were stem-group deuterostomes, allied to the Phylum Vetulicolia, to which they tentatively assigned them. Mallatt et al. (2003) discussed this paper and reasserted the craniate affinity for yunnanozoans based on the presence of gills and evidence for muscle fibers in the myomeres, a claim that Shu and Conway Morris (2003) rejected.

Mallatt and Chen (2003) provided a test of the hypothesis. They demonstrated the clear difference between cuticular folds and muscle fibers in *Haikouella*, illustrating strong evidence for the latter. More importantly they presented a detailed description and justification of a matrix of 40 characters that they used to perform a cladistic analysis of the relationships of *Haikouella* (i.e., yunnanozoans), protostomes, hemichordates, tunicates, cephalochordates, conodonts, and craniates. The result confirmed that of Holland and Chen (2001): *Haikouella*, and therefore the Yunnanozoa, is a sister group of vertebrates. A similar cladistic analysis to determine the position of vetulicolians (deuterostomes, arthropods, or other) remains to be done. Not to do so, once the evidence can be assembled, is to ignore the lessons of the debate that *Wonderful Life* stimulated.

Several new discoveries since 1989 in the Lower Cambrian Chengjiang fauna of southern China have extended the range of other groups that are also represented in the Middle Cambrian Burgess Shale. Two new genera of Ctenophora were described by Chen and Zhou (1997). A likely chaetognath from the Burgess Shale discovered by D. Collins (Royal Ontario Museum) awaits description (Briggs and Conway Morris 1986: p. 167). In the meantime a possible chaetognath (Fig. 3E) from

Kunming was reported by Chen and Huang (2002), extending the record of this group from the Carboniferous to the Lower Cambrian. The Phoronida are represented by a single genus from Chengjiang (Chen and Zhou 1997), the only body fossil of this group known. Sponges were reported from the Ediacara fauna of South Australia (Gehling and Rigby 1996).

Divergence Times of Major Clades and the Cambrian Explosion

The Chengjiang fauna (Atdabanian) demonstrates that, among fossilizable clades, several phylum-or class-level taxa extended back into the Early Cambrian, even allowing for the caveats about definitions of higher taxa discussed above. It is inescapable that the divergence of these groups from their common ancestors happened before the time represented by this early diverse fauna (Fortey et al. 1996). In spite of the discovery of earliest Cambrian (Nemakyt–Daldyn) embryos (Bengtson and Yue 1997) it remains debatable whether these derived groups were also present during the duration of this stage (ca. 10 Ma)—although the presence of plausible arthropod and mollusc trace fossils (Brasier et al. 1994) perhaps could be taken as evidence that some of the major groups had already appeared. The suggestion that the absence of Precambrian metazoan fossils might be attributed to small size (Fortey et al. 1996) has proved controversial. Budd and Jensen (2000) inferred that the ancestral bilaterian would have been large, but Ruiz-Trillo et al. (2002) have argued that the ancestor was a small, simple, probably direct-developing acoel, judging from a phylogenetic analysis of platyhelminthes that showed them to be polyphyletic (see Peterson et al. 2005).

The affinities of the Vendian Ediacaran organisms remain controversial. They have long been interpreted as "ancestral" or stem-group metazoans (e.g., Glaessner 1984; Gehling 1991). Gould (1989), however, favored the alternative interpretation (Seilacher 1989) of the Ediacaran organisms as the product of a Precambrian evolutionary experiment that generated unfamiliar creatures that existed alongside metazoans, the latter represented mainly by trace fossils (e.g., those that show burrowing by peristaltic locomotion). In following Seilacher's interpretation Gould (1989) emphasized the evolutionary importance of the Cambrian explosion—the fossil evidence indicates that some phyla, such as brachiopods and arthropods, may not have evolved by late Precambrian times. The affinities of the Ediacaran organisms continue to excite controversy. Seilacher extended his interpretation (Seilacher et al. 2003) to argue that many of the problematic Ediacaran organisms represent procaryotic biomats and giant rhizopodan protozoans (Xenophyophoria and Vendobionta) that dominated a late Precambrian world in which metazoans were rare (sponges, coelenterates, a mollusc, and an echinoderm) and relatively small. This view contrasts markedly with that of Fedonkin (2003), who regarded most of the Ediacaran animals as diploblastic or triploblastic metazoans. Important evidence for Fedonkin's interpretation is the discovery of several trace fossils associated with the Ediacaran organisms that made them, including *Yorgia*, *Kimberella*, and *Dickinsonia*.

Given that there is no consensus on the affinities of the Ediacaran body fossils, which are relatively abundant and well preserved, it is not surprising that the older trace fossil evidence for metazoan activities has proved even more controversial. Early records have been scrutinized carefully as a potential means of reconciling dates for the origin of groups based on molecular data with the evidence of the fossil record. Budd and Jensen (2000) argued, however, that none of the trace fossils reported to predate the last Proterozoic ice age (Marinoan or Varangerian) stand up to scrutiny—they can be attributed to sedimentary structures or metaphytes, or they have been misdated. The difficulty here is that of preservation and distortion over time. The earliest trace fossils that are attributed confidently to stem-group bilaterians are about 555 million years old (Droser et al. 2002).

Unequivocal fossils of identifiable metazoans older than the base of the Cambrian have not been forthcoming. The embryos from the Neoproterozoic Doushantuo Formation are preserved at too early a stage to allow them to be identified to specific metazoan taxa (Xiao and Knoll 2000). Thus dating of molecular trees is based inevitably on Cambrian fossil occurrences. At the time Gould wrote *Wonderful Life,* methods of determining from independent evidence the timing of the deeper branches of phylogeny were in their infancy—as, for example, the divergence between chordates and echinoderms, or between protostomes and deuterostomes. Estimates became a possibility as comparisons of genomic sequences, such as that of small subunit RNA, from a variety of organisms allowed models to be developed for converting sequence difference into a measure of time, using assumptions about the "molecular clock" (Smith and Peterson 2002).

A pioneering study by Wray et al. (1996) used data from 18S ribosomal RNA and seven genes to estimate divergence times between metazoan phyla. Divergence times in the deep Precambrian were inferred: for example about 1200 Ma for a protostome-deuterostome split, and 1000 Ma for the echinoderm-chordate split. Thus, according to this estimate, phylogenetic divergence long preceded the Cambrian "explosion." Other analyses soon followed. Wang et al. (1998), for example, identified the chordate arthropod split at 995 $\pm$ 79 Ma and noted that "basal animal phyla (Porifera, Cnidaria, Ctenophora) diverged between 1200–1500 Ma," and "six animal phyla originated deep in the Precambrian more than 400 Ma earlier than their first appearance in the fossil record." Ayala et al. (1998) used pairwise distance methods to estimate divergence times, and they obtained a proterostome-deuterostome split at 670 Ma and an echinoderm chordate split at 600 Ma. Hence within a few years it was already apparent that using different genes and different methods yielded strikingly different estimates of divergence times between the same clades.

Methodological limitations may have been responsible for this range in estimates of divergence times. Confidence limits (e.g., in Wang et al. 1999) were often calculated by taking the standard error of the mean divergence estimates as the total estimate error, thereby ignoring the assumed Poisson variance of the substitutions. Bromham et al. (1998) noted the difficulties of incorporating this variance into the calculations, which concatenate the error on estimates progressively with deeper divergences; they used a method based on maximum likelihood. This increased the length of confidence limits on divergence times considerably compared with those derived by using other methods; although the means of the proterostome-deuterostome and echinoderm-chordate splits were still well within the Precambrian, for example, the upper bounds on the confidence limits of the latter approached the lower Cambrian. Further problems with "molecular clock" estimations of divergence times were outlined by Cutler (2000), who pointed out that usual approaches did not take into account "overdispersed" loci—those that fail relative rate tests—and that the Poisson assumptions tend to underestimate the confidence limits for such loci. He developed a maximum likelihood method using more general stationary process assumptions, and was able to show that the Poisson-based model could be rejected for some loci. Cutler's method produced considerable error bars on deep divergences, but the origin of metazoans close to the base of the Cambrian could *not* be ruled out, although it was less likely than a significantly older divergence.

Methods for "relaxing" the molecular clock, allowing for different rates in different parts of the tree, have been developed in the last few years and are likely to have a positive impact on accuracy of estimates of divergence times (Sanderson 2002). Bayesian approaches (Huelsenbeck et al. 2001), using Markov Chain Monte Carlo methods, have become computationally feasible, although these methods have not escaped criticism (Shoemaker et al. 1999). At the same time, the comparative analysis of a wider range of single-copy nuclear genes is likely to identify more reliable loci for the analysis of deep divergences and will be preferable to the earlier reliance upon 18S or mitochondrial genes. Sequencing of a critical range of invertebrate organisms selected according to their phylogenetic position (Smith and Peterson 2002)—rather than their ready availability as laboratory material—should help tackle some of the long-branch problems that have also influenced the accuracy of published divergence estimates.

As yet there is no "definitive" answer to the

question regarding the length of the Precambrian phylogenetic fuse(s). If one trend is discernible in modern studies of divergence times, it is that the earlier papers tended to produce very deep dates that may, in fact, be partly artifacts. As methods have become more refined and critical, mean dates for divergences have become younger, but error ranges have also generally become longer. Nonetheless, some of the fundamental "splits" in the tree of life fall consistently well within the Precambrian, and it would, perhaps, be surprising if they were to "jump" considerably upwards as a result of future work. Within the Metazoa, for example, the proterostome-deuterostome divergence in the majority of studies falls within the range 750–1000 Ma (Fortey et al. 2003: Fig. 3.3)—the exception is the latest Precambrian date of Lynch (1999). This suggests a 200 Ma period to allow diversification in metazoan phyla before the base of the Cambrian. However, there is still much flexibility in the estimates of divergence times between pairs of phyla. For example, Hedges and Kumar (2003) suggest a divergence between arthropods and vertebrates at 993 Ma and vertebrates from hemichordates at 751 Ma (error ranges ignored). However, Peterson et al. (2004, 2005) suggested that previous molecular estimates of divergence times may have been influenced by calibrating on the basis of rates of genetic change in vertebrates, which are slower than those in invertebrates, a difference that would push back times of origin artificially. They estimated the last common ancestor of protostomes and deuterostomes (exclusive of acoelomorphs) at ca. 570 Ma and the last common ancestor of protostomes at ca. 560 Ma. This opens up the possibility again that there was, indeed, a period of "explosive" origination—a return, in effect, to Gould's original formulation (Conway Morris 2000). However, it is still early days in the methodology: for example, one might ask even whether rates in invertebrates are "one size fits all"?

The question of molecular dating of phyletic divergences remains a rapidly changing field, and it may be premature to prejudge the outcome as it relates to the base of the Cambrian. At this stage, it may be safe to conclude that the basal divisions, as between Ecdysozoa, protostomes, and deuterostomes, were accomplished well within the Precambrian and provide a long phylogenetic "fuse." However, the timing of origin of the phyla that appeared at, or after, the basal Cambrian is still not firmly established and varies widely according to methodology and clock assumptions. Taking molecular grounds alone, Gould's "explosive" scenario is still a feasible option. For example, a recent application of Bayesian methods (Aris Brosou and Yang 2003) using 22 genes, and permitting rates to vary in time, identified a major burst of molecular evolution late in Precambrian time, bringing molecular estimates of metazoan divergence closer to the base of the Cambrian than in all previous studies.

Contingency

Perhaps the most lasting image in *Wonderful Life* is the device that Gould used to explain the concept of contingency: "Wind the tape of life back to Burgess times, and let it play again. If *Pikaia* does not survive in the replay, we are wiped out of future history—all of us, from shark to robin to orangutan" (Gould 1989: p. 323). There is no doubt that contingency is an important factor in evolution, not least to the extent that the loss of many taxa during mass extinctions, particularly those attributed to asteroid impacts, is randomly determined. However, Conway Morris (1998, 2003a), in particular, has argued that convergent evolution lends an inevitability to the emergence of a creature with faculties similar to our own. The two antagonists summarized their viewpoints in *Natural History* (Conway Morris and Gould 1998). Convergence on this scale, however, awaits serious investigation. The degree of convergence within groups of organisms can by quantified by using parsimony programs (in terms of homoplasy). Mapping genomes will eventually determine how deep in the tree of life control genes constrain and determine potential outcomes. Computer simulations can determine the probability of sequences of events. In the meantime, to recast the history of life in terms of convergence places a singular emphasis on one aspect of a complex story. The repeated evolution of sabertooth morphology among predatory mammals is doubtless an extraordinary phenomenon that might suggest some sort of tendency to evolve in a particular direction. However, the fossil record has also revealed designs that have never been replicated. Graptolites, for example, colonial planktonic organisms with an extraordinary array of spiral, s-shaped, or branching rhabdosomes, have never had close homeomorphs in the 400 million years since their extinction. Is humankind a "sabertooth" or a "graptolite"? Unfortunately the probability that consciousness would have evolved even if the vertebrates had been eliminated by some late Cambrian extinction is unlikely to be susceptible to scientific enquiry.

Conclusions

Gould's *Wonderful Life* (1989) was an inspiration to students and served to set the agenda for 15 years of intensive research on the meaning of the Cambrian evolutionary "explosion." New sources of data have arisen from several quarters—particularly the discovery of major new soft-bodied faunas from the Paleozoic and the application of molecular techniques to questions of phylogeny and divergence times. Many of the issues that Gould raised are still under debate—a number have attracted the attention of zoologists, paleontologists, and molecular biologists and fostered interdisciplinary interest in a way that should be a model for the future. Although some of the ideas in *Wonderful Life* have not stood the test of time, the challenges set have served to move the science forward.

Claims about strikingly elevated numbers of "new phyla" in the Cambrian have been countered. Cladistic analysis has placed the majority of Cambrian "weird wonders" as stem taxa of known clades. New fossil discoveries have resolved the problematic status of several former enigmas. Although new major extinct taxa (e.g., Phylum Vetulicolia [Shu et al. 2001b]) occasionally still appear in the literature, their erection is controversial and their objective basis unclear. Nor have claims about a maximum of morphological disparity in the Cambrian stood up to analysis. At most, disparity of design was equal to that at the Recent. Its range was established early and its compass shifted through time in total morphospace.

Debates about the Cambrian radiation have stimulated a debate over how "phyla" might be defined at all. One view bases a definition of phylum (or class) on the crown group and excludes stem taxa as plesions. This has the advantage of objectivity and ensures that molecular data are available for the whole taxon except where primitive survivors pull fossil taxa into the crown clade. The crown-group definition of a phylum is arbitrary, however, depending on the most primitive taxon that happens to have survived to the present day. An alternative view defines higher taxa on the basal divergence (total group), which is logically consistent, and assigns stem taxa to a phylum. A disadvantage is that the placing of extremely plesiomorphic forms, particularly those with peculiar autapomorphies, remains difficult.

The question of the length of the phylogenetic "fuse" below the Cambrian—the generation time of the living phyla—remains unresolved. Molecular-based estimates of divergence times vary widely according to the methods used and are only now being supplied with realistic error ranges. Nonetheless, deep Precambrian divergence of "superphyla" (protostomes, Ecdysozoa, deuterostomes) is favored by the majority of studies. It is tempting to relate subsequent rapid evolutionary events in the origin of phyla to global scenarios of environmental change, such as "snowball earth" (Hoffman et al. 1998; see Peterson et al. 2005). Such an attractive hypothesis must not, however, discourage efforts to augment the fossil record in the late Proterozoic, because one good fossil at an unexpected time or place could still have the effect of overturning any preconceptions.

Acknowledgments

We are grateful to S. Bengtson, J. Mallatt, and F. R. Schram for comments on an earlier draft of the manuscript. We acknowledge the major contribution of M. A. Wills to enabling us to address some of the major issues raised in *Wonderful Life*. J.-Y. Chen and M. A. Wills provided illustrations, and S. Butts assisted with the preparation of figures.

Literature Cited

Ahlberg, P. E., ed. 2001. Major events in early vertebrate evolution. Systematics Association Special Volume 61. Taylor and Francis, London.

Aldridge, R. J., D. E. G. Briggs, E. N. K. Clarkson, and M. P. Smith. 1986. The affinities of conodonts—new evidence from the Carboniferous of Edinburgh, Scotland. Lethaia 19:279–291.

Allison, P. A., and D. E. G. Briggs. 1993. Exceptional fossil record: distribution of soft-tissue preservation through the Phanerozoic. Geology 21:527–530.

Aris-Brosou, S., and Z-H. Yang. 2003. Bayesian models of episodic evolution support a late Precambrian explosive divergence of the Metazoa. Molecular Biology and Evolution 20: 1947–1954.

Averof, M., and M. Akam. 1995. Insect-crustacean relationships: insights from comparative developmental and molecular studies. Philosophical Transactions of the Royal Society of London B 347:293–303.

Ayala, F. J., A. Rzhetsky, and F. J. Ayala. 1998. Origin of the metazoan phyla: molecular clocks confirm paleontological estimates. Proceedings of the National Academy of Sciences USA 95:606–12.

Bengtson, S., and Z. Yue. 1997. Fossilized metazoan embryos from the earliest Cambrian. Science 277:1645–1648.

Benton, M. J., ed. 1993. The fossil record 2. Chapman and Hall, London.

———. 2000. Stems, nodes, crown clades, and rank-free lists: is Linnaeus dead? Biological Reviews 75:633–648.

Bourlat, S. J., C. Nielsen, A. E. Lockyer, D. T. J. Littlewood, and M. J. Telford. 2003. *Xenoturbella* is a deuterostome that eats molluscs. Nature 424:925–928.

Brasier, M. J., J. W. Cowie, and M. E. Taylor. 1994. Decision on the Precambrian–Cambrian boundary. Episodes 17:3–8.

Briggs, D. E. G. 1978. The morphology, mode of life and affinities of *Canadaspis perfecta* (Crustacea: Phyllocarida), Middle Cambrian, Burgess Shale. Philosophical Transactions of the Royal Society of London B 281:439–487.

———. 1992a. Phylogenetic significance of the Burgess Shale crustacean *Canadaspis*. Acta Zoologica, Stockholm 73:293–300.

———. 1992b. Conodonts—a major extinct group added to the vertebrates. Science 256:1285–1286.

Briggs, D. E. G., and C. Bartels. 2001. New arthropods from the Lower Devonian Hunsrück Slate (Lower Emsian, Rhenish Massif, western Germany). Palaeontology 44:275–303.

Briggs, D. E. G., and D. Collins. 1988. A Middle Cambrian chelicerate from Mount Stephen, British Columbia. Palaeontology 31:779–798.

Briggs, D. E. G., and S. Conway Morris. 1986. Problematica from the Middle Cambrian Burgess Shale of British Columbia. Pp. 167–183 *in* A. Hoffman and M. H. Nitecki, eds. Problematic fossil taxa. Oxford University Press, Oxford.

Briggs, D. E. G., and R. A. Fortey. 1989. The early radiation and relationships of the major arthropod groups. Science 246:241–243.

Briggs, D. E. G., and H. B. Whittington. 1981. Relationships of arthropods from the Burgess Shale and other Cambrian sequences. Proceedings of the second International Symposium on the Cambrian System. U.S. Geological Survey Open-file Report 81-743:38–41.

Briggs, D. E. G., R. A. Fortey, and M. A. Wills. 1992a. Morphological disparity in the Cambrian. Science 256:1670–1673.

———. 1992b. Cambrian and Recent morphological disparity: response to Foote and Gould, and Lees. Science 258:1817–1818.

Briggs, D. E. G., D. J. Siveter, and D. J. Siveter. 1996. Soft-bodied fossils from a Silurian volcaniclastic deposit. Nature 382:248–250.

Bromham, L., A. Rambaut, R. A. Fortey, A. Cooper, and D. Penny. 1998. Testing the Cambrian explosion hypothesis using a molecular dating technique. Proceedings of the National Academy of Sciences USA 95:12386–9.

Budd, G. E. 1993. A Cambrian gilled lobopod from Greenland. Nature 364:709–711.

———. 1996. The morphology of *Opabinia regalis* and the reconstruction of the arthropod stem-group. Lethaia 29:1–14.

———. 1999. Does evolution in body patterning genes drive morphological change—or vice versa? BioEssays 21:326–332.

———. 2002. A palaeontological solution to the arthropod head problem. Nature 417:271–75.

———. 2003. The Cambrian fossil record and the origin of the phyla. Integrative and Comparative Biology 43:157–165.

Budd, G. E., and S. Jensen. 2000. A critical appraisal of the fossil record of the bilaterian phyla. Biological Reviews 75:253–295.

Butterfield, N. J. 2003. Exceptional fossil preservation and the Cambrian explosion. Integrative and Comparative Biology 43:166–177.

Carroll, S. B., S. Weatherbee, and J. Langeland. 1995. Homeotic genes and the regulation and evolution of insect wing number. Nature 375:58–61.

Chen, J.-Y., and D.-Y. Huang. 2002. A possible Lower Cambrian chaetognath (arrow worm). Science 298:187.

Chen, J.-Y., and C.-W. Li. 1997. Early Cambrian chordate from Chengjiang, China. Bulletin of the National Museum of Natural Science 10:257–273.

Chen, J.-Y., and G.-Q. Zhou. 1997. Biology of the Chengjiang fauna. Bulletin of the National Museum of Natural Science 10:11–105.

Chen, J.-Y., X.-G. Hou, and G.-X. Li. 1989. Early Cambrian netted scale-bearing worm-like sea animal. Acta Palaeontologica Sinica 29:402–414.

Chen, J.-Y., J. Dzik, G. E. Edgecombe, L. Ramsköld, and G.-Q. Zhou. 1995. A possible Early Cambrian chordate. Nature 377: 720–722.

Chen, J.-Y., D.-Y. Huang, and C.-W. Li. 1999. An early Cambrian craniate-like chordate. Nature 402:518–522.

Chen, J.-Y., D.-Y. Huang, Q.-Q. Peng, H.-M. Chi, X.-Q. Wang, and M. Feng. 2003. The first tunicate from the Early Cambrian of South China. Proceedings of the National Academy of Sciences USA 100:8314–8318.

Collins, D. 1996. The "evolution" of *Anomalocaris* and its classification in the arthropod class Dinocarida (nov.) and order Radiodonta (nov.). Journal of Paleontology 70:280–93.

Conway Morris, S. 1977a. A new metazoan from the Cambrian Burgess Shale, British Columbia. Palaeontology 20:623–640.

———. 1977b. Fossil priapulid worms. Special Papers in Palaeontology 20:1–98.

———. 1998. The crucible of creation: the Burgess Shale and the rise of animals. Oxford University Press, Oxford.

———. 2000. The Cambrian "explosion": slow fuse or megatonnage? Proceedings of the National Academy of Sciences USA 97:4426–9.

———. 2003a. Life's solution: inevitable humans in a lonely universe. Cambridge University Press, Cambridge.

———. 2003b. The Cambrian "explosion" of metazoans and molecular biology: would Darwin be satisfied? International Journal of Developmental Biology 47:505–515.

Conway Morris, S., and S. J. Gould. 1998. Showdown on the Burgess Shale. [The Challenge by Simon Conway Morris and the Reply by Stephen Jay Gould.] Natural History 107:48–55.

Conway Morris, S., and J. S. Peel. 1995. Articulated halkieriids from the Lower Cambrian of North Greenland and their role in early protostome evolution. Philosophical Transactions of the Royal Society of London B 347:304–358.

Cooper, A., and R. A. Fortey. 1998. Evolutionary explosions and the phylogenetic fuse. Trends in Ecology and Evolution 13: 151–6.
Cutler, D. J. 2000. Estimating divergence times in the presence of an overdispersed molecular clock. Molecular Biology and Evolution 17:1647–60.
Dewel, R. A., and W. C. Dewel. 1998. The place of tardigrades in arthropod evolution. Pp. 109–123 *in* R. A. Fortey and R. H. Thomas, eds. Arthropod relationships. Chapman and Hall, London.
Droser, M. L., S. Jensen, and J. G. Gehling. 2002. Trace fossils and substrates of the terminal Proterozoic-Cambrian transition: implications for the record of early bilaterians and sediment mixing. Proceedings of the National Academy of Sciences USA 99:12572–12576.
Dzik, J. 1995. *Yunnanozoon* and the ancestry of the chordates. Acta Palaeontologica Polonica 40:341–360.
Edgecombe, G. D., ed. 1998. Arthropod fossils and phylogeny. Columbia University Press, New York.
Fedonkin, M. A. 2003. The origin of the Metazoa in the light of the Proterozoic fossil record. Paleontological Research 7:9–41.
Foote, M., and S. J. Gould. 1992. Cambrian and Recent morphological disparity. Science 258:1816.
Fortey, R. A. 1989. The collection connection. [Review of *Wonderful Life*.] Nature 342:303.
Fortey, R. A., D. E. G. Briggs, and M. A. Wills. 1996. The Cambrian evolutionary 'explosion': decoupling cladogenesis from morphological disparity. Biological Journal of the Linnean Society 57:13–33.
Fortey, R. A., J. Jackson, and J. Strugnell. 2003. Phylogenetic "fuses" and evolutionary explosions: conflicting evidence and critical tests. Pp. 41–65 *in* P. Donoghue, ed. Molecular evolution. Chapman and Hall.
Gee, H. 2001. On being vetulicolian. Nature 414:407–409.
Gehling, J. G. 1991. The case for Ediacaran fossil roots to the Metazoan tree. Pp. 181–224 *in* B. P. Radhakrishna, ed. The world of Martin F. Glaessner. Geological Society of India Memoir No. 20. Bangalore.
Gehling, J. G., and J. K. Rigby. 1996. Long expected sponges from the Neoproterozoic Ediacara fauna of South Australia. Journal of Paleontology 70:185–195.
Giribet, G., G. D. Edgecombe, and W. C. Wheeler. 2001. Arthropod phylogeny based on eight molecular loci and morphology. Nature 413:157–161.
Glaessner, M. F. 1984. The dawn of animal life: a biohistorical study. Cambridge University Press, Cambridge.
Gould, S. J. 1989. Wonderful life: the Burgess Shale and the nature of history. Norton, New York.
———. 1991. The disparity of the Burgess Shale arthropod fauna and the limits of cladistic analysis: why we must strive to quantify morphospace. Paleobiology 17:411–423.
———. 2002. The structure of evolutionary theory. Belknap Press of Harvard University Press, Cambridge.
Hedges, S. B., and S. Kumar. 2003. Genomic clocks and evolutionary timescales. Trends in Genetics 19:200–206.
Hoffman, P. F., A. J. Kaufman, G. P. Halverson, and D. P. Schrag. 1998. A Neoproterozoic snowball earth. Science 281:1342–1346.
Holland, N. D., and J.-Y. Chen. 2001. Origin and early evolution of the vertebrates: new insights from advances in molecular biology, anatomy, and palaeontology. BioEssays 23:142–151.
Hou, X.-G. 1987. Early Cambrian large bivalved arthropods from Chengjiang, Eastern Yunnan. Acta Paleontological Sinica 26:286–297.
Hou, X.-G., and J. Bergström. 1997. Arthropods of the Chengjiang fauna, southwest China. Fossils and Strata 45:1–116.
Hou, X.-G., L. Ramsköld, and J. Bergström. 1991. Composition and preservation of the Chengjiang fauna—a Lower Cambrian soft-bodied biota. Zoologica Scripta 20:395–411.
Hou, X.-G., J. Bergström, and P. Ahlberg. 1995. *Anomalocaris* and other large animals in the Lower Cambrian Chengjiang Fauna of southwestern China. Geologiska Föreningens Förhandlingar 117:163–183.
Hou, X.-G., R. J. Aldridge, D. J. Siveter, D. J. Siveter, and F. Xianghong. 2002. New evidence on the anatomy and phylogeny of the earliest vertebrates. Proceedings of the Royal Society of London B 269:1865–1869.
Huelsenbeck, J. P., F. Ronquist, R. Neilsen, and J. P. Bollback. 2001. Bayesian inference of phylogeny and its impact on evolutionary biology. Science 294:2310–4.
Hwang, U. W., M. Friedrich, and D. Tautz. 2001. Mitochondrial protein joins myriapods with chelicerates. Nature 413:154–7.
Jefferies, R. P. S. 1979. The origin of chordates—a methodological essay. *In* M. R. House, ed. The origin of major invertebrate groups. Systematics Association Special Volume 12:443–477.Academic Press, London.
Lacalli, T. C. 2002. Vetulicolians—are they deuterostomes? Chordates? BioEssays 24:208–211.
Lee, M. S. Y. 1992. Cambrian and Recent morphological disparity. Science 258:1816–1817.
Levinton, J. S. 2001. Genetics, paleontology, and macroevolution, 2d ed. Cambridge University Press, Cambridge.
Lofgren, A. S., R. E. Plotnick, and P. J. Wagner. 2003. Morphological diversity of Carboniferous arthropods and insights on disparity patterns through the Phanerozoic. Paleobiology 29: 349–368.
Lynch, M. 1999. The age and relationships of the major animal phyla. Evolution 53:319–25.
Mallatt, J., and J.-Y. Chen. 2003. Fossil sister group of craniates: predicted and found. Journal of Morphology 258:1–31.
Mallatt, J., and J. Sullivan. 1998. 28S and 18S rDNA sequences support the monophyly of lampreys and hagfishes. Molecular Biology and Evolution 15:1706–1718.
Mallatt, J., J.-Y. Chen, and N. D. Holland. 2003. Comment on "A new species of Yunnanozoan with implications for deuterostome evolution." Science 300:1372c.
Müller, K. J., D. Walossek, and A. Zakharov. 1995. "Orsten" type phosphatized soft-integument preservation and a new record from the Middle Cambrian Kuonamka Formation in Siberia. Neues Jahrbuch für Geologie und Paläontologie, Abhandlungen 197:101–118.
Patterson, C. 1981. Significance of fossils in determining evolutionary relationships. Annual Review of Ecology and Systematics 12:195–223.
Pennisi, E. 2003. Modernizing the tree of life. Science 300:1692–1697.
Peterson, K. J., J. B. Lyons, K. S. Nowak, C, M. Takacs, M. J. Wargo, and M. A. McPeek. 2004. Estimating metazoan divergence times with a molecular clock. Proceedings of the National Academy of Sciences USA 101:6536–6541.
Peterson, K. J., M. A. McPeek, and D. A. D. Evans. 2005. Tempo and mode of early animal evolution: inferences from rocks, Hox, and molecular clocks. [This volume.]
Ramsköld, L. 1992. The second leg row of *Hallucigenia* discovered. Lethaia 25:221–224.
Ramsköld, L., and J.-Y. Chen. 1998. Cambrian lobopodians: morphology and phylogeny. Pp. 107–150 *in* Edgecombe 1998.
Ramsköld, L., and X.-G. Hou. 1991. New early Cambrian animal and onychophoran affinities of enigmatic metazoans. Nature 351:225–227.
Ronshaugen, M., N. McGinnis, and W. McGinnis. 2002. Hox protein mutation and macroevolution of the insect body plan. Nature 415:914–917.
Ruiz-Trillo, I., J. Paps, M. Loukota, C. Ribera, U. Jondelius, J. Baguna, and M. Riutort. 2002. A phylogenetic analysis of myo-

sin heavy chain type II sequences corroborates that Acoela and Nemertodermatida are basal bilaterians. Proceedings of the National Academy of Sciences USA 99:11246–11251.

Sanderson, M. J. 2002. Estimating absolute rates of molecular evolution and divergence times: a penalized likelihood approach. Molecular Biology and Evolution 19:101–109.

Seilacher, A. 1989. Vendozoa—organismic construction in the Proterozoic biosphere. Lethaia 22:229–239.

Seilacher, A., D. Grazhdankin, and A. Legouta. 2003. Ediacaran biota: the dawn of animal life in the shadow of giant protists. Paleontological Research 7:43–54.

Selden, P. A., and J. A. Dunlop. 1998. Fossil taxa and relationships of chelicerates. Pp. 303–331 *in* Edgecombe 1998.

Shoemaker, J. S., I. S. Painter, and B. S. Weir. 1999. Bayesian statistics in genetics. a guide for the uninitiated. Trends in Genetics 15:354–8.

Shu, D.-G. 2003. A paleontological perspective of vertebrate origin. Chinese Science Bulletin 48:725–735.

Shu, D.-G., and S. Conway Morris. 2003. Response to comment on "A new species of Yunnanozoan with implications for deuterostome evolution." Science 300:1372d.

Shu, D.-G., S. Conway Morris, and X.-L. Zhang. 1996a. A *Pikaia*-like chordate from the Lower Cambrian of China. Nature 384: 157–158.

Shu, D.-G., X. Zhang, and L. Chen. 1996b. Reinterpretation of *Yunnanozoon* as the earliest known hemichordate. Nature 380: 428–430.

Shu, D.-G., H.-L. Luo, S. Conway Morris, X.-L. Zhang, S.-X. Hu, L. Chen, J. Han, M. Zhu, Y. Li, and L.-Z. Chen. 1999. Lower Cambrian vertebrates from south China. Nature 402:42–46.

Shu, D.-G., L. Chen, J. Han, and X.-L. Zhang. 2001a. An Early Cambrian tunicate from China. Nature 411:472–473.

Shu, D.-G., S. Conway Morris, J. Han, L. Chen, X.-L. Zhang, Z.-F. Zhang, H.-Q. Liu, Y. Li, and J.-N. Liu. 2001b. Primitive deuterostomes from the Chengjiang Lagerstätte (Lower Cambrian, China). Nature 414:419–424.

Shu, D.-G., S. Conway Morris, J. Han, Z.-F. Zhang, K. Yasui, P. Janvier, L. Chen, X.-L. Zhang, J.-N. Liu, Y. Li, and H.-Q. Liu. 2003a. Head and backbone of the Early Cambrian vertebrate *Haikouichthys*. Nature 421:526–529.

Shu, D.-G., S. Conway Morris, Z.-F. Zhang, J.-N. Liu, L. Chen, X.-L. Zhang, K. Yasui, and Y. Li. 2003b. A new species of yunnanozoan with implications for deuterostome evolution. Science 299:1380–1384.

Simonetta, A. M., and E. Insom. 1993. New animals from the Burgess Shale (Middle Cambrian) and their possible significance for the understanding of the Bilateria. Bollettino Zoologica 60:97–107.

Siveter, D. J., M. Williams, and D. Waloszek. 2001. A phosphatocopid crustacean with appendages from the Lower Cambrian. Science 293:479–80.

Smith, A. B., and K. J. Peterson. 2002. Dating the time of origin of major clades: molecular clocks and the fossil record. Annual Review of Earth and Planetary Sciences 30:65–88.

Smith, M. P., I. J. Sansom, and K. D. Cochrane. 2001. The Cambrian origin of vertebrates. Pp. 67–84 *in* Ahlberg 2001.

Sutton, M. D., D. E. G. Briggs, D. J. Siveter, and D. J. Siveter. 2001. An exceptionally preserved vermiform mollusc from the Silurian of England. Nature 410:461–3.

Takezaki, N., F. Figueroa, Z. Zaleska-Rutczynska, and J. Klein. 2003. Molecular phylogeny of early vertebrates: monophyly of the agnathans as revealed by sequences of 35 genes. Molecular Biology and Evolution 20:287–292.

Walossek, D. 1999. On the Cambrian diversity of Crustacea. Pp. 3–27 *in* F. R. Schram and J. C. von Vaupel Klein, eds. Proceedings of the Fourth International Crustacean Congress, Amsterdam, The Netherlands, Vol. 1. Brill, Leiden.

Waloszek, D., and J. A. Dunlop. 2002. A larval sea spider (Arthropoda: Pycnogonida) from the Upper Cambrian "Orsten" of Sweden, and the phylogenetic position of pycnogonids. Palaeontology 45:421–446.

Wang, D. Y-C., S. Kumar, and S. B. Hedges. 1999. Divergence time estimates for the early history of animal phyla and the origin of plants, animals and fungi. Proceedings of the Royal Society of London B 266:163–71.

Wheeler, W. C. 1998. Molecular systematics and arthropods. Pp. 9–32 *in* Edgecombe 1998.

Whittington, H. B., and D. E. G. Briggs. 1985. The largest Cambrian animal, *Anomalocaris*, Burgess Shale, British Columbia. Philosophical Transactions of the Royal Society of London B 309:569–609.

Wills, M. A. 1998. Cambrian and recent disparity: the picture from priapulids. Paleobiology 24:177–199.

Wills, M. A., and R. A. Fortey. 2000. The shape of life: how much is written in stone? BioEssays 22:1142–1152.

Wills, M. A., D. E. G. Briggs, and R. A. Fortey. 1994. Disparity as an evolutionary index: a comparison of Cambrian and Recent arthropods. Paleobiology 20:93–130.

———. 1998. An arthropod phylogeny based on fossil and Recent taxa. Pp. 33–106 *in* Edgecombe 1998.

Winnepenninckx, B. M. H., T. Backeljau, and R. M. Kristensen. 1998. Relations of the new phylum Cycliophora. Nature 393: 636–637.

Wray, G. A., J. S. Levinton, and L. H. Shapiro. 1996. Molecular evidence for deep Precambrian divergences among metazoan phyla. Science 274:568–72.

Xiao, S., and A. H. Knoll. 2000. Phosphatized animal embryos from the Neoproterozoic Doushantuo Formation at Weng'an, Guizhou, South China. Journal of Paleontology 74:767–788.

Paleobiology, 31(2), 2005, pp. 113–121

Stephen Jay Gould on species selection: 30 years of insight

Bruce S. Lieberman and Elisabeth S. Vrba

Abstract.—Stephen Jay Gould made impressive contributions to macroevolutionary theory; one of the topics in this area that particularly interested him was how to define and recognize species selection. Here we explore how and why Gould's ideas on concepts related to species selection evolved over 30 years, from the punctuated equilibria paper of 1972 to his "Structure of Evolutionary Theory" magnum opus published in 2002. Throughout his career his ideas on species selection shifted between three phases. Initially, Gould favored a definition of species selection that was more descriptive. Later, he came to distinguish between species sorting, which he called species selection in the broad sense, and true species selection, which is tied to the concept of species-level aptations. Finally, he came to view species selection in a broader, more inclusive way, effectively merging the two earlier viewpoints. His ideas on species selection changed over the years because he was trying to square his views on complex concepts like adaptation, natural selection, emergence, and the independence of macroevolutionary theory. Gould's thoughts on species selection not only help to define the history of debate on the concept but also help set a course for the future.

Bruce S. Lieberman. Department of Geology, 1475 Jayhawk Boulevard, 120 Lindley Hall, University of Kansas, Lawrence, Kansas 66045. E-mail: blieber@ku.edu
Elisabeth S. Vrba. Department of Geology and Geophysics, Kline Geology Lab, Post Office Box 08109, Yale University, New Haven, Connecticut 06520-8109

Accepted: 22 April 2004

Introduction

Stephen Jay Gould had a tremendous impact on the fields of paleobiology and evolutionary biology and some of his most important contributions are in the area of macroevolutionary theory, which he helped build into an exciting, vibrant research area. One of the most contentious and interesting research topics in macroevolution has been the levels of selection debate, especially the role that species selection played in shaping evolution. Gould thought that this was a particularly salient topic, one to which he devoted much thought and energy. Indeed, he "long regarded species selection as the most challenging and interesting of macroevolutionary phenomena, and the most promising centerpiece for macroevolutionary theory" (Gould 2002: p. 731), while admitting that "no other subject in evolutionary theory has so engaged and confused me, throughout my career, as the definition and elucidation of species selection" (Gould 2002: p. 670). Although he was not the first nor the only scientist to publish on this topic, he did make important contributions.

Gould's published scientific legacy preserves 30 years of engaging, insightful commentary of relevance to species selection. Here, we explore how and why Gould's views on this topic evolved, and their relevance for understanding his scientific legacy. Although over the years Gould's opinion did change regarding the nature of species selection and how to define it, he always understood the complex scientific issues associated with the levels of selection debate; his positional shifts did not represent waffling but rather his attempt to grapple with issues like adaptation, natural selection, emergence, and the primacy of macroevolutionary theory.

Punctuated Equilibrium and the Early Development of a Macroevolutionary View of Species Selection

Species selection, or at least the related group selection concept, had been recognized as a possible though not necessarily significant evolutionary force by biologists midway through the twentieth century (e.g., Fisher 1958; Wynne-Edwards 1962; Williams 1966). The concept even sneaks into discussions in Darwin's writing (1859). It is safe to say, however, that the concept was not really legiti-

 0094-8373/05/3102-0008/$1.00

mized until the development of the theory of punctuated equilibrium by Eldredge (1971) and Eldredge and Gould (1972); this represented a significant turning point for the fortunes of the species selection concept. Punctuated equilibrium is a theory that has relevance to many paleobiological topics. One aspect of its significance is that it emphasized the reality of species (Eldredge 1979, 1985, 1989; Gould 1980, 1982a,b, 1990, 2002; Vrba 1980, 1984a, 1992); as documented by Eldredge and Gould (1972) species had previously been viewed as ephemeral entities. The theory of punctuated equilibrium makes it possible to individuate species in both space and time (Eldredge 1979, 1982; Vrba 1980; Gould 1982a). Indeed, it was Ghiselin's (1974) and especially Hull's (1980) emphasis on the fact that species were individuals that contributed in an important way to discussions about species selections (see also Hull 1988). Fixity and permanence, as opposed to evanescence, make it easier to view species as objects that could be selected (Lieberman 1995).

Punctuated equilibrium was also important to the topic of species selection because it suggested a revised ontology of trends, which are one of the most significant paleontological phenomena (Gould 1990). When Eldredge and Gould (1972) focused on the nature of trends in light of punctuated equilibrium, they argued that evolutionary trends might not be due to the gradual, anagenetic modification of evolutionary lineages; instead, trends would involve cladogenesis followed by the differential success of species exhibiting change in a particular direction. Stanley (1975) expanded this notion. He argued that in light of punctuated equilibrium the differential birth or death of groups of species should be called species selection because it implied the action of a process different from strict natural selection: in Stanley's (1975) concept, species selection favors some clades of species because their included organisms are better adapted than other organisms in related clades. In this view, trends produced by species selection actually represent adaptive success at the organism level. A similar concept was subsequently used by others, including Gould and Eldredge (1977), Stanley (1979), Gould (1980), and Arnold and Fristrup (1982). Punctuated equilibrium triggered the development of "an expanded hierarchical theory (that) would not be Darwinism as strictly defined, but it would capture, in abstract form, the fundamental feature of Darwin's vision—direction of evolution by selection at each level" (Gould 1982a: p. 381).

Species Sorting and Species Selection

Sorting is a neutral description of differential birth and death. It contains no statement about causes, and these could in fact range from random drift to selection. In Table 1, the subset of species sorting that is most relevant to our discussion about species selection is nonrandom sorting of heritable variation. We further subdivide this into three categories. Two of these are relatively clear-cut cases that either do not, or do, represent species selection, respectively: (A) where the only selected characters are those of organisms (although in this and all other cases of sorting in sexual species it is important to recognize that the emergent species-level characters of reproductive systems are relevant to speciation and extinction); and (B2) where there are selected characters that are indisputably emergent at the species level. The middle ground, case B1, has proved to be more difficult to interpret, as we discuss below.

A brief history of the debate about species selection and its definition starts with Stanley (1975) and other treatments of species selection up to 1980 (Table 1: 1), and we list three cases that have been descriptively termed species selection. A contrasting view of these early treatments was offered by Vrba (1980), who distinguished between the effect hypothesis of species sorting and true species selection. Effect species sorting occurs when lineages vary in aggregate organismal characters and selective regimes, and that variation directly effects, and explains, differences in speciation and extinction rates. In this view (Vrba 1980, 1982), species selection can be invoked only when true species-level characters interact with the environment to produce species sorting and trends (Table 1: 2).

In the ensuing debate there emerged a general agreement that cases A and B2 need to be

TABLE 1. A history and classification of terminology applied to three modes of nonrandom species sorting, involving heritable character differences among species and clades. The three modes differ in the kinds of species characters, and therefore the processes, involved. The terms used by GOULD and others are in the body of the table: SS = species selection; and EH = the effect hypothesis. The linkage between characters and terms used in cited references was either argued explicitly (shown as entries without parentheses), or implied but not directly stated (shown as entries in parentheses).

	Heritable character differences among species influence species sorting (namely, there is "emergent species' fitness" sensu Lloyd and Gould 1993):		
		B. Species differ in characters emergent at the species level:	
Authors	A. Species differ in aggregate characters (genotypes and phenotypes fixed in species)	B1. Degree of genetic/phenotypic variability of the species' gene pool	B2. Population structure and distribution within species
1. Eldredge and GOULD 1972; Stanley 1975, 1979; GOULD and Eldredge 1977; GOULD 1980	SS	(SS)	(SS)
2. Vrba 1980	EH	(SS)	(SS)
3. GOULD 1982b	?SS broad sense ?EH	?SS broad sense ?EH	SS narrow sense
4. Sober 1984	"SS of"	?	"SS for"
5. Eldredge 1985; Gilinsky 1986; Jablonski 1986; Kitchell et al. 1986; Vrba and GOULD 1986; Doolittle 1987; Werdelin 1987	EH	EH	SS
6. Vrba 1989	EH	SS	SS
7. GOULD 1990	?SS broad sense ?EH	?SS broad sense ?EH	SS narrow sense
8. Lloyd and GOULD 1993; GOULD 2002	SS broad sense ("emergent fitness SS")	SS broad sense ("emergent fitness SS")	SS narrow sense ("emergent character SS")
9. Lieberman and Vrba 1995; this paper	EH	SS	SS

distinguished, although the terms proposed for case A vary (Table 1: 3–9). Gould's early input in this debate was somewhat confusing. On the one hand, he acknowledged the distinction between effect species sorting, which is not based on species-level properties, and species selection. He wrote, "I strongly recommend that the term 'species selection' be confined to . . . selection among species based on species-level properties. I shall present an example of true species selection (p. 95) . . . ." (Gould 1982b: 94). On the other hand, in the same paper he coined the term "*species selection* in the broad sense" (our emphasis) and implied its equivalence with effect species sorting; further, his examples of "true species selection," namely "species selection in the narrow sense" (p. 95), included the case of organismal eurytopy and stenotopy for which the effect hypothesis was originally illustrated (Vrba 1980). Although Gould, and others, in subsequent treatments during the 1980s accepted the conceptual and terminological distinction between effect species sorting and species selection (Table 1: 5), a similar consensus did not emerge on the intermediate case, B1. For example, Vrba (1989: Table 2) grouped case B1 with B2 as species selection, and distinguished it from effect species sorting A (as we do in this paper, Table 1: 6, 9). By contrast, Lloyd and Gould (1993) and Gould (2002) treated A and B1 as species selection in the broad sense, and called it "emergent fitness species selection" (Table 1: 8).

In spite of these differences, on the whole the definition of the effect hypothesis led many to conclude that examples of species selection require emergent characters of species (see citations after 1980 in Table 1; also see Eldredge 1982, 1985, 1989; Vrba 1982, 1984a; Vrba and Eldredge 1984; Lieberman and Vrba 1995). This emphasis on emergent characters of species recognized the fundamental association between aptation and selection (Gould and Vrba 1982; Vrba 1984a, 1989; Eldredge 1985, 1989; Lloyd and Gould 1993; Gould 2002). In order to invoke species selection there must be a clade with a character that is a species-level aptation (Lieberman 1995).

The boundaries of the domain of emergent species characters, however, remain fuzzy. There is scientific consensus that emergent characters of species cannot be reduced to characters of the component organisms, and population size, structure, and distribution are such characters (e.g., Eldredge and Salthe 1984; Vrba and Eldredge 1984; Jablonski 1986, 1987; Vrba 1989: Table 2). There is debate, however, as to how simple differences in overall variability between species and clades (case B1 in Table 1) relate to species sorting.

One of us has argued (and we both agree on this) that "emergent characters that are candidates for aptation [at the species level include] rate of variation production (by mutation and recombination) [and] variation patterns of the gene pool itself, provided they can be transmitted to descendant species" (Vrba 1989: p. 131). Such characters can be significant because related species and clades commonly differ in their degree of variability. Most such differences are evanescent and either are not passed on to descendant species or have no causal bearing on species sorting. If, however, there are genetically based differences in gene pool variability between species and clades, and these differences persist for millions of years, such differences may be based on heritable among-organism dynamics, perhaps relating to sexual reproductive interactions. These differences would qualify as emergent, species-level characters; if they influenced species sorting they might be species-level aptations, and then such a case would involve species selection.

The most problematic case relevant to the definition of species selection is when differences in levels of organismal variability cause species sorting involving differential extinction in one of two sister groups. Take, for example, two sister clades, X and Y, where the species in X have, as a species-level character, monomorphism in one or more habitat-related characters; further, and as a result, clade X has a higher extinction rate during times of climatic change. By contrast, the species-level polymorphism in these characters in the species of clade Y decreases their extinction probability during times of climatic change. (This example is actually not far fetched. Brooks [2002] argued that high genetically based variability within populations of some guppy fish has dampened speciation rate in these lineages in spite of a history of environmental change.) One might be tempted to conclude in this case that differing organismal characters and selection regimes led to effect species sorting. However, because of the presence of an emergent, heritable character associated with high variability in the species of clade Y, there is heritable, nonrandom species sorting, and this qualifies as species selection. Although we currently lean toward an interpretation of species selection in our hypothetical case, we expect that the study of the origin and long-term maintenance of organismal variability in species, and its macroevolutionary consequences, will result in expansion and sharpening of this concept.

Lloyd and Gould (1993) and Gould (2002) also concluded that if the degree of variability within species promotes species sorting, this would act as a species aptation and qualify as a case of species selection; however, they arrived at this conclusion in a different way. They did not consider or even require that such characters be emergent at the species level. Rather, they argued that such sorting involves "emergent fitness" at the species level, and that this is sufficient to qualify as "species selection in the broad sense." We partly diverge from them in this interpretation and argue that in sexual species *all* nonrandom species sorting that is caused by heritable variation, including effect species sorting, involves emergent fitness. This is because the emergent fitness in effect species sorting arises as an incidental effect of the interaction of organismal selection with the boundaries imposed by closed gene pools.

In summary, the changes in the use of the term "species selection" (Table 1) are based on the fact that Eldredge and Gould (1972), and Stanley (1975) in more detail, had identified a legitimate and important pattern that had long been ignored by evolutionary biologists; however, this pattern, species sorting, did not necessarily imply that species were actually being selected (Vrba 1980; Eldredge 1982; Gould 1982b; Vrba and Gould 1986; Lieberman and Vrba 1995). Some of the cases that had been classified as species selection by Stanley (1975, 1979), Gould (1980), and Arnold and Fristrup (1982) but that no longer deserved such an appellation (Vrba 1980; Eldredge 1982) under the revised concept included trends caused by effects related to the direction of speciation, and differences in speciation and extinction rates that could be explained by the action of natural selection (Vrba 1980). An example of the latter includes diversity dynamics in two clades of African antelopes, the Aepycerotini and Alcelaphini, that were produced by the effect hypothesis rather than by species selection (Vrba 1984b, 1987).

Still, in the literature to date there may be several examples of species selection based on population structure and distribution (in the narrow sense sensu Gould 2002; B2 in Table 1); and we expect that many more will be found in the future as the number of macroevolutionary analyses increases and analytical methods improve. Some of the cases documented by Hansen (1978), Gilinsky (1981, 1986), Jablonski and Lutz (1983), and Jablonski (1986) may be included as examples. The mere existence, however, of an emergent population or species-level character does not necessarily prove the operation of species selection (Lieberman et al. 1993; Lieberman and Vrba 1995). For example, Hansen (1978), Jablonski and Lutz (1983), and others have documented how in gastropods a non-planktonic larval type produces a population structure that favors population fragmentation and speciation. They predicted that through time the number of species with a non-planktonic larval type will increase relative to the number of species with a planktonic larval type not because the organisms in these species were more fit but rather because they were more likely to speciate (Gould 1982b, 2002). In one group of gastropods, the turritellids, the number of species with a non-planktonic larval type does increase through the Cenozoic relative to the number of species with a planktonic larval type. However, phylogenetic analysis suggested that the primary reason for the trend in the turritellids was not species selection but instead the repeated conversion of planktonic to non-planktonic species with absent or minimal reversion (Lieberman et al. 1993; Lieberman 1995). Instead, mechanisms involving either development, called cell-lineage drive (Lieberman et al. 1993; Lieberman 1995) using Buss's (1987) ideas on germ-line sequestration, or organismal adaptation (Strathmann 1978), seemed implicated. Therefore, at least some of the trend pattern in the turritellids is compatible with species sorting, but the trend was not caused by species selection (Lieberman and Vrba 1995).

The distinction between species selection and other forms of species sorting, and between different kinds of species selection, is worth considering in greater detail because Gould (1982b) argued that "the inevitable confusion between (species selection in the) broad and narrow sense is most unfortunate especially since the existence of true group selection in some (but not all) trends is an important component of our (Gould and Eldredge's 1977) argument for the independence of macroevolution" (Gould 1982b: p. 94). We partly disagree with this interpretation: the independence of macroevolution is affirmed not only by species selection but also by other processes such as effect sorting among species (Vrba 1980: p. 81). (We here agree with Grantham's [1995: p. 309] conclusion on the effect hypothesis as illustrated by African mammal clades in Vrba 1984b, 1987: "The species-level sorting is merely an incidental effect of organismic selection. Although Vrba's explanation does not introduce a higher-level process, I would maintain that this explanation is not reducible.") Our view is more in line with Gould's recognition that "the key issue for the independence of macroevolution is not whether species selection operates in all trends (it does not), but whether the necessity, under punctuated equilibrium, of regarding trends as a higher-level sorting of species implies a new level in a hierarchy of evolutionary explanation" (Gould 1982b: p. 94). Thus, Gould agreed that it was important to recognize species selection as a special type of species sorting, and the mere existence of species sorting, the pattern emphasized by Stanley (1975), was enough to justify the importance of macroevolutionary theory. Macroevolution is given expanded meaning by punctuated equilibrium, which is a theory more about species and their reality and individuality (sensu Hull 1980) than about speciation (Lieberman 1995).

In the middle and late 1980s Gould's views on species selection restricted the term to what he called species selection in the strong sense, and he used a definition requiring the presence of characters emergent at the species level that interacted with the environment to produce differential speciation rates (i.e., the definition of Vrba 1980, 1984a, 1989; and Eldredge 1982, 1985). As we shall see, however, Gould came to feel that this definition differed from how natural selection was defined. We do not agree with his revised interpretation. Traditionally, natural selection was always acknowledged to involve the interaction between the environment and genetically based phenotypes—namely, emergent characters of organisms—with fitness consequences emergent at the organismal level. Gould felt that the emergent character definition unnecessarily constrained the purview of macroevolution, making it a field more about documenting patterns of species sorting than identifying novel examples of species selection. (Again, we differ because we suspect the domain of species selection [as circumscribed by cases B1 and B2 in Table 1] is large; there remains the challenge of distinguishing between the various causal processes of species sorting irrespective of the terminology used.) By 1990 Gould had come to "vacillate between a strict definition (of species selection) based on emergent characters and a more inclusive construction" (Gould 1990: p. 19) (Table 1: 7), and three years later he had come to embrace the more inclusive construction (Table 1: 8).

Species Selection and Emergent Fitness

The shift in Gould's thoughts on species selection was first thoroughly documented in a paper by Lloyd and Gould (1993) where the concept of emergent fitness was introduced (Table 1). This they distinguished from the more narrowly circumscribed species selection in the strict sense, which they referred to as the emergent character definition. They argued that under the emergent fitness definition a character emergent at the species level is no longer required; instead, some differential pattern of speciation or extinction rates is necessary, and this would have to be correlated with a trait emergent at any hierarchical level. This emergent fitness definition greatly expands the amount of evolution that would be due to species selection. Still included would be the emergent character type of species selection, but also other examples that had been treated as nonselective species sorting by Vrba (1980, 1984) and Vrba and Gould (1986) would now qualify as valid examples of species selection (Table 1: 8). Further, Lloyd and Gould (1993) argued that species-level variability, a character that they did not necessarily believe was emergent at the species level, is crucial to the expanded vision of the emergent fitness criterion of species selection. Lloyd and Gould (1993) postulated that variability within species might be heritable and could promote extinction resistance; conceivably, such characters might also promote increased speciation rates, though Lloyd and Gould (1993) did not discuss this.

Gould (2002) offered three reasons that motivated his development of, and preference for, the emergent fitness concept of species selection as opposed to the emergent character approach. Although we do not necessarily agree with him, they are the following: (1) he believed that the emergent fitness definition is more in line with the way selection is identified at the organism level; (2) he felt that the emergent character approach limits species selection to a small number of cases (he considered that only category B2 in Table 1 qualifies; as noted above, Vrba's [1989] emergent character concept and our present one is actually more inclusive); and (3) he felt that convincingly demonstrating that characters are truly emergent at the species level is problematic. At least two crucial issues are raised by Lloyd and Gould's (1993) and Gould's (2002) emergent fitness approach: the concept of adaptation or exaptation and its relationship to species selection, and how to treat characters such as species-level variability.

Adaptation and Exaptation at the Species Level.—As defined by Gould and Vrba (1982) adaptations are characters that are currently enhancing fitness and that were constructed by natural selection to function in that particular role; thus, their selection context has not varied historically. By contrast, exaptations are characters that now perform a current function that is subject to selection, but they initially either were not shaped by selection at all or were shaped by selection for a different role. Lloyd and Gould (1993) and Gould (2002) argued that whereas the emergent character approach to studying species selection requires the identification of adaptations at the species level that interact with the environment and produce differential survival (extinction) and especially differential birth (speciation) rates, the emergent fitness approach only requires the identification of exaptations of species that arise at the organismal level and pass upwards as effects to the species level. We do not, however, see this as a valid distinction between the strict and broad species selection concepts. As argued by Vrba (1989: pp. 135–136), when using the strict concept of species selection, most species aptations are likely to be exaptations, while true species adaptations, if they exist at all, must be much rarer.

Species-Level Variability and Species Selection.—The primary example that Lloyd and Gould (1993) and Gould (2002) invoked to demonstrate a case of their version of species selection involved two hypothetical species: one with little variation that is well adapted to a particular narrow environment, and another with abundant variation that is moderately adapted to several environments. If there is an environmental perturbation, it is the second more poorly adapted, variable species that is most likely to survive, whereas the better adapted but less variable species is more likely to go extinct. Lloyd and Gould (1993) argued that this example represents emergent fitness without emergent species properties. Their example is closely related to the one we described above of sister taxa X and Y, differing in levels of organismal variability and therefore in extinction rate, which we explored above as a case of species selection. In discussing this variability-based example of "emergent fitness species selection," Gould concluded that "Vrba's solution . . . requires . . . that we interpret such cases as upward causation from the traditional organismal level . . . [consistent with the] 'effect hypothesis'" (Gould 2002: p. 658; see also Lloyd and Gould 1993). However, this is erroneous and based on a misunderstanding of Vrba (1989), who included "variation patterns of the gene pool itself, provided they can be transmitted to descendant species" (p. 131) as emergent species characters relevant to species selection.

Whatever one calls this example identified by Lloyd and Gould (1993) and Gould (2002), it is clear that it involves a type of pattern that cannot be explained by recourse to the traditional neo-Darwinian world view that prevailed before the development of punctuated equilibrium and the demonstration of the stability, reality, and individuality of species (see Hull 1980, 1988). Their hypothetical example involving species-level variability cannot be explained by natural selection operating in a world where species are evanescent. To understand it, organisms must instead be viewed in the context of species; Lloyd and Gould have presented a classic example of nonselective species sorting. Their example demonstrates how the existence of species can powerfully shape the course of the evolutionary process. Although we disagree with Lloyd and Gould (1993) and Gould (2002) in the particulars about their definition of species selection, preferring the concept Gould embraced earlier (e.g., Vrba and Gould 1986), we share strong concordance with them that species provide the fundamental context-dependence for organisms that influences patterns and processes in the history of life. Even if selection never operated above the level of individual organisms (which we do not believe), still, the geometry of evolution will be very different if species are real and relatively stable entities, compared with a world where species are not real and stable through time (Lieberman and Vrba 1995).

Conclusions

The issues that Stephen Jay Gould raised throughout his career regarding the prevalence and nature of species selection are insightful and important. Indeed, this topic, along with the related topic that species are real entities, stable for most of their history, which derives from his and Eldredge's theory of punctuated equilibrium, permeates many of his scientific writings. His work in this area helped develop macroevolutionary theory and expand current visions of evolutionary biology. He did this not by focusing primarily on the complex adaptations that organisms possess. Although a legitimate area of research, this has been thoroughly explored by Darwin and others (and also considered by Gould in many of his publications). Instead, with this work Gould focused on the related and equally interesting issue of what promotes trends within clades whose organisms possess these adaptations, and how clades wax and wane over geological time. The latter issues are primarily within the direct purview of paleontology. The emphasis on the stability of species provided by punctuated equilibrium suggests that characters that increase organismic fitness do not necessarily enhance speciation or prevent extinction, and that an important part of evolutionary theory is the causes and consequences of species sorting (Eldredge 1979, 1982; Vrba 1980; Gould 1982b, 2002). In effect, then, the discontinuity of species in space championed by Dobzhansky (1937) and Mayr (1942), and also their stability through time, first championed by Eldredge (1971) and Eldredge and Gould (1972), implied that the causes for how adaptive diversity is distributed within and among clades might reside not simply at the organismic level (mediated by natural selection) but also at the species level (species sorting and species selection). Future debates in macroevolutionary theory will continue to address the nature and various causes of species sorting. In any event, it is clear that Stephen Jay Gould was at the vanguard of many macroevolutionary topics and helped define not only the present but also the future of the debate about the nature of macroevolutionary theory in general and species selection in particular. In closing, it is worth recognizing that one topic that Gould championed throughout his career (e.g., Gould 1989, 1996, 2002) was contingency, for "in contingency lies the power of each person . . . to make a difference . . . spelling . . . vast improvement" (Gould 2002: p. 1346). This principle of contingency, and the impact a single person can have on a scientific field, is wonderfully illustrated by Stephen Jay Gould's contributions to paleobiology.

Acknowledgments

We thank D. Ackerley, N. Eldredge, R. Kaesler, and one anonymous reviewer for comments on earlier versions of this paper. B.S.L.'s research was supported by National Science Foundation grant EAR 0106885, a Self Faculty Fellowship, and NASA Astrobiology grant NNG04GM41G.

Literature Cited

Arnold, A. J., and K. Fristrup. 1982. The theory of evolution by natural selection: a hierarchical expansion. Paleobiology 8: 113–129.

Brooks, R. 2002. Variation in female mate choice within guppy populations: population divergence, multiple ornaments and the maintenance of polymorphism. Genetica 116:343–358.

Buss, L. W. 1987. The evolution of individuality. Princeton University Press, Princeton, N.J.

Darwin, C. 1859. On the origin of species. (Facsimile of first edition.) Harvard University Press, Cambridge.

Dobzhansky, T. 1937. Genetics and the origin of species. Columbia University Press, New York.

Doolittle, W. F. 1987. The origin and function of intervening sequences: a review. American Naturalist 130:55–85.

Eldredge, N. 1971. The allopatric model and phylogeny in Paleozoic invertebrates. Evolution 25:156–167.

———. 1979. Alternative approaches to evolutionary theory. Bulletin of the Carnegie Museum of Natural History 13:7–19.

———. 1982. Phenomenological levels and evolutionary rates. Systematic Zoology 31:338–347.

———. 1985. Unfinished synthesis. Oxford University Press, New York.

———. 1989. Macroevolutionary dynamics. McGraw Hill, New York.
Eldredge, N., and S. J. Gould. 1972. Punctuated equilibria: an alternative to phyletic gradualism. Pp. 82–115 *in* T. J. M. Schopf, ed. Models in paleobiology. Freeman, Cooper, San Francisco.
Eldredge, N., and S. N. Salthe. 1984. Hierarchy and evolution. Oxford Surveys in Evolutionary Biology 1:184–208.
Fisher, R. A. 1958. The genetical theory of natural selection, 2d ed. Dover, New York.
Ghiselin, M. T. 1974. A radical solution to the species problem. Systematic Zoology 23:536–544.
Gilinsky, N. L. 1981. Stabilizing species selection in the Archaeogastropoda. Paleobiology 7:316–331.
———. 1986. Species selection as a causal process. Evolutionary Biology 20:248–273.
Gould, S. J. 1980. Is a new and general theory of evolution emerging? Paleobiology 6:119–130.
———. 1982a. Darwinism and the expansion of evolutionary theory. Science 216:380–387.
———. 1982b. The meaning of punctuated equilibrium and its role in validating a hierarchical approach to macroevolution. Pp. 83–104 *in* R. Milkman, ed. Perspectives on evolution. Sinauer, Sunderland, Mass.
———. 1989. Wonderful life. Norton, New York.
———. 1990. Speciation and sorting as the source of evolutionary trends, or 'things are seldom what they seem.' Pp. 3–27 *in* K. J. McNamara, ed. Evolutionary trends. Belhaven Press, London.
———. 1996. Full house. Harmony Books, New York.
———. 2002. The structure of evolutionary thought. Harvard University Press, Cambridge.
Gould, S. J., and N. Eldredge. 1977. Punctuated equilibria: the tempo and mode of evolution reconsidered. Paleobiology 3: 115–151.
Gould, S. J., and E. S. Vrba. 1982. Exaptation—a missing term in the science of form. Paleobiology 8:4–15.
Grantham, T. A. 1995. Hierarchical approaches to macroevolution: recent work on species selection and the "effect hypothesis." Annual Review of Ecology and Systematics 26:301–321.
Hansen, T. A. 1978. Larval dispersal and species longevity in Lower Tertiary gastropods. Science 199:885–887.
Hull, David L. 1980. Individuality and selection. Annual Review of Ecology and Systematics 11:311–332.
———. 1988. Science as a process. University of Chicago Press, Chicago.
Jablonski, D. 1986. Larval ecology and macroevolution in marine invertebrates. Bulletins of Marine Science 39:565–587.
———. 1987. Heritability at the species level: analysis of geographic ranges of Cretaceous mollusks. Science 238:360–363.
Jablonski, D., and R. A. Lutz. 1983. Larval ecology of marine benthic invertebrates: paleobiological implications. Biological Reviews 58:21–89.
Kitchell, J. A., D. L. Clark, and A. M. Gombos. 1986. The selectivity of mass extinction: causal dependency between life history and survivorship. Palaios 1:504–511.
Lieberman, B. S. 1995. Phylogenetic trends and speciation: analyzing macroevolutionary processes and levels of selection. Pp. 316–337 *in* D. H. Erwin and R. L. Anstey, eds. New approaches to speciation in the fossil record. Columbia University Press, New York.
Lieberman, B. S., and E. S. Vrba. 1995. Hierarchy theory, selection, and sorting. BioScience 45:394–399.
Lieberman, B. S., W. D. Allmon, and N. Eldredge. 1993. Levels of selection and macroevolutionary patterns in the turritellid gastropods. Paleobiology 19:205–215.
Lloyd, E. A., and S. J. Gould. 1993. Species selection on variability. Proceedings of the National Academy of Sciences USA 90:595–599.
Mayr, E. 1942. Systematics and the origin of species. Columbia University Press, New York.
Sober, E. 1984. The nature of selection. MIT Press, Cambridge.
Stanley, S. M. 1975. A theory of evolution above the species level. Proceedings of the National Academy of Sciences USA 72: 646–650.
———. 1979. Macroevolution: pattern and process. W. H. Freeman, San Francisco.
Strathmann, R. R. 1978. The evolution and loss of feeding larval stages in marine invertebrates. Evolution 32:894–906.
Vrba, E. S. 1980. Evolution, species and fossils: how does life evolve? South African Journal of Science 76:61–84.
———. 1982. The evolution of trends. Pp. 239–246 *in* J. Chaline, ed. Modalités, rythmes, et mécanismes de l'évolution biologique (Colloques internationaux du centre national de la recherche scientifique, No. 330). Editions du Centre National de la Recherche Scientifique, Paris.
———. 1984a. What is species selection? Systematic Zoology 33: 318–328.
———. 1984b. Evolutionary pattern and process in the sister-group Alcelaphini-Aepycerotini (Mammalia: Bovidae). Pp. 62–79 *in* N. Eldredge and S. M. Stanley, eds. Living fossils. Springer, New York.
———. 1987. Ecology in relation to speciation rates: some case histories of Miocene-Recent mammal clades. Evolutionary Ecology 1:283–300.
———. 1989. Levels of selection and sorting with special reference to the species level. Oxford Surveys of Evolutionary Biology 6:111–168.
———. 1992. Mammals as a key to evolutionary theory. Journal of Mammalogy 73:1–28.
Vrba, E. S., and N. Eldredge. 1984. Individuals, hierarchies and processes: towards a more complete evolutionary theory. Paleobiology 10:146–171.
Vrba, E. S., and S. J. Gould. 1986. The hierarchical expansion of sorting and selection: sorting and selection cannot be equated. Paleobiology 12:217–228.
Werdelin, L. 1987. Jaw geometry and molar morphology in marsupial carnivores: analyses of a constraint and its macroevolutionary consequences. Paleobiology 13:342–350.
Williams, G. C. 1966. Adaptation and natural selection. Oxford University Press, New York.
Wynne-Edwards, V. C. 1962. Animal dispersion in relation to social behaviour. Oliver and Boyd, Edinburgh.

Paleobiology, 31(2), 2005, pp. 122–132

The neutral theory of biodiversity and biogeography and Stephen Jay Gould

Stephen P. Hubbell

Abstract.—Neutral theory in ecology is based on the symmetry assumption that ecologically similar species in a community can be treated as demographically equivalent on a per capita basis—equivalent in birth and death rates, in rates of dispersal, and even in the probability of speciating. Although only a first approximation, the symmetry assumption allows the development of a quantitative neutral theory of relative species abundance and dynamic null hypotheses for the assembly of communities in ecological time and for phylogeny and phylogeography in evolutionary time. Although Steve Gould was not a neutralist, he made use of ideas of symmetry and of null models in his science, both of which are fundamental to neutral theory in ecology. Here I give a brief overview of the current status of neural theory in ecology and phylogeny and, where relevant, connect these newer ideas to Gould's work. In particular, I focus on modes of speciation under neutrality, particularly peripheral isolate speciation, and their implications for relative species abundance and species life spans. Gould was one of the pioneers in the study of neutral models of phylogeny, but the modern theory suggests that at least some of the conclusions from these early neutral models were premature. Modern neutral theory is a remarkably rich source of new ideas to test in ecology and paleobiology, the potential of which has only begun to be realized.

Stephen P. Hubbell. Department of Plant Biology, University of Georgia, Athens, Georgia 30602, and Smithsonian Tropical Research Institute, Unit 0948, APO AA 34002-0948. E-mail: shubbell@plantbio.uga.edu

Accepted: 12 September 2004

Introduction

Many of Gould's ideas generated controversy, but one for which he was never criticized was being a neutralist, which he most definitely was not. On the contrary, Gould argued that patterns of punctuated equilibria in phylogeny can be explained only if natural selection operates not just among individuals within populations, but also among species and higher taxon levels (Gould 2002). Selection operating at levels above the individual would of course imply that species and higher taxa differ in fundamental ways that affect their relative fitnesses on geological timescales, which in turn affect their relative life spans in the fossil record.

Although Gould was not a neutralist, there are at least two deep philosophical connections between Gould and current neutral theory. The first is his recognition of the importance of null models in science, and more specifically Gould's pioneering work using neutral models to study phylogeny. The unified neutral theory of biodiversity and biogeography is a recent example of such models, and it generates a set of formal null hypotheses for the origin, maintenance, and loss of species in ecological communities or in phylogenies (Hubbell 2001a). Gould understood the critical importance of null models in his own science (Gould et al. 1977), and in science in general. He also understood their importance as a control on one's own preconceptions and biases, and what can happen when such controls are absent or abused, as in the racist theories of human evolution, which Gould vigorously refuted (Gould 1981). In the 1970s, Gould and his collaborators used null models of phylogeny as a means to evaluate whether the patterns generated solely by random birth-death processes were consistent with the punctuated equilibrium hypothesis of evolution, with busts of speciation interspersed with long periods of relative quietude (Raup et al. 1973; Gould et al. 1977). They found that randomly generated patterns were not consistent with observed patterns, and so Gould and company concluded that non-neutral selection processes must be at work in the phylogenies he studied. As it turns out, however, these conclusions were somewhat premature, be-

0094-8373/05/3102-0009/$1.00

cause there were several problems with the formulations of the models that were not immediately apparent at the time.

The other philosophical connection of neutral theory to Gould is through the concept of symmetry and symmetry breaking. In the neutral theory of biodiversity, all species are treated as symmetric, which means that, to a first approximation, all species are assumed to be demographically identical on a per capita basis. The principal use of the neutral theory is to evaluate when, and to what degree, asymmetries among species are required to explain the assembly of observed ecological communities. Although Gould did not use this terminology, nevertheless the concepts he wrote about were the same. Paleontology could be described as the study of how morphological symmetries are transmitted, broken, and reestablished over evolutionary time. In phylogenies, more closely related species are more similar (symmetric) than less related species (Harvey and Pagel 1991). At least for the living species, the new tools of genomics and evolutionary developmental biology promise to answer many if not all of the recalcitrant classic questions in ontogeny and phylogeny (Gould 1977). These tools are revealing that the symmetries of life run far deeper than anyone ever supposed, as illustrated, for example, by the discovery of ancient and phylogenetically pervasive homeobox master regulatory genes (Carroll et al. 2001; Davidson 2001).

In this paper, I first present a brief synopsis of the neutral theory (Hubbell 2001a; Volkov et al. 2003). The current theory is best developed for ecological scales of time and space. However, the unified neutral theory embodies more of a macroecological and deep-time perspective than do most contemporary theories in ecology. This is because it is one of very few theories in ecology to incorporate a process of speciation explicitly. Here I focus mainly on the results for speciation, particularly peripheral isolate speciation, and its implications for biodiversity, speciation rates, and species life spans. I conclude with a brief prospect for the future of symmetric neutral theory in ecology and paleobiology, and how new models with symmetry broken in various ways promise to move us forward.

The Neural Theory and Relative Species Abundance

The origins of modern neutral theory in ecology can be traced back to the theory of island biogeography (MacArthur and Wilson 1963, 1967). The theory of island biogeography hypothesizes that ecological communities are assembled purely by dispersal. This and other dispersal assembly theories assert that the species richness on islands or in local communities represents a dynamic equilibrium between the rates of immigration of species into the community and the rate of their subsequent local extinction. Thus, such theories assert that communities are in taxonomic nonequilibrium with continual species turnover. The theory of island biogeography is neutral because it assumes that species are identical (symmetric) in their probabilities of arrival and survival. This theory was, and remains, a radical departure from most contemporary theory in ecology, which says that ecological nature is fundamentally asymmetric, that communities are equilibrium or near-equilibrium assemblages of niche-differentiated species, each of which is the best competitor in its own ecological niche (Chase and Leibold 2003). There has been a persistent theoretical tension in ecology between these two conflicting worldviews. Both perspectives have strong elements of truth, although typically on very different spatial and temporal scales (Hubbell 2001a). The theoretical quest has been the search for ways to reconcile and combine these divergent perspectives into a single seamless theory for ecology.

The unified neutral theory of biodiversity begins to build a theoretical bridge between these two perspectives by incorporating a dynamic theory of relative species abundance into the theory of island biogeography (Hubbell 2001a; Bell 2000, 2001). As in the original theory, the unified neutral theory treats species as identical (symmetric) in their per capita vital rates of birth, death, and migration. Unlike the theory of island biogeography, however, the unified neutral theory makes the neutrality assumption at the individual level, not the species level, a change that allows species to differentiate in relative abundance through ecological drift (demographic stochasticity). The persistence times of species under drift are then dictated by their abundances, so that the extinction rate is a genuine prediction of the theory, not a free parameter as it was in the original theory of island biogeography. In the unified neutral theory, the "metacommunity" replaces the mainland source area concept of the theory of island biogeography. The metacommunity is the phylogeographic unit within which most member species spend their entire evolutionary lifetimes. The neutral theory also generates a natural length scale—the biogeographic correlation length—that measures the size of metacommunities. In the theory of island biogeography, however, the size of the source area is not defined.

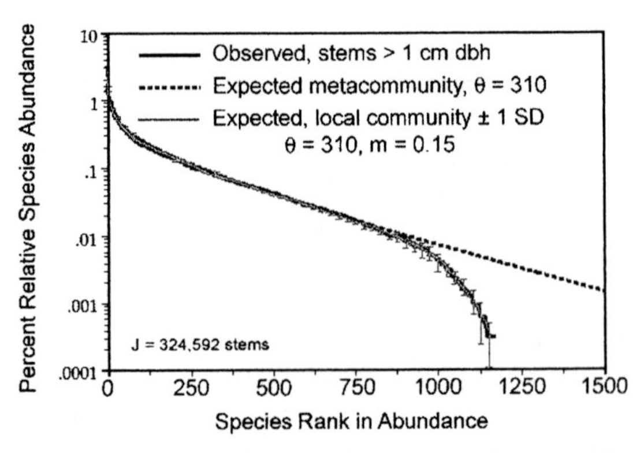

FIGURE 1. The fit of the unified neutral theory to the dominance-diversity curve for tropical tree species in a 50-ha permanent plot of rainforest in Lambir Hills National Park, Sarawak, Borneo. The dotted line extending diagonally down to the right is the best-fit metacommunity curve for $\theta = 310$, assuming no dispersal limitation ($m = 1$). The relative abundance for the 50-ha plot was best fit with $\theta = 310$ and $m = 0.18$. The error bars are $\pm$ one standard deviation. The heavier solid line is the observed dominance-diversity curve. Note the excellent fit even for very rare species. This fit to 1197 species is achieved with just three parameters, θ, m, and local community size J, the latter of which is known from the plot census data ($J = 324{,}592$).

Previous theories of relative species abundance have been largely static, phenomenological models (e.g., Preston 1948, 1960) and involve fitting generic statistical distributions whose parameters are not clearly derivable from first principles in population biology (Hubbell 2005). Because the previous models are not dynamic or mechanistic, they do not generate hypotheses about how basic demographic processes affect species richness and the distribution of relative species abundance. In contrast, the parameters of the neutral theory of relative species abundance all have straightforward biological interpretations, such as per capita birth and death rates, dispersal rates, and rates of speciation. Incorporating a process of speciation was especially key to developing a neutral theory of relative species abundance. Lacking a speciation mechanism for generating new diversity, the older phenomenological models have been generally unable to make predictions about the expected patterns of relative species abundance on large biogeographic spatial scales and evolutionary timescales. Fits of the neutral theory to relative abundance data are often quite good, especially in species-rich communities such as tropical rain forests (Fig. 1).

The importance of studying relative species abundance, especially on large landscape scales, cannot be overstated. Apart from species richness, no other general attribute of ecological communities has received more theoretical and empirical attention in ecology than relative species abundance, attention that is fully justified. One of the most important insights gained from the unified neutral theory is that speciation rates and patterns of relative species abundance on large spatio-temporal scales are inextricably and causally linked. At large biogeographic scales and between punctuational events, the steady-state diversity and distribution of relative species abundance are set by the balance between speciation and extinction rates. Relative species abundance is directly involved in this steady state because species extinctions are not drawn at random with respect to the abundances of species. Both theoretically and empirically, we know that rare species are more extinction prone (Richter-Dyn and Goel 1972; Lande et al. 1993). Although high global abundance appears to offer little or no protection from the agents of mass extinction (Jablonski 2001, 2002), there is evidence that abundant and geographically widespread taxa are more persistent during "normal" times (Jablonski 1995; Jackson 1995).

The correlation between global abundance and taxon longevity implies that during "normal'' times, the diversity steady state is maintained principally by the balance between the origination of new species and the extinction of mostly *rare* species. This conclusion is important in considering tests of the apparent dynamic quiescence of diversity between punctuational events. If rare species are harder to find in the fossil record than common species, there will be a built-in sampling bias that will underestimate taxon turnover rates during these periods. As data sets improve for fossil assemblages to include ever rarer taxa, I predict that estimated turnover rates for periods between punctuational events will steadily rise. In studies of paleocommunities, relative species abundance has played a less prominent role than species richness, but this situation is changing as improved data sets become available (e.g., Kidwell 2001). However, there are already data to support high rates of taxon turnover during periods of diversity steady states in some taxa, even in the absence of analyses of relative abundance data (e.g., Patzkowsky and Holland 1997).

Because relative species abundance is fundamental to any discussion of speciation under the neutral theory, it is important to review briefly the formal connection between the two subjects. Since publication of my book (Hubbell 2001a), there have been many significant changes and improvements in the mathematical framework of the theory (Houchmandzadeh and Vallade 2003; Vallade and Houchmandzadeh 2003; Volkov et al. 2003; Etienne and Olff 2004; Hubbell 2004; McKane et al. 2004). One of the advantages of the new framework is that several important problems in the symmetric neutral theory that were addressed only by simulations (numerical experiments) in my book are now tractable to analytical solutions. In particular, we have made substantial progress in understanding the relationship of neutral theory to the two most celebrated statistical distributions used to describe the distribution of relative species abundance: the logseries (Fisher et al. 1943), and the lognormal (Preston 1948). The neutral theory clarifies the situation and shows that the logseries is the fundamental distribution of relative species abundance expected at large spatiotemporal scales (the ''metacommunity'') under the simplest mode of speciation (''point mutation'' speciation; see below). However, this fundamental distribution becomes modified on local scales under dispersal limitation and on large spatial scales under different modes of speciation, and these modified distributions are more lognormal-like (Hubbell 2001a; Volkov et al. 2003; Hubbell and Borda-de-Água 2004).

The formal connection between speciation and relative species abundance in the metacommunity can be shown by simultaneously deriving the distribution of relative species abundance and the speciation rate from the fundamental dynamical equations of population growth under neutrality (see the Appendix for details). Under the metacommunity distribution of relative species abundance, which is the log-series, the expected mean number of species with n individuals $\langle \phi_n \rangle$ is given by

$$\langle \phi_n \rangle = \alpha \frac{x^n}{n},$$

where parameter x is a positive number <1, and α is the diversity parameter, known as Fisher's α. Fisher's α (Fisher et al. 1943) is the oldest, most famous, and most widely used measure of species diversity in ecology (Magurran 1988). One reason it is used so widely is that it is remarkably stable in the face of increasing sample sizes of relative abundance data from communities. Until the unified neutral theory, however, there has been no clear theoretical explanation for the stability and universality of Fisher's α nor any biological interpretation of parameter x of the logseries.

Neutral theory explains that the diversity parameter, Fisher's α, is so stable and universal because it is a linear function of the average speciation rate across the entire metacommunity as well as of the size of the metacommunity, defined as the sum of the population sizes of all species in the metacommunity. Fisher's α–called θ by Hubbell (2001a)—is a dimensionless, fundamental biodiversity number that crops up over and over again throughout the neutral theory. Neutral theory also demonstrates that parameter x of the logseries is equal to the ratio of the average per capita birth rate to the average per capita death rate of all species in the metacommunity (Volkov et al. 2003) (see Appendix). This ratio is very slightly less than unity when the distribution of relative species abundance in the metacommunity is in steady state. This means that at equilibrium diversity, there is a minute excess of deaths over births, and this small deficit in births is exactly balanced by the slow rate of introduction of new species into the metacommunity. Thus, under the framework of the unified neutral theory, the fundamental distribution of relative species abundance at large landscape scales is directly derivable from the speciation rate, the size of the metacommunity, and the average rates of birth and death in the metacommunity.

The Neutral Theory and Speciation

Coyne and Orr (2004) have recently reviewed current evidence in favor of various modes of speciation. They favor the biological species concept, which leads them naturally to a focus on the origin and maintenance of reproductive isolation. They conclude that, despite modest evidence for the hybrid origin of some species or for other mechanisms of sympatric speciation, the vast majority of species arise through allopatric speciation, following closely the now classical model of Mayr (1963). The biological species concept is, of course, not very useful in paleontology because tests of reproductive isolation are not possible. However, from the perspective of neutral theory, the only question about speciation that matters is how the mode of speciation affects the mean size of the founding population of new species. This is the critical question because initial population size determines not only the mean life span of a species, but also steady-state species richness and relative species abundance in the metacommunity.

In my book, I studied two modes of speciation, "point mutation" speciation and "random fission" speciation (Hubbell 2001a). I chose to study these two modes because they represent the end extremes of a speciation continuum in terms of the size of founding populations and the predicted life spans of new species. Mean species life spans are very short under "point mutation" speciation, because under this mode, new species arise as lineages founded by single individuals, and most of these lineages go extinct quickly. "Random-fission" speciation creates new species by the random uniform partition of an ancestral species into two daughter species. Species life spans are much longer under "random fission" speciation because this mode produces the largest average population size of new species. Large founding population sizes buffer species from rapid extinction, which, in turn, increases the steady state species richness in the metacommunity. In my book I suggested that random fission speciation is the analog to Mayr's allopatric speciation model (Hubbell 2001a). However, we now have a more complete and general formulation of the problem, in a mode of speciation we call "peripheral isolate" speciation (Hubbell and Lake 2003), which is discussed below.

Ricklefs (2003) has recently argued that these two modes of speciation are unrealistic because "point mutation" speciation produces too many short-lived species, whereas "random fission" speciation leads to overly long-lived species. In my response, (Hubbell 2003), I argued that the problems with "point mutation" speciation were easily resolved if one viewed this mode as actually a model of the fate of all lineages, most of which die out rapidly, and only a very few of which survive and become numerous enough to be discovered and sufficiently reproductively isolated to be called "good species." The "point mutation" mode is the only known speciation mechanism that gives rise to Fisher's α and the logseries distribution for the metacommunity. Whenever metacommunity relative species abundance distributions are observed to be consistent with the logseries, such a finding necessarily implies that the population sizes at origination must be small to very small. Species that arise by sudden changes in ploidy number or by hybridization are good candidates for origins by "point mutation" speciation.

Regarding "random fission" speciation, however, I think Ricklefs's point is well taken, and certainly most current data seem to be inconsistent with the "random fission" mode. In response to Ricklefs, Jeff Lake and I have explored the consequences of a third and intermediate mode of speciation, dubbed "peripheral isolate" speciation (Hubbell 2003; Hubbell and Lake 2003). Under this mode, founding populations are not as small as singleton-founded lineages, nor as large as those under "random fission," but nevertheless are fairly modest in size. "Peripheral isolate" speciation is probably commonplace. Most species are distributed as discontinuous metapopulations, and it seems likely that new species arise from one or more of the local isolated demes of metapopulations. This mode of speciation does indeed produce species having intermediate life spans and equilibrium metacommunities with intermediate species richness and relative species abundance distributions (Hubbell and Lake 2003). If the incipient species originate in small populations, however, empirically they may be difficult to distinguish from "point mutation" speciation events.

In my book, only the "point mutation" mode was solved analytically (Hubbell 2001a). Now, however, all three modes of speciation, including "peripheral isolate" speciation, have been solved analytically (I. Volkov personal communication 2004). Although the analytical results will be reported elsewhere, we can make three general statements about the findings here. The first conclusion is that the general case is "peripheral isolate" speciation; the other two modes are special cases of this general mode. The second conclusion confirms the simulation results of Hubbell (2001a) and Hubbell and Lake (2003) that increasing the size of the founding population greatly increases the steady-state species richness in the metacommunity. Figure 2 shows this effect for a value of Fisher's α (Hubbell's θ) of 10, for a birth-death ratio of 0.9999, and for various values of the size of the peripheral isolate at the point of speciation.

The third conclusion addresses one of the main concerns of Ricklefs (2003) about speciation rates that are too high under "point mutation" speciation. Because of the slower rate of extinction of new species under "peripheral isolate" speciation, a given level of species richness in the metacommunity can be explained by a much slower rate of speciation.

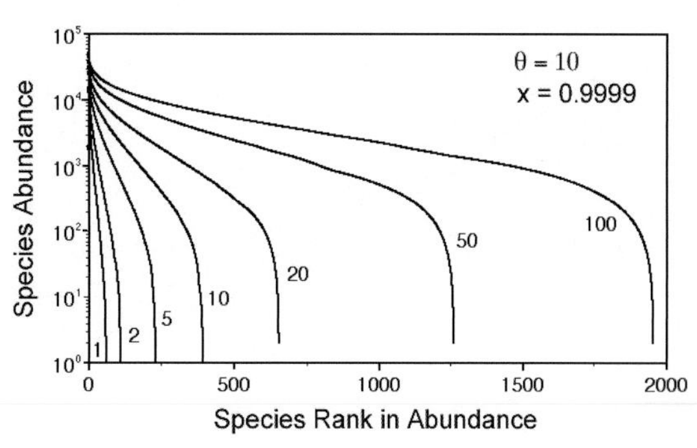

FIGURE 2. As the initial size of species populations at origination increases, more species are present at steady state in the metacommunity, for a fixed speciation rate. The figure shows an example of the effect of varying population size at origination on the steady-state metacommunity species richness and distribution of relative species abundance ("dominance-diversity" curves), for a fixed value of the fundamental biodiversity number θ (Fisher's α) and for a metacommunity average per capita birth/death ratio of $x = 0.9999$. The numbers beside each dominance-diversity curve are the initial population sizes at the origination of new species. The steepest dominance-diversity curve for a founding population size of unity corresponds to the logseries distribution for the limiting case of "point mutation" speciation.

Thus, one will consistently overestimate the speciation rate by fitting the "point mutation" speciation equations to relative abundance data. By how much the speciation rate is overestimated will depend on the mean size of peripheral isolate populations at origination. Figure 3 shows that speciation rates under "peripheral isolate" speciation are orders of magnitude smaller than those expected under "point mutation" speciation. The curves represent the ratio of the speciation rate under "peripheral isolate" speciation to the speciation rate under "point mutation" speciation that yields an equivalent metacommunity species richness, as a function of the population size of the isolate at speciation. Thus, the intercept for all curves is a ratio of 1.0 for an initial population size of unity, which corresponds to the "point mutation" limiting case. A smaller and smaller speciation rate is required to achieve the same metacommunity diversity the larger the founding peripheral isolate population becomes, and the closer the average metacommunity birth/death rate ratio approaches unity.

Testing these predictions about speciation and relative species abundance in metacom-

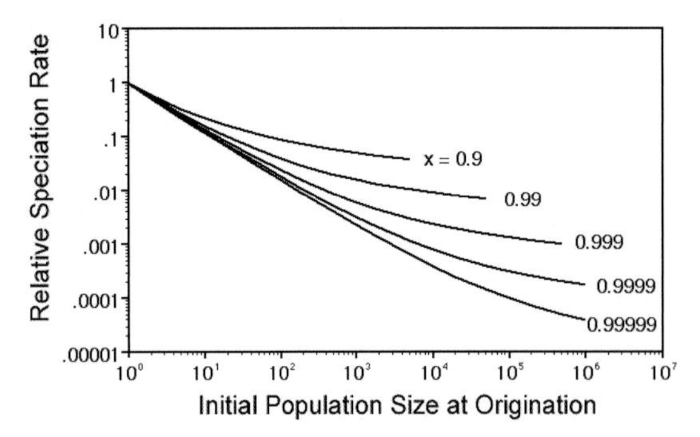

FIGURE 3. Holding species richness in the metacommunity constant, we can calculate the speciation rate under "peripheral isolate" speciation that is necessary to produce the same species richness relative to the speciation rate required under "point mutation" speciation (initial size = 1). These relative speciation rates are independent of the starting value of the fundamental biodiversity number θ (Fisher's α) under "point mutation" speciation, but they do depend on the mean per capita birth rate to death rate ratio, x. For large metacommunities, x is expected to be extremely close to unity, so that even moderate-sized peripheral isolates will result in a reduction of the effective speciation rate by several to many orders of magnitude over than required under "point mutation" speciation to produce the same metacommunity diversity.

munities will be a considerable challenge. The best tests will involve independent measures of speciation rates and metacommunity sizes, so that we have independently derived estimates of the fundamental biodiversity number θ. However, there are genuine empirical difficulties. One is that, if "peripheral isolate" speciation is the dominant mode but the average size of the founding populations is small, then ecologists, and especially paleontologists, may have considerable difficulty in finding and recognizing these nascent species, leading us to underestimate the true speciation rates, potentially seriously. One possible approach to testing the theory relies on the fact that the structure of phylogenetic trees is expected to be fractal with a single scaling domain under "point mutation" speciation, whereas it is expected to be compound fractal under "peripheral isolate" speciation (Hubbell 2001b). In the former case, the neutral theory says that we should expect a linear relationship between θ and the fractal dimension of the phylogeny (Hubbell 2001b). Because the deep-time, higher taxonomic divisions of phylogenetic clades leading to modern species are better sampled and known than the most recent species-level taxa, an estimate of θ from the deep-time structure of the phylogeny may be more accurate, and this in turn may help us assess how much modern diversity we are still missing. Ironically, it may well be that fossil assemblages can be used to test these ideas better than living taxa can, providing a much deeper time perspective.

Gould's Contribution to Neutral Theory and a Modern Update

As mentioned, Gould and a number of his colleagues made a pioneering contribution to neutral theory in their early studies of neutral phylogenies. After the publication of Eldredge and Gould (1972) on the theory punctuated equilibrium, one of the questions that arose was whether randomly generated phylogenies would produce patterns similar to those seen in the fossil record, which, depending on the taxon, often exhibited sudden, episodic increases in diversity, separated by longer periods of relatively calm and steady diversity levels. In part to try to answer this question, Raup et al. (1973) and Gould et al. (1977) took a "demographic approach" to phylogeny, in which they modeled monophyletic clades evolving as a stochastic birth-death branching process, picking up from the much earlier work of Yule (1925). In these models, lineages were assigned "birth" and "death" rates. When the birth rate exceeded the death rate, the general outcome of these models was exponential growth in the number of descendant lineages, and somewhat slower growth if all extinct lineages were pruned out. In no cases did the models yield the punctuational pattern postulated by Eldredge and Gould. In the last decade, in a series of elegant papers, Nee and his collaborators have provided analytical solutions to these models to study the process of "phylogenetic reconstruction" (e.g., Nee et al. 1994). Nee argued that observed phylogenies exhibit clades that are too "bushy," with some subclades containing too many species relative to the predictions of the null models, and argued that this was strong evidence for nonrandom processes in clade evolution.

These models meshed very well with Gould's concepts about species, which he has long argued can be treated as analogous to individuals when it comes to species-level selection (Gould 2001). Indeed, in the unified neutral theory under "point mutation" speciation, the hypothesis that new species arise from individually founded lineages blurs the distinction between species and individuals. This said, there is a fundamental difference between species and individuals that led the original neutral models of cladogenesis of Raup et al. (1973) and Gould et al. (1977)—and their analytical counterparts (Nee et al. 1994)—astray. The essential problem lies in the failure to take the relative abundance of lineages into account: populations have abundances, individuals do not. In the models of Raup, Gould, and Nee the evolutionary unit is the lineage, and lineages have assigned probabilities of speciating or going extinct. However, in the unified neutral theory, the evolutionary unit is the individual, and lineages per se have no preassigned speciation and extinction rates. Instead, the probability of speciating or going extinct is determined by lineage abundance, which is dictated in turn by the fundamental biodiversity number θ (or Fisher's α). In the models of Gould and Raup, lineage abundances are ignored completely, yet the abundance of a lineage (i.e., a species) will have a profound effect on the time to extinction of the lineage. Models that are pure birth-death branding processes trace the fate of lineages as if they had equal probabilistic fates, but this is not true because the fate of globally abundant lineages is very different on average from the fate of rare and local endemics.

Including lineage abundance changes many of the conclusions of the original neutral theory of phylogeny and suggests a series of testable hypotheses. First, globally abundant species are expected to be much older on average than rare species, a result consistent with the now classical theory of stochastic extinction (Richter-Dyn and Goel 1972). However, there is no such expectation under Gould's theory because lineage abundance is not considered. Second, these globally abundant and widespread metacommunity species are expected to be the ancestors of many more modern species than are rare and local species. This is a consequence not only of their much longer expected life spans, but also of their much higher total birth rate per unit time (opportunities for speciation) than in the case of rare species. This effect will make certain subclades much more speciose than expected under Gould-type neutral models of phylogeny. Third, the unified neutral theory produces a genuine diversity steady state at equilibrium between speciation and extinction. As mentioned, there is increasing evidence that these diversity equilibria between punctuational events (Eldredge and Gould 1972, 1988) are dynamic steady states with continual species turnover (Patzkowsky and Holland 1997). In contrast, Gould's neutral theory does not produce a diversity steady state, but instead produces exponential growth in the number of surviving lineages (Nee et al. 1994). This is because in Gould's theory, lineages have preassigned birth and death rates that do not change with lineage abundance.

Current evidence gives equivocal support to these various predictions of the unified neutral theory. As mentioned, there is evidence that globally abundant taxa do indeed have longer evolutionary life spans, at least during "normal" extinction times (Jablonski 1995; Jackson 1995), but this pattern breaks down during mass extinctions (Jablonski 2001, 2002). In molluscs and foraminifera, there are well-established relationships between dispersal ability and global abundance. However, there is also evidence that dispersal ability and global abundance are negatively correlated with rates of speciation, suggesting that gene flow is a major cohesive force in maintaining the integrity of species (Jablonski and Roy 2003). It should be noted that increased dispersal is expected to reduce the slope of the species-area relationship (Hubbell 2001a). This occurs for two reasons. One reason is simply more complete mixing. The other reason is extinctions of rare species caused by increased dispersal of common species and the overwhelming mass effect of their greater absolute birth rates (even when per capita birth rates are the same). Thus, even in a fully neutral model at a per capita level, increasing dispersal can cause a reduction of metacommunity diversity through increasing the extinction rate of rare species (Hubbell 2001a).

Conclusions

Although the neutral theory is simple, it nevertheless fits many macroecological patterns as well as or better than current niche theory in ecology. Perhaps the deepest question raised by neutral theory is why it performs so well despite its symmetry assumptions. I expect that some of the best and most rigorous tests will come from paleobiology. One of the most important conclusions from neutral theory is that processes of speciation and macroecological patterns of species richness and relative species abundance are inextricably and causally linked. This finding suggests that understanding relative species abundance in fossil communities better will provide further insights into both speciation and extinction processes. There are encouraging signs that major improvements in data on patterns of relative species abundance in fossil assemblages are possible (Kidwell 2001), and this would be a major boon in testing the predictions of the unified neutral theory with fossil data. It is also extremely important to obtain improved spatial data on the geographic range of fossil communities.

Symmetric neutral theory will be a rich source of hypotheses and tests about community assembly rules. I predict that one of the most productive uses of the unified neutral theory will be in testing when, how, and to what extent symmetry is broken in actual ecological communities. The theory is still in its infancy; and there are many exciting, unresolved theoretical challenges in and beyond the unified neutral theory to tackle for years to come (Chave 2004). I also anticipate that the legacy of the exciting, challenging, and still unanswered questions left by Steve Gould in paleobiology will continue to inspire major new contributions to our understanding of the assembly of ecological communities, past and present.

Acknowledgments

I thank the National Science Foundation, the John D. and Catherine T. MacArthur Foundation, the Pew Charitable Trusts, the John Simon Guggenheim Foundation, and many other donors—individuals and private organizations—who have supported my ecological research over the past 25 years, and the development of the neutral theory over the past ten years. I thank the Center for Tropical Forest Science of the Smithsonian Tropical Research Institute for permission to analyze the relative species abundance data for tree species in the 52-hectare plot in Lambir Hills National Park, Sarawak (Fig. 1).

Literature Cited

Bell, G. 2000. The distribution of abundance in neutral communities. American Naturalist 155:606–617.

———. 2001. Neutral macroecology. Science 201:2413–2417.

Carroll, S. B., J. K. Grenier, and S. D. Weatherbee. 2001. From DNA to diversity: molecular genetics and the evolution of animal design. Blackwell Scientific, Oxford.

Chase, J. M., and M. A. Leibold. 2003. Ecological niches: lining classical and contemporary approaches. University of Chicago Press, Chicago.

Chave, J. 2004. Neutral theory and community ecology. Ecology Letters 7:241–253.

Coyne, J. A., and H. A. Orr. 2004. Speciation. Sinauer, Sunderland, Mass.

Davidson, E. 2001. Genomic regulatory systems. Academic Press, New York.

Eldredge, N., and S. J. Gould. 1972. Punctuated equilibria: an alternative to phyletic gradualism. Pp. 82–115 *in* T. J. M. Schopf, ed. Models in paleobiology. Freeman, Cooper, San Francisco.

———. 1988. Punctuated equilibrium prevails. Nature 332:211–212.

Etienne, R. S., and H. Olff. 2004. A novel genealogical approach to neutral biodiversity theory. Ecology Letters 7:170–175.

Fisher, R. A., A. S. Corbet, and C. B. Williams. 1943. The relation between the number of species and the number of individuals in a random sample of an animal population. Journal of Animal Ecology 12:42–58.

Gould, S. J. 1977. Ontogeny and phylogeny. Harvard University Press, Cambridge.

———. 1981. The mismeasure of man. Norton, New York.

———. 2002. The structure of evolutionary theory. Belknap Press of Harvard University Press, Cambridge.

Gould, S. J., D. M. Raup, J. J. Sepkoski Jr., T. J. M. Schopf, and D. S. Simberloff. 1977. The shape of evolution: a comparison of real and random clades. Paleobiology 3:23–40.

Harvey, P. H., and M. D. Pagel. 1991. The comparative method in evolutionary biology. Oxford University Press, Oxford.

Houchmandzadeh, B., and M. Vallade. 2003. Clustering in neutral ecology. Physical Review E 68: Art. No. 061912.

Hubbell, S. P. 2001a. The unified neutral theory of biodiversity and biogeography. Princeton University Press, Princeton, N.J.

———. 2001b. The unified neutral theory of biodiversity and biogeography: a synopsis of the theory and some challenges ahead. Pp. 393–411 *in* J. Silvertown and J. Antonovics, eds. Integrating ecology and evolution in a spatial context. Blackwell Scientific, Oxford.

———. 2003. Modes of speciation and the lifespans of species under neutrality: a response to the comment of Robert E. Ricklefs. Oikos 100:193–199.

———. 2005. Neutral theory in ecology and the evolution of functional equivalence. Ecology (in press).

Hubbell, S. P., and J. Lake. 2003. The neutral theory of biogeography and biodiversity: and beyond. Pp. 45–63 *in* T. Black-

burn and K. Gaston, eds. Macroecology: concepts and consequences. Blackwell Scientific, Oxford.
Hubbell, S. P., and L. Borda-de-Água. 2004. The unified neutral theory of biogeography and biogeography: reply. Ecology 85: 3175–3178.
Jablonski, D. 1995. Extinctions in the fossil record. Pp. 25–44 *in* J. H. Lawton and R. M. May, eds. Extinction rates. Oxford University Press, Oxford.
———. 2001. Lesson from the past: evolutionary impacts of mass extinction. Proceedings of the National Academy of Sciences USA 98:5393–5398.
———. 2002. Survival without recovery after mass extinctions. Proceedings of the National Academy of Sciences USA 99: 8139–8144.
Jablonski, D., and K. Roy. 2003. Geographical ranges and speciation in fossil and living molluscs. Proceedings of the Royal Society of London B 270:401–406.
Jackson, J. B. C. 1995. Constancy and change in the life of the sea. Pp. 45–54 *in* J. H. Lawton and R. M. May, eds. Extinction rates. Oxford University Press, Oxford.
Kidwell, S. M. 2001. Preservation of species abundance in marine death assemblages. Science 294:1091–1094.
MacArthur, R. H., and E. O. Wilson. 1963. An equilibrium theory of insular zoogeography. Evolution 17:373–387.
———. 1967. The theory of island biogeography. Princeton University Press, Princeton, N.J.
Magurran, A. E. 1988. Ecological diversity and its measurement. Princeton University Press, Princeton, N.J.
Mayr, E. 1963. Animal species and evolution. Belknap Press of Harvard University Press, Cambridge.
McKane, A. J., D. Alonso, and R. V. Sole. 2004. Analytic solution of Hubbell's model of local community dynamics. Theoretical Population Biology 65:67–73.
Nee, S., R. M. May, and P. H. Harvey. 1994. The reconstructed evolutionary process. Philosophical Transactions of the Royal Society of London B 344:305–311.
Patzkowsky, M. F., and S. M. Holland. 1997. Patterns of turnover in Middle to Upper Ordovician brachiopods of the eastern United States: a test of coordinated stasis. Paleobiology 23: 420–443.
Preston, F. W. 1948. The commonness, and rarity, of species. Ecology 29:254–283.
———. 1960. Time and space variation of species. Ecology 41: 611–627.
Raup, D. M., S. J. Gould, T. J. M. Schopf, and D. S. Simberloff. 1973. Stochastic models of phylogeny and the evolution of diversity. Journal of Geology 81:525–542.
Richter-Dyn, N., and S. S. Goel. 1972. On the extinction of a colonizing species. Theoretical Population Biology 3:406–433.
Ricklefs, R. E. 2003. A comment on Hubbell's zero-sum ecological drift model. Oikos 100:187–193.
Yule, G. U. 1925. A mathematical theory of evolution, based on the conclusions of Dr. J. C. Willis, F. R. S. Philosophical Transactions of the Royal Society of London B 213:21–87.
Vallade, M., and B. Houchmandzadeh. 2003. Analytical solution of a neutral model of biodiversity. Physical Review E 68: Art. No. 061902.
Volkov, I., J. R. Banavar, S. P. Hubbell, and A. Maritan. 2003. Neutral theory and relative species abundance in ecology. Nature 424:1035–1037.

Appendix

A full ecological explanation of Fisher's logseries in terms of population dynamics is now available from new theoretical results on the neutral theory of relative species abundance. This explanation emerges from a more general theoretical formulation of the equations of demographic stochasticity that underpin population dynamics under the unified neutral theory (Volkov et al. 2003). This new formulation incorporates birth and death rates explicitly. Let $b_{n,k}$ and $d_{n,k}$ be the probabilities of birth and death of an arbitrary species k at abundance n. Let $P_{n,k}(t)$ be the probability that species k is at abundance n at time t. Then the rate of change of this probability is given by

$$\frac{dp_{n,k}(t)}{dt} = p_{n+1,k}(t)d_{n+1,k} + p_{n-1,k}(t)b_{n-1,k} - p_{n,k}(t)(b_{n,k} + d_{n,k}). \quad \text{(A1)}$$

This equation is straightforward and easy to understand. The first term on the right represents the transition from abundance $n + 1$ to n, due to a death. The second term is the transition from abundance $n - 1$ to n due to a birth. The last two terms are losses to $P_{n,k}(t)$ because they are transitions away from abundance n to either $n + 1$ or $n - 1$ through a birth or death, respectively. On first consideration, equation (1) appears to be little more than a bookkeeping exercise, but it is actually much more. Because it is written as a recursive function on abundance, it allows an equilibrium solution to be found for species of arbitrary abundance n. When all derivatives at all abundances are set equal to zero, then the solution is said to satisfy "detailed balance," which means that each abundance transition is in equilibrium. Now let $P_{n,k}$ denote this equilibrium. Then $P_{n,k} = P_{n-1,k}{:}(b_{n-1,k}/d_{n,k})$, and more generally, this corresponds to a global equilibrium solution for the metacommunity:

$$P_{n,k} = P_{0,k} \prod_{i=0}^{n-1} \frac{b_{i,k}}{d_{i+1,k}}. \quad \text{(A2)}$$

Note that the probability of being at abundance n is a function of the product of the ratios of birth rate to death rate over all abundances below n. Because the $P_{n,k}$'s must sum to unity, we can find the value of $P_{0,k}$ from this sum, and therefore all other terms as well.

Now consider a symmetric neutral community of S species that are all alike on a demographic basis, such that they all have the same birth rates and death rates; that is, $b_{n,k} \equiv b_n$ and $d_{n,k} \equiv d_n$ (i.e., the species identifier k doesn't matter, and we denote the probabilities by P_n). We can introduce speciation by recognizing a special "birth rate" in this general metacommunity solution; i.e., $b_0 = \nu$, the speciation rate. The mean number of species with n individuals, $\langle\phi_n\rangle$, in a community of S identical species is simply proportional to P_n:

$$\langle\phi_n\rangle = SP_0 \prod_{i=0}^{n-1} \frac{b_i}{d_{i+1}}. \quad \text{(A3)}$$

From equation (3) we are now in a position to derive Fisher's logseries under density independence. Density independence means that the birth rate of a species of current abundance n is simply n times the birth rate of a species with abundance 1; i.e., $b_n = nb_1$, or density independence. Similarly, suppose that the death rates are density independent, $d_n = nd_1$. Substituting these expressions into equation (3), we immediately obtain Fisher's logseries:

$$\langle\phi_n\rangle_M = S_M P_0 \frac{b_0 b_1 \cdots b_{n-1}}{d_1 d_2 \cdots d_n} = \theta\frac{x^n}{n}, \quad \text{(A4)}$$

where the subscript M refers to the metacommunity, $x = b_n/d_n = b_1/d_1 = b/d$, $b_0 = \nu$, and $\theta = \alpha = S_m P_0 \nu/b$ of Fisher's logseries. The derivation of equation (4) reveals that the mysterious parameter x of the logseries is now biologically interpretable: x is the ratio of the density-independent, per capita birth rate to per capita death rate. Note that when one introduces speciation, parameter x must be slightly less than 1 to maintain a finite metacommunity size. At very large spatial scales, the total birth and death rates must be nearly in material balance, resulting in a metacommunity b/d ratio only infinitesimally less than unity. The very slight deficit in birth rates versus death rates at the metacommunity biodiversity equilibrium is made up by the very slow input of new species.

Thus, we now have a complete derivation of the logseries and its parameters θ and x, from the neutral theory. It is interesting that Hubbell (2001a) derived θ following a completely different route from the one taken by Volkov et al. (2003), so we now have further insights into parametric relationships under the unified neutral theory. The greatest significance of this result, however, is demonstrating that the logseries relative species abundance distribution arises at the metacommunity speciation-extinction equilibrium when the birth and death rates are density independent and the metacommunity is symmetric (all species exhibit the same mean per capita rates).

The expected relative species abundance distribution on islands under dispersal limitation (the classical problem in the theory of island biogeography) is not the logseries, however, but is a distribution that resembles a skewed lognormal (Hubbell 2001a). The second advance in the unified neutral theory is the discovery of an analytical solution for the relative species abundance distribution in a local community under immigration from the metacommunity (Volkov et al. 2003), previously available only by simulation (Hubbell 2001a). Once again, let $\langle \phi_n \rangle$ be the mean number of species with n individuals. Then

$$\langle \phi_n \rangle = \theta \frac{J!}{n!(J-n)!} \frac{\Gamma(\gamma)}{\Gamma(J+\gamma)} \times \int_0^{\gamma} \frac{\Gamma(n+y)}{\Gamma(1+y)} \frac{\Gamma(J-n+\gamma-y)}{\Gamma(\gamma-y)} \exp\left(\frac{-y\theta}{\gamma}\right) dy \qquad \text{(A5)}$$

where $\Gamma(z) = \int_0^\infty t^{z-1} e^{-t}\, dt$ which is equal to $z - 1)!$ for integer z, and $\gamma = [m(J-1)]/(1-m)$. As in Hubbell 2001a, parameter m is the immigration rate. Equation (5) was derived by making the following functional substitutions for the per capita birth and death rates in equation (2):

$$b_{n,k} = (1-m)\left(\frac{n}{J}\right)\left(\frac{J-n}{J-1}\right) + m\left(\frac{\mu_k}{J_M}\right)\left(1-\frac{n}{J}\right) \quad \text{and}$$

$$d_{n,k} = (1-m)\left(\frac{n}{J}\right)\left(\frac{J-n}{J-1}\right) + m\left(1-\frac{\mu_k}{J_M}\right)\left(\frac{n}{J}\right) \qquad \text{(A6)}$$

where μ_k is the abundance of the kth species in the metacommunity under the logseries, and J_M is the size of the metacommunity. The first (second) term of $b_{n,k}$ and $d_{n,k}$ is the probability of an increase or decrease by one individual, the kth species in the local community, as a function of whether an immigration event occurred (did not occur), respectively.

The expression in equation (5) can be solved numerically quite accurately. Programs in C are attached electronically to the paper by Hubbell and Borda-de-Água (2004). As the immigration rate m decreases, the relative species abundance distribution in the local community given by equation (5) becomes progressively more skewed. Thus, as islands or local communities become more isolated, rare species become rarer and common species become commoner. The degree of skewness of the relative species abundance distribution is also a function of local community size (Hubbell and Borda-de-Água 2004).

Paleobiology, 31(2), 2005, pp. 133–145

The dynamics of evolutionary stasis

Niles Eldredge, John N. Thompson, Paul M. Brakefield, Sergey Gavrilets, David Jablonski, Jeremy B. C. Jackson, Richard E. Lenski, Bruce S. Lieberman, Mark A. McPeek, and William Miller III

Abstract.—The fossil record displays remarkable stasis in many species over long time periods, yet studies of extant populations often reveal rapid phenotypic evolution and genetic differentiation among populations. Recent advances in our understanding of the fossil record and in population genetics and evolutionary ecology point to the complex geographic structure of species being fundamental to resolution of how taxa can commonly exhibit both short-term evolutionary dynamics and long-term stasis.

Niles Eldredge. Division of Paleontology, American Museum of Natural History, Central Park West at Seventy-ninth Street, New York, New York 10024. E-mail: epunkeek@amnh.org
John N. Thompson. Department of Ecology and Evolutionary Biology, A316 Earth and Marine Sciences Building, University of California, Santa Cruz, California 95060. E-mail: thompson@biology.ucsc.edu
Paul M. Brakefield. Institute of Biology, Leiden University, Post Office Box 9516, 2300 RA Leiden, The Netherlands. E-mail: brakefield@rulsfb.leidenuniv.nl
Sergey Gavrilets. Department of Ecology and Evolutionary Biology and Department of Mathematics, University of Tennessee, Knoxville, Tennessee 37996. E-mail: gavrila@tiem.utk.edu
David Jablonski. Department of Geophysical Sciences, 5734 South Ellis Avenue, University of Chicago, Chicago, Illinois 60637. E-mail: djablons@midway.uchicago.edu
Jeremy B. C. Jackson. Scripps Institution of Oceanography, University of California, San Diego, La Jolla, California 92039. E-mail: jbjackson@ucsd.edu
Richard E. Lenski. Center for Microbial Ecology, Michigan State University, East Lansing, Michigan 48824. E-mail: lenski@pilot.msu.edu
Bruce S. Lieberman. Departments of Geology and Ecology and Evolutionary Biology, University of Kansas, 120 Lindley Hall, Lawrence, Kansas 66045. E-mail: blieber@ku.edu
Mark A. McPeek. Department of Biological Sciences, Dartmouth College, Hanover, New Hampshire 03755. E-mail: mark.mcpeek@dartmouth.edu
William Miller III. Department of Geology, Humboldt State University, 1 Harpst Street, Arcata, California 95521. E-mail: wm1@axe.humboldt.edu

Accepted: 17 April 2004

Introduction

The pronounced morphological stability displayed by many fossil species (Eldredge 1971; Eldredge and Gould 1972; Gould and Eldredge 1977; Stanley and Yang 1987; Jackson and Cheetham 1999; Jablonski 2000), often for millions of years, contrasts sharply with the rapid, often adaptive, evolutionary changes documented in many extant species (Reznick et al. 1997; Thompson 1998; Huey et al. 2000; Thomas et al. 2001). If evolutionary change occurs frequently within populations, why is it that in some species so little of it is conserved and translated through time as net change? In this paper we examine what paleobiologists, population geneticists, and evolutionary ecologists have learned about stasis and rapid evolution over the past decade as new approaches have been adopted and results obtained in all these fields. Our basic conclusion—that stasis derives from the geographic structure and partitioning of genetic information within widespread species—is derived from a consideration of all known population genetic processes that promote (or conversely hinder) genetic change, as well as from analysis of data from the fossil record.

Stasis is generally defined as little or no net accrued species-wide morphological change during a species-lineage's existence up to millions of years—instantly begging the question of the precise meaning of "little or no" net evolutionary change. All well-analyzed fossil species lineages, as would be expected, display variation within and among populations, but the distribution of this variation typically remains much the same even in samples sep-

0094-8373/05/3102-0010/$1.00

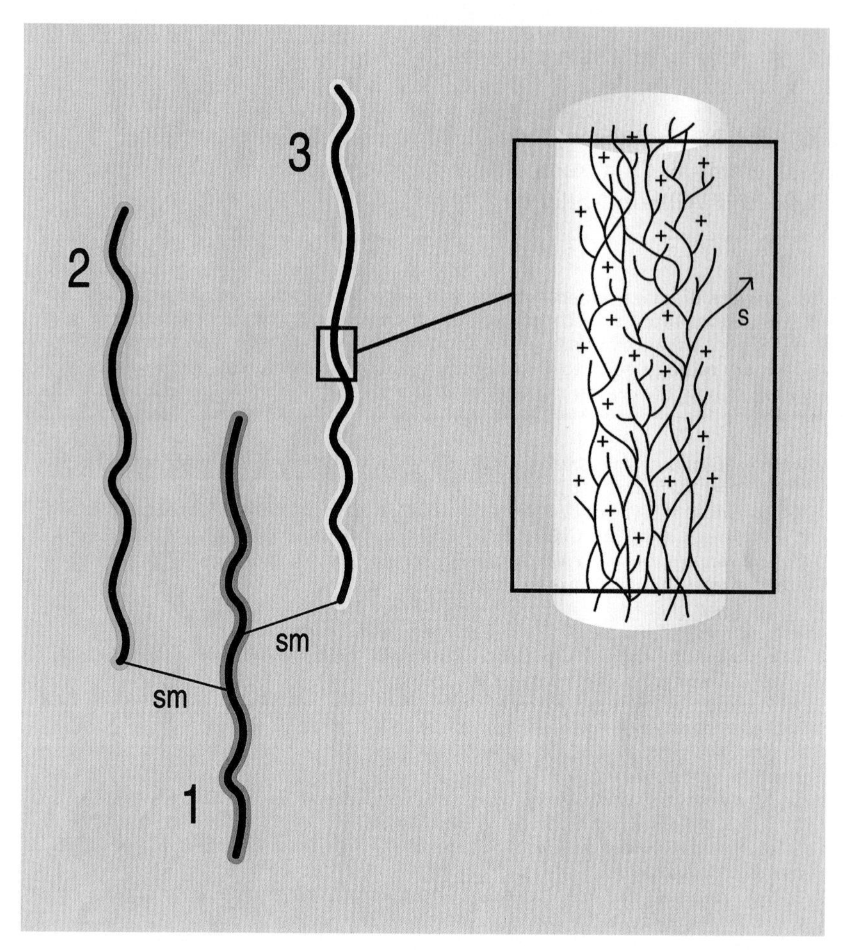

FIGURE 1. Species stasis in the face of ongoing population-level evolution. Species (lineages 1, 2, and 3 on the left) exhibit negligible net phenotypic changes, while their component population systems (on the right) continually differentiate, fuse, or go extinct. Stasis is occasionally broken by establishment and spread of novel phenotypes (s); when this is matched with ecological opportunity, highly differentiated new lineages (sm) may be formed that eventually develop internal (population) dynamics and geographic structure resulting, again, in stasis. (In this view, species-lineages consist of anastomosing population systems and, at the same time, belong to clades composed of similar lineages).

arated by millions of years (Fig. 1). This view of fossil variation has been reinforced over the past decade as paleontological studies have applied higher sampling intensities in time and space, improvements in both relative and absolute stratigraphic dating, more comprehensive use of multivariate statistical analysis, and better controls for sampling biases.

Although it is now clear that some fossil species lineages do indeed accrue morphological change through time (Geary 1995), it is also now evident that many do not. Well-documented examples of stasis range from Paleozoic brachiopods (Lieberman et al. 1995) to late Cenozoic bivalves (Stanley and Yang 1987) and bryozoans (Jackson and Cheetham 1999). Inventories of evolutionary tempo and mode across entire clades are sparse, but Jackson and Cheetham's (1999) survey of well-documented case studies in the Neogene fossil record found 52 instances of stasis and only two instances of anagenesis in nine benthic macroinvertebrate clades, and eight instances of stasis as opposed to 10–12 instances of anagenesis in marine microplankton. Anagenesis occurs in only eight of 88 trilobite lineages in the Ordovician of Spitsbergen, and in but one of 34 scallop lineages in the northern European Jurassic (Jablonski 2000).

Studies of extant taxa with rich fossil records provide mounting evidence that morphologically defined species-level lineages recognized in fossil sequences often correspond to genetically defined species in the modern biota (Jablonski 2000). Such studies are crucial to the demonstration that patterns of stasis in the fossil record constitute a genuine problem for evolutionary theory. Perhaps the most rigorous and detailed of such studies (and one that has proven compelling to population geneticists) are those on tropical American Neogene cheilostome bryozoans (Jackson and Cheetham 1999). Cheilostomes are small, clonal marine animals that grow in plantlike shapes by budding modules (zooids) to form a colony. They are abundant in Recent seas and in the fossil record. In the tropical American genera *Metrarabdotos* and *Stylopoma*, all long-ranging species (11 in each genus) persisted essentially morphologically unchanged for 2–16 Myr. New species appear abruptly in the fossil record, with morphological change occurring within the limits of stratigraphic resolution of sampling (approximately 150,000 years). Studies of extant species in these genera indicate that morphological stasis also reflects stasis in key life history traits, with occasional rapid change. For example, the size of larval brood chambers, which is correlated with larval size, differs by up to twofold among closely related species, and entirely arborescent species have given rise to entirely encrusting species. Such examples show that stasis can include reproductive and behavioral characteristics in addition to pure morphology. We find this example and other such case studies compelling evidence that morphological stasis is a common pattern in the fossil record, which thus requires an examination of how evolutionary and ecological processes can account for it.

If many, perhaps even most, species accrue little morphological change during their lifetimes, then a corollary is immediately raised: the possibility that much of the morphological change accrued within evolutionary lineages over time is concentrated in relatively brief episodes of speciation. Mayr (1954) suggested a link between speciation and evolutionary change, and stasis+morphological change concentrated at speciation events is the core of punctuated equilibria (Eldredge 1971; Eldredge and Gould 1972). Recently, Webster, Payne, and Pagel (2003), in their analysis of speciation events and underlying genetic change in 56 phylogenies, concluded that "rapid genetic evolution frequently attends speciation," and that their results provide a "genetic component" to the pattern of stasis and change of morphological traits seen so commonly in the fossil record.

Our purpose here is to explore further the dynamics generating such patterns, particularly insofar as stasis itself is concerned. Given patterns of changes in heritable phenotypic variation and genetic variation commonly seen in local populations, what factors prevent such change from becoming species-wide? Do novelties arise only at speciation, or do they arise but are typically not conserved throughout the history of species—perhaps further suggesting that speciation conserves rather than prompts the generation of novelty? Previous authors (Darwin 1871; Ohta 1972; Futuyma 1987; Eldredge 1989; Lieberman et al. 1995) have discussed the difficulties inherent in conserving evolutionary novelties arising in local populations and their spread over the entire range of a far-flung, heterogeneous species. Futuyma (1987) in particular has discussed the closely related corollary that speciation may be the key to the phylogenetic conservation of such novelties. More recently, the geographic mosaic of ongoing local adaptation has become the very foundation for new views of how coevolving interactions between species persist over long periods of time in a constantly changing world (Thompson 1994, 1999a,b).

What, then, constrains the species-wide spread of evolutionary change when experimental and field data clearly show that the potential for rapid change within populations is nearly always present? We divide the question into three stages related to the establishment of evolutionary change in a geographically heterogeneous world: origin, local population establishment, and species-wide spread. Our analysis of studies from the past decade, including examples drawn from our own work, suggests that patterns and processes related to geographic structure contribute importantly to the maintenance of stasis.

Origin, Local Population Establishment, and Species-wide Spread

To be preserved in the fossil record with any reasonable likelihood, a novel genotype must originate, become established in a local population, and then spread and increase in numbers across a large geographic area. Failure to complete all three of these stages will result in stasis in the fossil record. Consequently, if we are to understand the evolutionary dynamics of stasis, we need to understand where most failures occur along this sequence of origin and spread of novelty. Many earlier attempts to reconcile our understanding of the evolutionary dynamics of extant species with the paleobiological evidence for stasis focused on the role of genetic constraints and stabilizing selection in preventing the origin and establishment of novelty within local populations (Charlesworth et al. 1982; Van Valen 1982; Levinton 1983; Maynard Smith 1983; Wake et al. 1983; Williamson 1987). More recent mathematical and empirical studies have refined our understanding of the roles of these evolutionary forces, and they have shown that the spatial structure of species strongly influences the pattern of establishment of novel types.

Constraints on the Origin and Local Establishment of Novelty.—From a theoretical perspective, the origin of novel genotypes involves a set of processes (mutation and recombination) distinct from those processes that determine the local fate of the variants that are produced (drift and selection). From an empirical perspective, however, the actual rate of production of novel variants is very rarely observed directly. Instead, the failure to produce novelty, on the one hand, versus the failure of novelties to become locally established, on the other hand, must often be inferred indirectly from the dynamics of experimental systems. Therefore, we combine our analysis of these two dynamical stages in the section that follows.

The simplest potential explanations for stasis are exhaustion of standing genetic variation or the limited production of useful novelties within populations. Even when genetic variation is present, however, evolutionary potential is not equal in all traits, and the origination of useful novelties may depend upon mutations appearing in a particular sequence (Mani and Clarke 1990). Antagonistic pleiotropy (leading to negative genetic correlations), epistasis, and linkage disequilibrium can all constrain the generation of novel genotypes, even when standing genetic variation is not limited by population size or the previous history of selection (Barton and Partridge 2000). Some artificial-selection experiments have shown that rates of phenotypic change may decelerate during prolonged directional selection (Falconer and Mackay 1996). This pattern has often been attributed to the depletion of the genetic variation for the selected trait that was present in the founding population or, alternatively, depletion of variation in fitness more generally, such as when selection on some other aspect of organismal performance opposes the response to artificial selection (Barton and Partridge 2000; Falconer and Mackay 1996). Consistent with these explanations, response to selection can be accelerated by increasing population size, which both increases the overall level of genetic variation and opens new permissible directions ("ridges") available to selection in the multidimensional adaptive landscape (Weber 1996).

Pronounced decelerations in rates of phenotypic evolution have also been observed over thousands of generations in asexual populations of *Escherichia coli* founded from a single cell (Cooper and Lenski 2000). In these populations, new mutations provide the only source of genetic variation, and this mutational source continues indefinitely. In this case, stasis cannot derive from depletion of preexisting variation, nor from exhaustion of genetic variation more generally. In fact, the amount of genetic variation increased in these populations even as the rate of phenotypic evolution declined (Sniegowski et al. 1997). Instead, these populations evidently approached a local adaptive peak or plateau, at which point most potential (i.e., genetically accessible) beneficial mutations were fixed. Consistent with this explanation, the rate of adaptive evolution was re-accelerated by perturbing populations from their proximity to an adaptive peak, either by changing the environment (Travisano et al. 1995) or by introducing deleterious mutations (Moore et al. 2000).

These studies show that relative stasis can arise fairly quickly following periods of rapid adaptive evolution. They also indicate that the exhaustion of beneficial variants—whether preexisting or potentially accessible by mutation—can contribute to stasis. However, the depletion of standing variation is relevant only in small populations, which contribute very little to the fossil record. Species-wide depletion of accessible beneficial mutations requires a degree of environmental constancy that is not typical of the earth's history (Lambeck and Chappell 2001; Zachos et al. 2001).

More likely, genetic and developmental correlations among traits can also influence both the direction and extent of change in local populations, and advances in evolutionary developmental biology are suggesting the extent to which these genetic interactions may influence stasis. For example, some experiments on butterfly wing patterns show that multiple eyespots are made by the same developmental pathway and, consequently, there exist strong genetic correlations among them. Selection to increase the size of the posterior eyespot on the forewing of *Bicyclus anynana*, in the absence of any selection on the anterior eyespot (Beldade et al. 2002), will typically increase the size of both eyespots. Nonetheless, selection can readily uncouple the two eyespots to produce highly divergent morphologies in all directions of morphological space (Fig. 2). Indeed, novel patterns not seen in any related species can be obtained after 25 or so generations. These results indicate that genetic and developmental processes can produce genetic correlations and favor evolution along paths of least resistance, but they need not absolutely constrain the process of adaptive radiation (Brakefield et al. 2003).

In fact, recent studies have revealed a variety of genetic mechanisms that may overcome constraints imposed by gene interaction. Epistatic components of genetic variance may be converted into additive variance, promoting evolutionary change in small, perturbed populations (Wade and Goodnight 1998). Environmental stress may disrupt developmental stability sufficiently to uncover latent genetic variance that can promote evolvability (Rutherford and Lindquist 1998). Yet other processes—including gene or genome duplication, polyploidy, hybridization, and horizontal gene transfer—can further promote novel paths of evolution (Rieseberg 1997; Soltis and Soltis 1999; Lynch and Force 2000; Sandstrom et al. 2001). Consequently, it increasingly seems that neither an absence of genetic variation nor genetic constraints are sufficient to account for long-term stasis.

Expression of advantageous genetic variation in highly variable environments, however, may constrain the breaking of stasis within local populations. Recent theoretical studies of multidimensional genotype space have demonstrated the possibility of prolonged phenotypic change within local populations by a chain of substitutions that are nearly neutral with respect to overall fitness in the absence of a highly variable environment (Gavrilets 1997). Only a small proportion of mutations with significant phenotypic effects are expected to be advantageous or even neutral. The more variable the environment over time, the more restricted the range of these genotypes with equal or higher fitness, because each genotype must function under a wide range of environmental conditions.

When stasis breaks down, it may do so either in large or in small populations. Considering both the production of mutations and their subsequent fate, advantageous mutations will become established more often in larger than in smaller populations. An environmental change, by redefining the "optimum" phenotype, may result in increasing the probability of mutations being conditionally advantageous or neutral, thereby promoting evolutionary change. On the other hand, decreasing population size will increase the role of stochastic fluctuations, creating an opportunity to overcome stabilizing selection or incumbency effects (Barton and Charlesworth 1984) and facilitating evolution along an adaptive ridge of genotypes that are nearly equal in fitness (Gavrilets 1999). Strong competition for a resource may potentially lead to sympatric or-

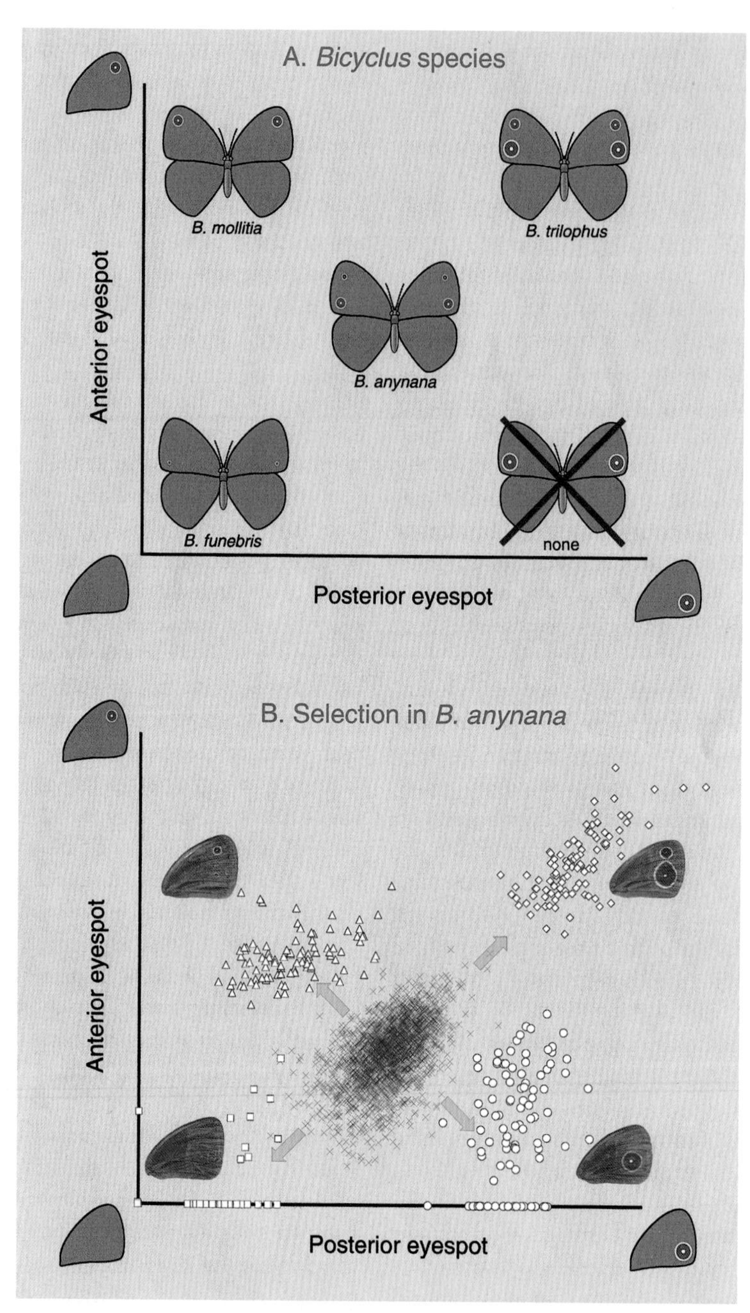

FIGURE 2. Analysis of a potential evolutionary constraint. A, Occupation by species of the butterfly genus *Bicyclus* of morphological space for the pattern of the forewing eyespot size. Names of representatives from among the 80 or so species are given. B, Responses obtained over 25 generations of artificial selection in replicate lines of *B. anynana*. Results show that butterflies similar to each corner pattern were produced from standing genetic variation in a single laboratory stock, including one morphology not seen in any extant species. Crosses indicate butterflies from the base population, and open symbols show samples from generation 25 in each direction of selection (green arrow) together with a representative forewing. Redrawn from Beldade et al. 2002.

igin and within-deme establishment of genetic novelties (Kawata 2002). More complete cessation of gene flow can result in rapid evolutionary change in a population experiencing a novel environment (Garcia-Ramos and Kirkpatrick 1997).

These theoretical expectations on the conditions allowing occasional breakdown of stasis receive support from experimental studies during the past decade. Besides the deceleration in phenotypic evolution found during the long-term experiments in *E. coli* (Lenski and Travisano 1994), both performance and morphology show a stair-step dynamic over shorter periods. Most of the changes in the first 3000 generations were concentrated in a few episodes that appeared instantaneous at a 100-generation sampling interval. These episodes have a simple explanation: each step in performance reflects a selective fixation of a beneficial mutation, and the morphological changes are pleiotropic effects of these mutations.

Rapid diversification of a lineage may therefore often involve the invasion of a new selective environment by one or a few local populations. The breakdown of stasis occurs as a local population adapts rapidly to an initially inhospitable habitat before it would otherwise be driven extinct (Gomulkiewicz and Holt 1995). Rapid, pulsed diversification of some phytophagous insects as they colonize new host taxa in local populations has long been a working model for studies of plant-insect interactions (Ehrlich and Raven 1964). Molecular phylogenetic analyses of insect taxa during the past decade have provided evidence for such bouts of rapid diversification at the bases of clades, as species colonize new host lineages (Pellmyr et al. 1998). Similarly, the occasional invasion by *E. coli* of thermally stressful environments, beyond the tolerance limits of ancestral populations, fits this model (Mongold et al. 2001).

One well-studied example of invasion of a novel environment leading to the breakdown of stasis—and the generation of evolutionary novelty correlated with speciation—is found in the diversification of damselflies. *Enallagma* damselflies have diversified in North America into permanent ponds and lakes with either fish or large dragonflies as the top predators (McPeek and Brown 2000). *Enallagma* species differ in their vulnerability to these predators and are thus capable of living with only one of them. Species that coexist with fish use crypsis to avoid predators, whereas species that coexist with large dragonflies are more active and swim away from attacking predators (McPeek 1998). Moreover, several *Enallagma* species are found in each lake type, and co-occurring species are phenotypically very similar (McPeek 2000). Lakes with fish are the ancestral habitat for the genus, and at least two independent invasions by damselflies into the dragonfly lake environment have occurred (McPeek and Brown 2000). These habitat shifts have been accompanied by rapid evolution in a number of morphological, behavioral, and biochemical characters that enhance burst swimming speed because of selection imposed by dragonfly predators in the new environment (McPeek 2000; McPeek and Brown 2000). It may have taken the invading lineages only a few hundred years to gain a high degree of local adaptation to their new environment (McPeek 1997).

Such rapid evolutionary change would appear saltatory in the fossil record. In contrast, rates of evolution in these characters within the fish lake environment are very slow. Millions of years of evolution within the fish lake environment have produced few or no differences among species in many other characters that are important in determining their ecological performance (McPeek 2000; McPeek and Brown 2000). Importantly, shifts to dragonfly lakes and accompanying rapid evolution have been rare events, occurring in only one of the two primary clades of *Enallagma*. That clade has a number of phenotypic characters that are already similar to phenotypes favored by selection in dragonfly lakes (McPeek 2000). Hence, there appears to be a fundamental niche conservatism that dooms shifts by most populations to failure, thereby contributing to stasis within many *Enallagma*.

Recent studies therefore suggest that the absence of useful novelties, or their failure to become established within local populations, may contribute to stasis in certain limited cases. But more generally, the field of evolutionary ecology has clearly shown the ability of local populations to evolve rapidly under changing conditions. Consequently, species-wide stasis would seem to require additional constraints acting above the level of local populations.

Species-wide Spread.—A key change in population genetic theory and evolutionary ecology over the past decade has been the increasing incorporation of geographic structure into our understanding of the evolutionary dynamics of species. We now know from various genetic modeling approaches that spatial structure can decrease the likelihood of regional extinction, maintain genetic polymorphisms across populations, and shape evolutionary and coevolutionary trajectories (Gandon et al. 1996; Thrall and Burdon 1997; Gomulkiewicz et al. 2000; Nuismer et al. 2000). At the same time, a burgeoning number of studies in molecular ecology and evolutionary ecology reveal even more widespread genetic differentiation among populations than was apparent from earlier studies that often underestimated spatial genetic structure. These modeling and empirical results together suggest that the geographic genetic structure of species must be a central component of any overall theory resolving the discrepancy between short-term dynamics and long-term stasis.

Novel forms must spread beyond their site of origin if they are to have a reasonable chance of being preserved in the fossil record. If a local population is already reproductively isolated from its neighbors, then novel forms must successfully expand beyond their initial geographic limits. Alternatively, if the local population is still genetically connected to other populations, then a novel form must be able to spread across those other populations if it is to become sufficiently widespread to leave a record of the change. In both cases the key problem is expansion of geographic range (Kirkpatrick and Barton 1997; Thomas et al. 2001). We now know that, even once established locally, novel forms may face large hurdles in spreading beyond their site of origin.

These spatially induced hurdles may be the most potent evolutionary forces maintaining stasis. Established species often have an intrinsic advantage over invaders because they occur at high relative frequency or density (Gomulkiewicz and Holt 1995; McPeek 2000). The incumbency of established species can be further maintained through effects on hybrids. Recent mathematical models show that if hybrids between the novel and incumbent forms have reduced fitness, then the chance of spread of the novel form is further reduced (Gavrilets 1996; Coyne et al. 1997). Moreover, through asymmetric gene flow most hybrids are likely to occur within the population where the novel genotype originated, because the absolute numbers will often be less than surrounding populations. This asymmetric gene flow will therefore minimize the chance that a novel form will rise to high frequencies elsewhere.

The development of metapopulation theory (Hanski and Gilpin 1997) has provided yet additional insights into the problem of spread in novel forms (Lande 1985; Tachida and Ilizuka 1991). Some current models suggest that high population turnover rates can reduce the chances of establishment and spread of novel genotypes, unless those genotypes are favored by their very rarity through negative frequency-dependent selection as occurs in gene-for-gene coevolution between some plants and pathogens (Burdon and Thrall 1999; Gandon et al. 1996). This kind of negative frequency-dependent selection, which maintains polymorphisms by favoring rare genotypes within and among populations, may also maintain stasis within a species rather than lead to diversification. When metapopulation structure is coupled with heterogeneous selection across landscapes, it may become even more difficult for novel genotypes to spread.

Paleobiologists have argued that widespread species are expected to exhibit slower rates of species-wide evolution than species with small ranges, because natural selection will not be consistently directional across space and time (Eldredge 2003; Jablonski 2000; Lieberman et al. 1995)—e.g., ecological conditions acting on local populations of American robins in the southwestern United States are clearly different from those present in the deep woodlands of the Northeast. That overall expectation is supported by population genetic theory, which suggests that it is difficult for a mutant to be advantageous under all conditions required by a highly heterogeneous environment (Ohta 1972). Consistent with the expectation that most mutations that are locally adaptive would not be globally advantageous, lines of *E. coli* adapted to a glucose-containing environment for 20,000 generations tend to have reduced performance on a range of other substrates (Cooper and Lenski 2000).

The developing mathematical theory of species ranges provides additional indications that the spatial structure of habitats and heterogeneous selection may be important sources of stasis. Gene flow from the center of a species range can impede novel adaptation at the periphery and prevent the range from expanding outward (Kirkpatrick and Barton 1997). The problems of spatial structure and heterogeneous selection may therefore contribute to the kind of sustained habitat tracking found in the fossil record (Eldredge 2003). Data from several paleontological studies on Pleistocene plants (Davis 1983), beetles (Coope 1979), foraminifera (Bennett 1990), and mollusks (Valentine and Jablonski 1993) have demonstrated little morphological response to protracted climate change. Instead, geographic distributions changed. Species tended to survive, usually with little or no discernible morphological change, as long as recognizable habitats could be tracked. That does not mean that natural selection is not acting (Davis and Shaw 2001; Hoekstra et al. 2001), as the data from population and evolutionary genetics show that populations are constantly under selection. Rather, it means that selection often acts in ways that favor populations that are evolutionarily conservative at the species level.

The geographic mosaic of coevolution may also contribute to species-wide stasis, even though coevolution is one of the evolutionary forces most commonly thought to generate novelty. Studies over the past decade have indicated that selection mosaics, coevolutionary hotspots, and gene flow can combine to create extensive coevolutionary dynamics (Thompson 1994, 1997, 1999a). This ongoing coevolution creates local novelty and is undoubtedly important to the ecological dynamics of species. Moreover, it may be crucial for keeping coevolving species in the evolutionary game as one species or the other temporarily gains the upper hand in different environments. But most of these dynamics may not result in much net change at the species level. Recent mathematical models have indicated that geographically structured coevolution can actually constrain the escalation of antagonistic arms races. These interactions may continually recycle defenses and counterdefenses through frequency-dependent selection, because geographic structure may maintain the polymorphisms on which frequency-dependent selection depends (Gandon et al. 1996; Gomulkiewicz et al. 2000; Nuismer et al. 2000). Long-term studies of gene-for-gene coevolution within natural populations support this mathematical prediction (Burdon and Thrall 2000). Similarly, geographic structure may stabilize some kinds of coevolved mutualisms, by maintaining previously fixed traits in the face of moderate gene flow (Nuismer et al. 2000). As a result, the geographic mosaic of coevolution may often create ongoing genetic dynamics embedded within longer-term stasis, with populations only rarely breaking through in fundamentally novel directions.

Increasingly detailed paleontological studies corroborate the potential importance of spatial structure in maintaining stasis. Moreover, the great strength of paleontological data is that within-population variation can be compared over time as well as space, allowing analysis of the importance of the spatial structuring of species throughout a species' history. Analysis of two broadly distributed species lineages of Devonian brachiopods highlights the significance of spatial structuring within species to the generation of patterns of stasis and change (Lieberman et al. 1995). Statistical analysis revealed no discernible net change in the morphology of either species over their respective five-million-year histories. However, within any single environment, large morphological shifts did occur—larger, in fact, than the net morphological change across the entire environmental distribution of the species over the same time period (Fig. 3). As these changes were in different directions in different en-

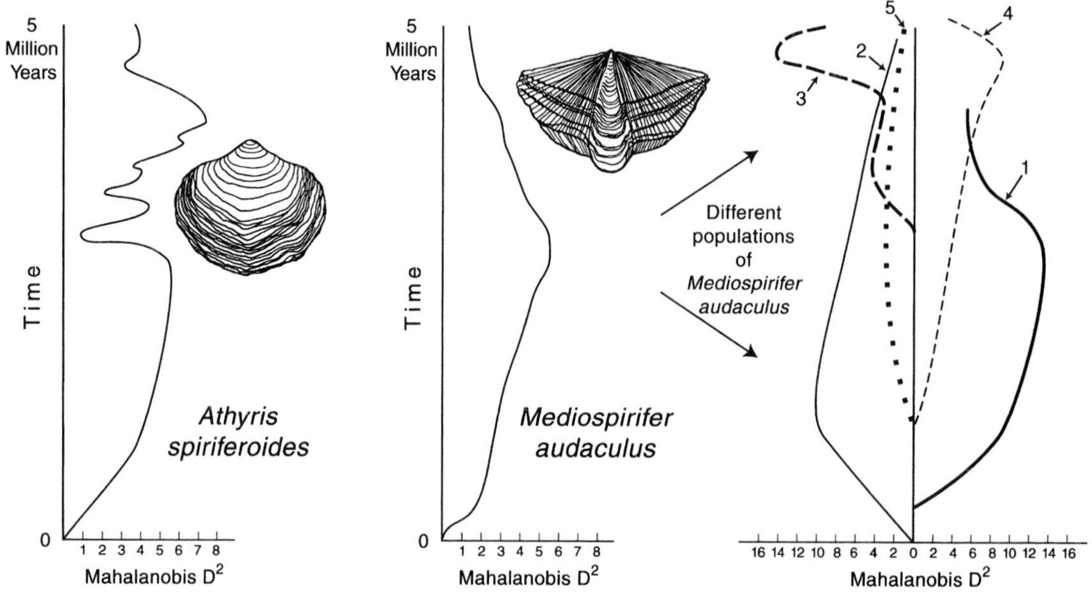

FIGURE 3. Schematic diagram showing temporal and environmental (spatial) patterns of morphological change in two species of Middle Devonian brachiopods, measured as Mahalanobis D^2 values from canonical discriminant analysis of morphometric data. Each of these species occurred in five distinct environments over a period of 5 Myr. Note the oscillatory nature of morphological change in each species (left and middle panels). The morphological changes of *Mediospirifer audaculus* sampled from the five distinct environments (far right panel) are also oscillatory, but have larger D^2 distance excursions than when samples of the species are lumped as a whole (see middle panel). Moreover, changes within individual environments tend to cancel out, leading to negligible net change for the species as a whole.

vironments, they tended to cancel out, resulting in no net change: stasis resulted at least in part from a species' presence in several distinct environments (Lieberman et al. 1995; Lieberman and Dudgeon 1996).

These results agree with the expectation that spatial structuring of widespread species will, as a rule, lead to stasis—but that local populations, under certain conditions, can be expected to develop more substantial amounts of morphological change in the short term. They further suggest that the patterns of generally fluctuating change documented by Gingerich (1976) in Eocene mammals and Sheldon (1987) in Ordovician trilobites reflect the evolutionary histories of geographically localized populations of these species. Gingerich's data involved meticulously collected time series from the Bighorn Basin of Wyoming, a localized subset of the regions over which the *Hyopsodus* and other species have been documented to have lived. Likewise, Sheldon's study of eight trilobite lineages from the Builth Inlier of Wales did not include substantial geographic sampling. Both of these data sets contain examples of short-term evolutionary change that is repeatedly reversed over longer timescales (Gingerich 1983)—much like the fluctuations in beak morphology in Galapagos finches (Grant 1986) and in floral color of desert plants (Schemske and Bierzychudek 2001). Thus, the entire spatiotemporal history of a species can reveal less net change than what is documented in temporal or geographic subsets of a species lineage.

Conclusions

Both theoretical and empirical studies of the past decade suggest that the complex pattern of selection imposed on geographically structured populations by heterogeneous environments and coevolution can paradoxically maintain stasis at the species level over long periods of time. By contrast, neither lack of genetic variation nor genetic and developmental constraint is probably sufficient in and of itself to account for species-wide stasis.

Further resolution of our understanding of the dynamics of evolutionary stasis will require novel integration of modeling and empirical analyses. Comparison of rates of gradual change in widespread versus endemic species will help us better test our conclusion that geographic range shapes stasis. Such analyses of the genetic and geographic structure of species when placed within a phylogenetic context will help us further test the relative contributions of geographic structure, underlying genetic variation, and development to the ongoing dynamics of stasis.

Remaining issues include finer resolution of the issue of conservation versus generation of novelty in short bursts of speciation, and the possibility that many such bursts of speciation are spatiotemporally correlated among sympatric lineages in regional ecological settings. Such bursts could reflect "turnovers" (Vrba 1985) or the events between episodes of "coordinated stasis" (Brett and Baird 1995), reflecting spatiotemporal scales intermediate to local ecological succession, on the one hand, and the well-documented evolutionary responses to episodes of global mass extinctions on the other (Eldredge 2003). And what dynamic processes underlie the emergence of stable species (Miller 2003)? The solution to these and related problems will demand further integration of the fields of evolutionary ecology and evolutionary developmental biology into evolutionary genetic and paleontological approaches.

Acknowledgments

This work was conducted as part of the Ecological Processes and Evolutionary Rates Working Group supported by National Center for Ecological Analysis and Synthesis at University of California, Santa Barbara (NSF DEB-94-21535); and by National Science Foundation support to J.N.T., D.J., R.E.L., B.S.L., M.A.M., S.G.; Human Frontiers Science Program support to P.M.B.; and National Institutes of Health support to S.G. We thank D. J. Futuyma for discussion, and C. Thomas, J. Valentine, and anonymous reviewers for comments on the manuscript.

Literature Cited

Barton, N. H., and B. Charlesworth. 1984. Genetic revolutions, founder effects, and speciation. Annual Review of Ecology and Systematics 15:133–164.

Barton, N. H., and L. Partridge. 2000. Limits to natural selection. BioEssays 22:1075–1084.

Beldade, P., K. Koops, and P. M. Brakefield. 2002. Developmental constraints versus flexibility in morphological evolution. Nature 416:844–847.

Bennett, K. D. 1990. Milankovitch cycles and their effects on species in ecological and evolutionary time. Paleobiology 16:11–21.

Brakefield, P. M., V. French, and B. J. Zwaan. 2003. Development and the genetics of evolutionary change within insect species. Annual Review of Ecology, Evolution and Systematics 34:633–660.

Brett, C. E., and G. Baird. 1995. Coordinated stasis and evolutionary ecology of Silurian to Middle Devonian faunas in the Appalachian Basin. Pp. 285–315 *in* R. Anstey and D. H. Erwin, eds. Speciation in the fossil record. Columbia University Press, New York.

Burdon, J. J., and P. H. Thrall. 1999. Spatial and temporal patterns in coevolving plant and pathogen associations. American Naturalist 153:S15–S33.

———. 2000. Coevolution at multiple spatial scales: *Linum marginale-Melampsora lini*—from the individual to the species. Evolutionary Ecology 14:261–281.

Charlesworth, B., R. Lande, and M. Slatkin. 1982. A neo-Darwinian commentary on macroevolution. Evolution 36:474–498.

Coope, G. R. 1979. Late Cenozoic fossil Coleoptera: evolution, biogeography and ecology. Annual Review of Ecology and Systematics 10:247–267.

Cooper, V. S., and R. E. Lenski. 2000. The population genetics of ecological specialization in evolving *E. coli* populations. Nature 407:736–739.

Coyne, J. A., N. H. Barton, and M. Turelli. 1997. A critique of Sewall Wright's shifting balance theory of evolution. Evolution 51:643–671.

Darwin, C. D. 1871. The descent of man, and selection in relation to sex. John Murray, London.

Davis, M. 1983. Quaternary history of deciduous forests of eastern North America and Europe. Annals of the Missouri Botanical Garden 20:550–563.

Davis, M. B., and R. G. Shaw. 2001. Range shifts and adaptive responses to Quaternary climate change. Science 292:673–679.

Ehrlich, P. R., and P. H. Raven. 1964. Butterflies and plants: a study in coevolution. Evolution 18:586–608.

Eldredge, N. 1971. The allopatric model and phylogeny in Paleozoic invertebrates. Evolution 25:156–167.

———. 1989. Macroevolutionary dynamics: species, niches and adaptive peaks. McGraw-Hill, New York.

———. 2003. The sloshing bucket: how the physical realm controls evolution. Pp. 3–32 *in* J. Crutchfield and P. Schuster, eds. Evolutionary dynamics: exploring the interplay of selection, accident, neutrality, and function (SFI Studies in the Sciences of Complexity Series). Oxford University Press, New York.

Eldredge, N., and S. J. Gould. 1972. Punctuated equilibrium: an alternative to phyletic gradualism. Pp. 82–115 *in* T. J. M. Schopf, ed. Models in paleobiology. Freeman, Cooper, San Francisco.

Falconer, D. S., and T. Mackay. 1996. Introduction to quantitative genetics, 4th ed. Longman, London.

Futuyma, D. J. 1987. On the role of species in anagenesis. American Naturalist 130:465–473.

Gandon, S., Y. Capowiez, Y. Dubois, Y. Michalakis, and I. Olivieri. 1996. Local adaptation and gene-for-gene coevolution in

a metapopulation model. Proceedings of the Royal Society of London B 263:1003–1009.
Garcia-Ramos, G., and M. Kirkpatrick. 1997. Genetic models of adaptation and gene flow in peripheral populations. Evolution 51:21–28.
Gavrilets, S. 1996. On phase three of the shifting-balance theory. Evolution 50:1034–1041.
———. 1997. Evolution and speciation on holey adaptive landscapes. Trends in Ecology and Evolution 13:307–312.
———. 1999. A dynamical theory of speciation on holey adaptive landscapes. American Naturalist 154:1–22.
Geary, D. H. 1995. The importance of gradual change in species-level transitions. Pp. 67–86 *in* D. H. Erwin and R. L. Anstey, eds. New approaches to speciation in the fossil record. Columbia University Press, New York.
Gingerich, P. D. 1976. Paleontology and phylogeny: patterns of evolution at the species level in Early Tertiary mammals. American Journal of Science 276:1–28.
———. 1983. Rates of evolution: effects of time and temporal scaling. Science 222:159–161.
Gomulkiewicz, R., and R. D. Holt. 1995. When does evolution by natural selection prevent extinction? Evolution 49:201–207.
Gomulkiewicz, R., J. N. Thompson, R. D. Holt, S. L. Nuismer, and M. E. Hochberg. 2000. Hot spots, cold spots, and the geographic mosaic theory of coevolution. American Naturalist 156:156–174.
Gould, S. J., and N. Eldredge. 1977. Punctuated equilibrium: the tempo and mode of evolution reconsidered. Paleobiology 3: 115–151.
Grant, P. R. 1986. Ecology and evolution of Darwin's finches. Princeton University Press, Princeton, N.J.
Hanski, I., and M. E. Gilpin. 1997. Metapopulation biology: ecology, genetics, and evolution. Academic Press, San Diego.
Hoekstra, H. E., J. M. Hoekstra, D. Berrigan, S. N. Vignieri, A. Hoang, C. E. Hill, P. Beerli, et al. 2001. Strength and tempo of directional selection in the wild. Proceedings of the National Academy of Sciences USA 98:9157–9160.
Huey, R. B., G. W. Gilchrist, M. L. Carlson, D. Berrigan, and L. Serra. 2000. Rapid evolution of a geographic cline in size in an introduced fly. Science 287:308–309.
Jablonski, D. 2000. Micro- and macroevolution scale and hierarchy in evolutionary biology and paleobiology. *In* D. H. Erwin and S. L. Wing, eds. Deep time: *Paleobiology*'s perspective. Paleobiology 26(Suppl. to No. 4):15–52.
Jackson, J. B. C., and A. H. Cheetham. 1999. Tempo and mode of speciation in the sea. Trends in Ecology and Evolution 14: 72–77.
Kawata, M. 2002. Invasion of vacant niches and subsequent sympatric speciation. Proceedings of the Royal Society of London B 269:55–63.
Kirkpatrick, M., and N. H. Barton. 1997. Evolution of a species' range. American Naturalist 150:1–23.
Lambeck, K., and J. Chappell. 2001. Sea level change through the last glacial cycle. Science 292:679–686.
Lande, R. 1985. The fixation of chromosomal rearrangements in a subdivided population with local extinction and colonization. Heredity 54:323–332.
Lenski, R. E., and M. Travisano. 1994. Dynamics of adaptation and diversification: a 10,000- generation experiment with bacterial populations. Proceedings of the National Academy of Sciences USA 91:6808–6814.
Levinton, J. S. 1983. Stasis in progress: the empirical basis of macroevolution. Annual Review of Ecology of Systematics 14: 103–137.
Lieberman, B. S., and S. Dudgeon. 1996. An evaluation of stabilizing selection as a mechanism for stasis. Palaeogeography, Palaeoclimatology and Palaeoecology 127:229–238.
Lieberman, B. S., C. E. Brett, and N. Eldredge. 1995. A study of stasis and change in two species lineages from the Middle Devonian of New York State. Paleobiology 21:15–27.
Lynch, M., and A. Force. 2000. The probability of duplicate gene preservation by subfunctionalization. Genetics 154:459–473.
Mani, G. S., and B. C. C. Clarke. 1990. Mutational order: a major stochastic process in evolution. Proceedings of the Royal Society of London B 240:29–37.
Maynard Smith, J. 1983. The genetics of stasis and punctuation. Annual Review of Genetics 17:11–25.
Mayr, E. 1954. Change of genetic environment and evolution. Pp. 157–180 *in* J. Huxley, A.C. Hardy, and E. B. Ford, eds. Evolution as a process. Allen and Unwin, London.
McPeek, M. A. 1997. Measuring phenotypic selection on an adaptation: lamellae of damselflies experiencing dragonfly predation. Evolution 51:459–466.
———. 1998. The consequences of changing the top predator in a food web: a comparative experimental approach. Ecological Monographs 68:1–23.
———. 2000. Predisposed to adapt: clade-level differences in characters affecting swimming performance in damselflies. Evolution 54:2072–2080.
McPeek, M. A., and J. M. Brown. 2000. Building a regional species pool: diversification of the *Enallagma* damselflies of eastern North American waters. Ecology 81:904–920.
Miller, W., III. 2003. A place for phyletic evolution within the theory of punctuated equilibria: Eldredge pathways. Neues Jahrbuch für Geologie und Paläontologie Monatshefte 2003: 463–476.
Mongold, J. A., A. F. Bennett, and R. E. Lenski. 2001. Evolutionary adaptation to temperature. VII. Extension of the upper thermal limit of *Escherichia coli*. Evolution 53:386–394.
Moore, F. B.-G., D. E. Rozen, and R. E. Lenski. 2000. Pervasive compensatory adaptation in *Escherichia coli*. Proceedings of the Royal Society of London B 267:515–522.
Nuismer, S. L., J. N. Thompson, and R. Gomulkiewicz. 2000. Coevolutionary clines across selection mosaics. Evolution 54: 1102–1115.
Ohta, T. 1972. Population size and rate of evolution. Journal of Molecular Evolution 1:305–314.
Pellmyr, O., J. Leebens-Mack, and J. N. Thompson. 1998. Herbivores and molecular clocks as tools in plant biogeography. Biological Journal of the Linnean Society 63:367–378.
Reznick, D. N., F. H. Shaw, F. H. Rodd, and R. G. Shaw. 1997. Evaluation of the rate of evolution in natural populations of guppies (*Poecilia reticulata*). Science 275:1934–1936.
Rieseberg, L. H. 1997. Hybrid origins of plant species. Annual Review of Ecology and Systematics 28:359–389.
Rutherford, S. L., and S. Lindquist. 1998. Hsp90 as a capacitor for morphological evolution. Nature 396:336–342.
Sandstrom, J. P., J. A. Russell, J. P. White, and N. A. Moran. 2001. Independent origins and horizontal transfer of bacterial symbionts of aphids. Molecular Ecology 10:217–228.
Schemske, D. W., and P. Bierzychudek. 2001. Evolution of flower color in the desert annual *Linanthus parryae*: Wright revisited. Evolution 55:1269–1282.
Sheldon, P. R. 1987. Parallel gradualistic evolution of Ordovician trilobites. Nature 330:561–563.
Sniegowski, P. D., P. J. Gerrish, and R. E. Lenski. 1997. Evolution of high mutation rates in experimental populations of *Escherichia coli*. Nature 387:703–705.
Soltis, D. E., and P. S. Soltis. 1999. Polyploidy: recurrent formation and genome evolution. Trends in Ecology and Evolution 14:348–352.
Stanley, S. M., and X. Yang. 1987. Approximate evolutionary stasis for bivalve morphology over millions of years: a multivariate, multilineage study. Paleobiology 13:113–1119.
Tachida, H., and M. Iizuka. 1991. Fixation probability in spatially changing environments. Genetical Research 58:243–251.

Thomas, C. D., E. J. Bodsworth, R. J. Wilson, A. D. Simmons, Z. G. Davies, M. Musche, and L. Conradt. 2001. Ecological and evolutionary processes at expanding range margins. Nature 411:577–581.

Thompson, J. N. 1994. The coevolutionary process. University of Chicago Press, Chicago.

———. 1997. Evaluating the dynamics of coevolution among geographically structured populations. Ecology 78:1619–1623.

———. 1998. Rapid evolution as an ecological process. Trends in Ecology and Evolution 13:329–332.

———. 1999a. The evolution of species interactions. Science 284: 2116–2118.

———. 1999b. Coevolution and escalation: are ongoing coevolutionary meanderings important? American Naturalist 153: S92–S93.

Thrall, P. H., and J. J. Burdon. 1997. Host-pathogen dynamics in a metapopulation context: the ecological and evolutionary consequences of being spatial. Journal of Ecology 85:743–753.

Travisano, M., J. A. Mongold, A. F. Bennet, and R. E. Lenski. 1995. Experimental tests of the roles of adaptation, chance, and history in evolution. Science 267:87–90.

Valentine, J. W., and D. Jablonski. 1993. Fossil communities: compositional variation at many time scales. Pp. 341–349 *in* R. E. Ricklefs and D. Schluter, eds. Species diversity in ecological communities. University of Chicago Press, Chicago.

Van Valen, L. M. 1982. Integration of species: stasis and biogeography. Evolutionary Theory 6:99–112.

Vrba, E. S. 1985. Environment and evolution: alternative causes of the temporal distribution of evolutionary events. South African Journal of Science 81:229–236.

Wade, M. J., and C. J. Goodnight. 1998. The theories of Fisher and Wright in the context of metapopulations: when nature does many small experiments. Evolution 52:1537–1553.

Wake, D. B., G. Roth, and M. H. Wake. 1983. On the problem of stasis in organismal evolution. Journal of Theoretical Biology 101:211–224.

Weber, K. E. 1996. Large genetic change at small fitness cost in large populations of *Drosophila melanogaster* selection for wind tunnel flight: rethinking fitness surfaces. Genetics 144:205–213.

Webster, A. J., R. J. H. Payne, and M. Pagel. 2003. Molecular phylogenies link rates of evolution and speciation. Science 301: 478.

Williamson, P. G. 1987. Selection or constraint? A proposal on the mechanism for stasis. Pp. 129–142 *in* K. S. W. Campbell and M. F. Day, eds. Rates of evolution. Allen and Unwin, London.

Zachos, J., M. Pagani, L. Sloan, E. Thomas, and K. Billups. 2001. Trend, rhythms, and aberrations in global climates 65 Ma to present. Science 292:686–693.

Paleobiology, 31(2), 2005, pp. 146–156

The evolution of complexity without natural selection, a possible large-scale trend of the fourth kind

Daniel W. McShea

Abstract.—A simple principle predicts a tendency, or vector, toward increasing organismal complexity in the history of life: As the parts of an organism accumulate variations in evolution, they should tend to become more different from each other. In other words, the variance among the parts, or what I call the "internal variance" of the organism, will tend to increase spontaneously. Internal variance is complexity, I argue, albeit complexity in a purely structural sense, divorced from any notion of function. If the principle is correct, this tendency should exist in all lineages, and the resulting trend (if there is one) will be driven, or more precisely, driven by constraint (as opposed to selection). The existence of a trend is uncertain, because the internal-variance principle predicts only that the range of options offered up to selection will be increasingly complex, on average. And it is unclear whether selection will enhance this vector, act neutrally, or oppose it, perhaps negating it. The vector might also be negated if variations producing certain kinds of developmental truncations are especially common in evolution.

Constraint-driven trends—or what I call large-scale trends of the fourth kind—have been in bad odor in evolutionary studies since the Modern Synthesis. Indeed, one such trend, orthogenesis, is famous for having been discredited. In Stephen Jay Gould's last book, *The Structure of Evolutionary Thought,* he tried to rehabilitate this category (although not orthogenesis), showing how constraint-driven trends could be produced by processes well within the mainstream of contemporary evolutionary theory. The internal-variance principle contributes to Gould's project by adding another candidate trend to this category.

Daniel W. McShea. Department of Biology, Duke University, Box 90338, Durham, North Carolina 27708-0338. E-mail: dmcshea@duke.edu

Accepted: 18 August 2004

There may be no marvel greater than the complex functionality of organisms: the precise coordination of their parts, their reliability in the face of environmental upset, and their ability to act appropriately in their own interests. And then there's the sheer number of their parts and part types, across a range of physical scales, most obvious in the larger multicellular organisms. These properties together have been aptly called "adaptive complexity" (Ruse 2003), and the apparent increase in the adaptive complexity of organisms over the history of life is one of the great mysteries of biology.

The mystery is half solved by breaking the notion of adaptive complexity into its two components, adaptation and complexity, and thinking about them separately. Notions like "coordination," "reliability," and "appropriate action" (from the first sentence) are all aspects of function, or adaptation. And function or adaptation in organisms is no mystery at all. Since Darwin, we have had a perfectly adequate explanation for it, natural selection.

Complexity is another matter. Suppose we do a thought experiment in which we subtract all aspects of adaptation from our understanding of an organism. In other words, we put aside everything we know about how the organism functions. What's left is an assemblage of parts, diverse in their sizes, compositions, shapes, and orientations with respect to each other. To capture that diversity, we will call the organism complex, although keep in mind that this is what you might call "pure complexity," divorced from any notion of function. For complexity in this sense, biology has no satisfactory explanation. Consider just complexity in the sense of number of part types. Why should organisms have so many? And why would complexity in this sense increase in evolution (as the common view suggests it does)? In other words, why would modern organisms be more complex than ancient ones?

One possibility is that natural selection has favored complexity along with functionality. Perhaps functional improvement requires

0094-8373/05/3102-0011/$1.00

greater division of labor, which in turn requires more part types, in other words, greater complexity. Or maybe evolutionary increases in body size—also favored by selection perhaps—demand greater complexity for functional reasons (Bonner 1988). (For example, very small organisms can breathe by diffusion but larger ones need to add a circulatory system.) These arguments are plausible, and others equally plausible can be imagined. But they may not be necessary.

Here I argue that organisms are expected to accumulate variations spontaneously as they evolve, with the result that their internal parts become more differentiated. In other words, there is a spontaneous tendency for what I call their "internal variance" to increase. I further argue that internal variance is an aspect of complexity, and therefore the internal-variance principle generates a vector in evolution toward increasing complexity. The internal-variance principle is quite general. It explains, for example, the increase in differentiation of a newly painted picket fence as the paint on each picket acquires its own unique pattern of wear, and the increase in differentiation of the surface of the moon as it accumulates impact craters.

The vector is a generative tendency, or a bias in the production of variants, and therefore it predicts a trend, at least prior to any consideration of selection. But I call it a vector rather than a trend, because the evolutionary resultant is unknown. Selection could reinforce the internal-variance vector, act neutrally, or oppose it. If it acts in opposition, selection might even overwhelm the vector, either precisely canceling it or producing a net vector toward decrease.

In the following discussion, I do not challenge the conventional wisdom that a long-term trend in mean complexity—understood as differentiation of parts—in fact occurred, at least in metazoans (although later I will neutrally consider what it would mean for the internal-variance principle if we were to discover that no trend in the mean had occurred). It is worth noting in this context that some evidence for a metazoan trend exists (e.g., Valentine et al. 1994) but that doubts have been raised, especially for a trend after the Cambrian explosion (e.g., Simpson 1967; McShea 1996).

In a classification of causes of evolutionary trends, the internal-variance principle falls in with some heretical company. As will be seen, if the principle is correct, the vector it predicts should be pervasive, affecting all lineages in all times and all places in the history of life, and therefore would constitute a kind of evolutionary "drive" (sensu McShea 1994; Gould 2002). And any trend resulting must be considered driven, not by selection but by a kind of mathematical or statistical constraint.

Constraint-driven trends—when they occur in higher taxa over substantial time spans—fall into a category that I call large-scale trends of the fourth kind (explained later in a discussion of a four-part trend-classification scheme). This category includes some, such as orthogenesis, that are notorious in the history of biology for having been discredited. In his last book and magnum opus, Stephen Jay Gould (2002) tried to rehabilitate constraint-driven trends (although not orthogenesis itself), offering a number of real and hypothetical examples and showing that they could be produced by causes falling well within the present theoretical mainstream. The internal-variance principle contributes to this rehabilitation, adding one—and I believe an important one, at the highest level of generality—to Gould's list of examples.

The internal-variance principle has a long and diverse ancestry. It comes mainly from Herbert Spencer's (1900, 1904) metaphysics, especially his notion of "the instability of the homogeneous" (see discussion in McShea 1991), which he invoked to explain organismal complexity specifically. Thus my treatment here can be understood as an attempt to revive, update, and further explore Spencer's notion. The principle can also be found in other guises. It is closely related to another old notion in biology, the duplication and differentiation of parts (e.g., Gregory 1935; Weiss 1990). And it is implicit in modern treatments of morphological evolution as a diffusive or Markov process (e.g., Raup 1977). Finally, what is essentially the same principle has been recognized in a related context, the evolution of differences among species (rather than as here, differences among parts in an organism). In particular, it underlies the intuition that the degree of morphological differentiation among species in a group, or their "disparity" (Foote 1997), should tend to increase spontaneously (Ciampaglio et al. 2001). And it emerges in the writings of one of the thermodynamic schools of thought in evolutionary studies (e.g., Wicken 1987; Brooks and Wiley 1988), where it has been invoked mainly to explain species diversity.

The Internal-Variance Principle

Figure 1 illustrates the principle with a simple model. Five parts of an organism are plotted on an axis representing some dimension, say, their length (actually log-length). Initially all five parts are identical (e.g., serial homologs, perhaps), with their lengths set at the same value, arbitrarily, 10 millimeters. In the first time step, random heritable variation is added to each part, so that its length increases or decreases by the same factor, divided by 0.9 for increases and multiplied by 0.9 for decreases, with the direction of change chosen at random (probability of increase = probability of decrease = 0.5). Each part is treated independently. And in each subsequent time step, more variation is introduced, and new values calculated for each of the five parts based on their values in the previous time step, so that change is cumulative.

The top box in Figure 1 shows the starting distribution of logged part-lengths in a single run of the model. The parts have the same lengths so the points fall precisely on top of each other. The lower boxes show the distributions after 1, 2, 5, 10, and 20 time steps. The figure also shows the increase in internal variance quantitatively. The term "variance" in this phrase is intended in a general sense, to mean something like "amount of variation" or "degree of differentiation," not in a formal statistical sense to refer to a sum of squared deviations from a mean. However, statistical variance is one possible measure of internal variance. In Figure 1, internal variance is measured as the square root of the statistical variance, i.e., the standard deviation.

Figure 2A shows the trajectory of the standard deviation over the entire run, and Figure

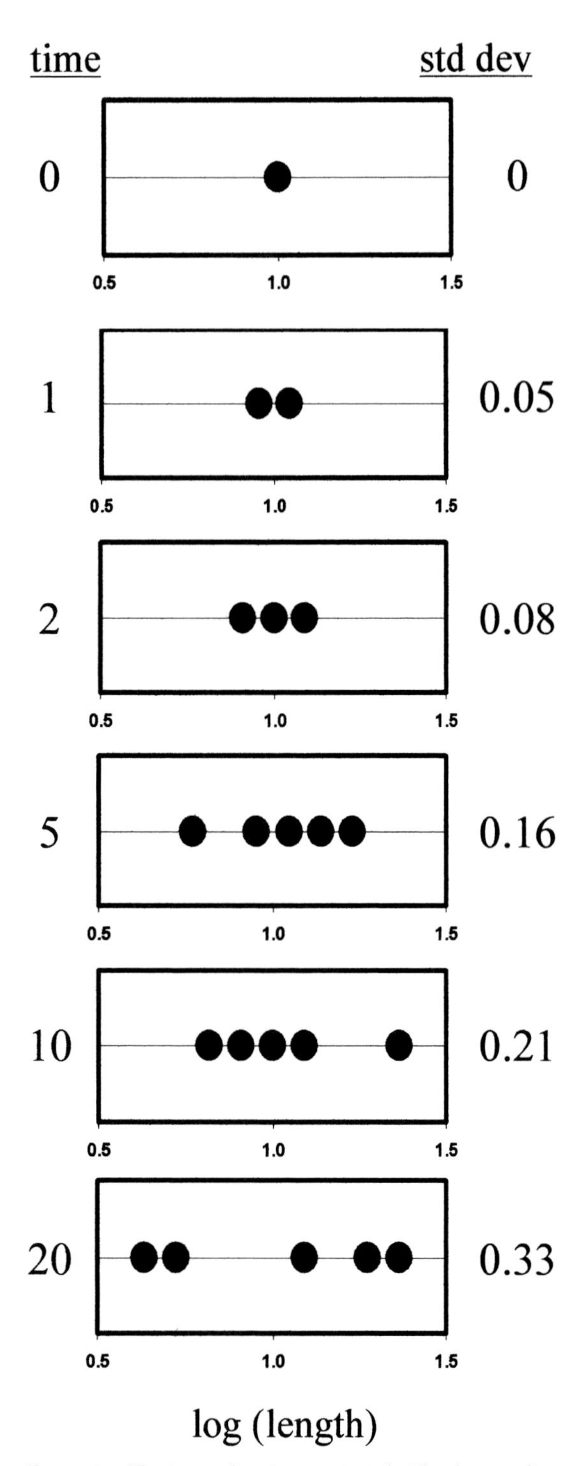

FIGURE 1. The internal-variance principle. The dots are five initially identical parts (overlapping in the top frame). In each time step, a variation introduced to each part changes its length by a factor of 0.9 (length is multiplied by 0.9 for decreases and divided by 0.9 for increases), with increases and decreases occurring with equal probability (0.5). Internal variance is measured as the standard deviation among log-lengths. The numbers on the right show the internal variance at time 0 (top frame), 1, 2, 5, 10, and 20 (bottom frame).

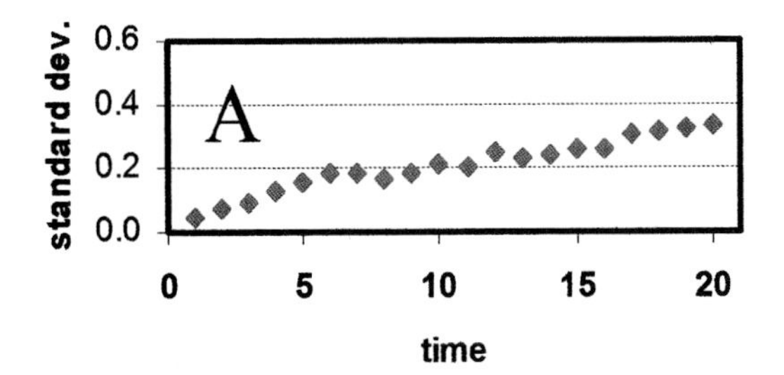

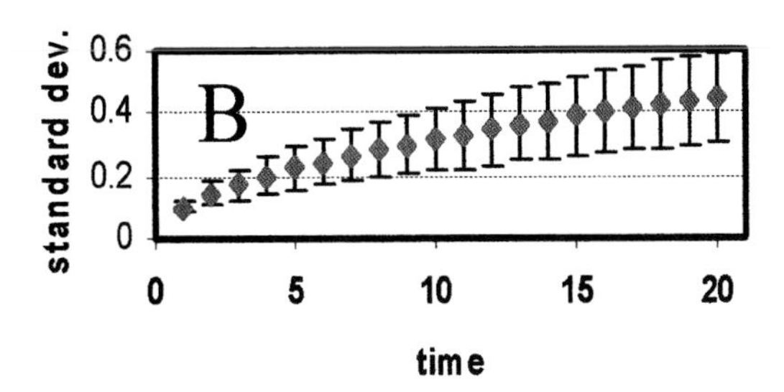

FIGURE 2. A, The trajectory of the standard deviation for a single run of the model in Figure 1 over 20 time intervals. B, The trajectory averaged over 1000 runs. Error bars show one standard deviation of the internal variance (i.e., the standard deviation of the standard deviation).

2B shows the mean standard deviation over 1000 runs, with error bars (i.e., the standard deviation of the standard deviation). Analytically, it can be shown that the internal variance rises as the square root of the number of time steps. Notice that the internal variance rises even after parts have become highly differentiated. That is, even when complexity is high, the expectation is further increase, without limit.

The explanation for the trend should be obvious. In a very small number of cases, the introduced random variations may by chance make the parts more similar to each other, but in the vast majority of cases, they will cause the parts to diverge, to differentiate. In other words, there are many more ways for internal variance to increase than for it to decrease, and therefore most combinations of random variations will produce increases.

This example is a paradigmatic case. But the principle should be far more general. In particular, it should apply in the following cases: (1) To any organism with some degree of modular construction, i.e., any organism with parts (although an analogous principle could be developed for a hypothetical organism without any clear compartmentalization). (2) To any set of parts within an organism sharing a common dimension, not just initially identical parts as in the example above. Thus, the variance should increase even when the parts in the set are disparate and seemingly incomparable, such as the set: foot, kidney, ear, and aorta (perhaps in a mammal). All of these share a dimension, length for example, and therefore the variance in that dimension should increase over time. (3) To any dimension or set of dimensions. For those four mammalian parts, the variances among their lengths, widths, permeabilities, color, metabolic rates, etc. should all increase over time. And if internal variance is measured in a multidimensional space, the overall rate of increase should be higher with greater dimensionality. (4) For all intuitively reasonable measures of internal variance. (5) To cases where discrete part types are recognized. To see this, imagine that the simulation above were run with the length axis in Figure 1 divided into equal-sized bins, with each bin corresponding to a discrete part type. (6) To individuals in all lineages, over the history of life.

(7) For most types of random variation. In the example above, variation was introduced to each part independently, using a simple two-state function. However, the vast majority of arbitrary functions that could be introduced to represent variation should also produce increases in internal variance, on average, even the very complex or highly structured ones. In two other papers (McShea 1992, in press), I experimented with real and hypothetical homologous series (representing vertebral columns), to study the effect of introducing highly structured variations in which changes among adjacent parts are correlated (e.g., a sine curve), such as might be caused by introduced fields or gradients. The thinking was that such variations were at least somewhat realistic and might represent counterexamples, that is, they might tend to produce decreases in internal variance. In fact, some functions could produce decreases, but only occasionally, with certain parameter settings. Increases still predominated.

Notice that the internal-variance principle is different from another trend mechanism, what has been called variously a passive mechanism, an increase in variance, or diffusion in the presence of a lower bound (Stanley 1973; Fisher 1986; McShea 1994; Gould 1996). The internal-variance principle does invoke boundaries in the sense that variances are bounded at zero. And it invokes diffusion in the sense that the morphologies of parts within an organism diffuse in morphospace. But at the lineage level, the principle predicts a pervasive increasing tendency. It predicts that parts should vary so that differences among them increase in essentially all lineages, from the simplest to the most complex. In other words, the trend in complexity should be driven (McShea 1994; Gould 2002). In contrast, a passive mechanism predicts equal numbers of lineages increasing and decreasing.

Internal Variance Is a Type of Complexity

I cannot give the full argument here for this claim (but see McShea 1996, in press). But briefly, the argument is based on a view of complexity as the amount of differentiation among parts, where variation is continuous, or as number of part types, where variation is discrete.

Some objections will arise. First, in this view, complexity is purely about structure, but colloquially the word includes some notion of function as well. We commonly think of complex systems not just as differentiated, but also capable. Now this compound usage may be useful in certain contexts, but it confounds the two central issues involved in adaptive complexity: adaptation (i.e., function), for which we have an almost universally acknowledged explanation, and structural complexity, for which we don't. And to address them separately, it is essential to use conceptually nonoverlapping terms. The case for conceptual separation is bolstered by the ease with which we can imagine objects with any of the four combinations of properties: (1) functional and structurally simple (e.g., virus, screwdriver); (2) functional and structurally complex (*Amoeba*, vacuum cleaner); (3) non-functional and simple (grain of sand, drop of water); and (4) non-functional and complex (dead *Amoeba*, surface of the moon).

A second objection is that even when used in a purely structural sense, complexity sometimes refers to more than just internal variance. There is also complexity in the sense of degree of hierarchical structure (number of levels of parts within wholes), irregularity of spatial configuration, and so on. The argument here is that the internal-variance principle predicts a rise in complexity only in the sense of differentiation among parts (or number of part types), what I elsewhere call "non-hierarchical object complexity" (McShea 1996). However, the argument is extendable to make analogous predictions at least for irregularity of spatial configuration, and perhaps for other senses of complexity as well. (At present, though, I do not see how it could predict hierarchy.)

Third, this view might seem to equate complexity with entropy, and indeed entropy is the term some have chosen for this same principle in other contexts (e.g., Brooks and Wiley 1988). Entropy in its information-theoretical interpretation refers to a relationship between microstates and macrostate, namely the more entropic macrostate is the one with the greater number of microstates corresponding to it. This is what makes the more entropic macrostate the more probable one and the one toward which the system spontaneously evolves. My house is more likely to become messier than neater, because many more configurations of my stuff (microstates) correspond to messier (a macrostate) and relatively few correspond to neater (a macrostate). Similarly, my argument here is that many more character state distributions correspond to greater internal variance than correspond to less, so greater internal variance is more probable, and it is toward greater internal variance that the organism spontaneously evolves (considerations of selection aside). So entropy would seem to be the right word.

Confusion arises however, because the word in common usage also connotes disorganization and loss of function. We think of systems as falling apart, and failing to work, as they become more entropic. But this connotation runs at a right angle to the technical meaning, which is completely silent on such matters. My messy house may be quite functional, for my purposes, even if it has high entropy. Perhaps various messy distributions of items allow me to find and use them efficiently. Likewise, an organism may have high entropy, with many disparate parts, irregularly configured, and yet it may function quite well. That is, it may be quite fit. Technically, nothing in the notion of entropy conflicts with the notion of functionality, or even with increase in functionality. But given the friction between technical meaning and common-usage connotation, I have opted against entropy in favor of a more neutral term, internal variance.

To conclude this section, I understand internal variance as a type or subcategory of complexity. However, in what follows, I will risk causing some confusion by using the terms interchangeably (in effect ignoring the many other senses of complexity). The point is to subtly reinforce the connection between two notions that might formerly have seemed unconnected—one a variable that is unproblematic and even measurable, internal variance, and the other a concept that has historically been very troublesome, complexity.

The Effect of Selection

The internal-variance principle describes a tendency for the accumulation of variation to produce greater complexity, prior to any consideration of selection. When we do take selection into account, there are four ways in which it could act: (1) A variation producing a complexity increase could have a greater probability of being favored than one producing a decrease, and in that case selection will just reinforce the internal-variance vector. (2) Selection could be neutral with respect to complexity. That is, suppose that every introduced variation is favored or rejected depending on its effect on adaptedness, but that there is no correlation between adaptedness and complexity. In that case, the few variations that pass the selective filter will be a representative sample of those generated, and therefore will be more complex, on average. It is as if selection were picking from a deck that has been stacked with black cards, where the black cards represent high complexity and the red represent low. If selection has no color preference (i.e., if it is neutral with respect to complexity) and if the aces are the well-adapted forms, then most of the aces selected will be black. In other words, most of the well-adapted forms will be complex.

(3) Selection could oppose (but not overwhelm) the internal-variance vector, meaning that it could disproportionately favor the rare variants producing decreases in complexity (or disproportionately oppose the many variants producing increases). There are at least two ways it might do this. First, selection might impose boundaries on variation, so that increase in the length, permeability, etc. of parts beyond some upper limit, and decrease below some lower limit, is disfavored. Such boundaries must exist in that all variation has functional limits. But notice that for boundaries to limit internal variance, the distribution of parts would have to be already maximal, filling the entire functionally permissible range. Also, maximal dispersion must have already been reached in all dimensions. If dispersion can continue in even one dimension, then overall internal variance should still increase.

Alternatively, it could be that simpler species have disproportionately more ecological opportunities, on average, than complex species. In other words, the generative bias produces more increases in internal variance than decreases, but this vector is partly offset by the disproportionate frequency of ecological opportunities, of niches or places in the economy of nature, for simpler species. This might seem implausible if ecological opportunities for increased complexity spring to mind more readily than those for decrease. For example, specialization is probably a common route to adaptation, and specialization might require greater complexity, on average. Arguably, however, opportunities for simplicity should also be common. Obvious candidates include the opportunities associated with the evolution of parasitism or of small size generally (as in the evolution of the so-called interstitial fauna or meiofauna). But the range is really much broader. For example, consider the evolution of the mammalian vertebral column in the transitions from terrestrial to aquatic living. Terrestrial ancestors typically had complex columns suitable for quadrupedal locomotion on land (e.g., with special modifications of the column for attachment of the hind legs). But their aquatic descendants, such as whales, tended to have simpler, more uniform, fishlike columns, suitable for undulatory propulsion in the water (McShea 1991). Selection favored simplicity, not for any reason having to do with simplicity itself, but on account of an accidental association between simplicity and a particular functional mode. Such associations must be quite common.

(4) Finally, selection might oppose the internal-variance vector—by either of the routes in case 3 above—strongly enough to cancel it or even to produce a reverse vector. Notice that for cancellation, selection must allow increases and decreases in equal numbers, not equal proportions (which would be equivalent to selective neutrality).

Cases 1, 2, and 3 predict a driven trend in mean complexity, produced either partly (case 1) or entirely (cases 2 and 3) by the internal-variance principle. Case 4 predicts no trend in the mean if selection precisely cancels internal variance, at least no driven trend (the mean could still increase by a passive mechanism). And it predicts a decreasing trend if selection is strong enough to produce reversal.

Unfortunately, we do not know the fact of the matter, for either the mean or the mechanism, for life as a whole or for any kingdom-level clade within it. Several cases have been studied within the metazoans, and both the passive and driven mechanisms have been found (McShea 1993; Saunders et al. 1999; Sidor 2001). The best that can be done at this point is to frame a conclusion in the conditional. In a given group, at some specified temporal scale, if no trend in the mean occurred, or if a trend occurred and is ultimately discovered to have been passive, then presumably selection opposes and cancels the internal-variance vector. But if mean complexity increases, and if the trend is ultimately found to be largely driven, then the internal-variance principle must account for it, at least partly. It would still be impossible to say whether the trend was occurring with the support, neutrality, or partial opposition of selection (cases 1, 2, and 3, above). To infer that, we would need to know the magnitude of the internal-variance vector.

One other possible selective route needs to be considered, selection above the level of the organism, such as species selection. For example, it might be that species with complex morphologies are more specialized and therefore more extinction prone. Again, however, nothing conclusive can be said. Schopf et al. (1975) investigated this in metazoans and found a positive correlation between complexity and extinction probability but interpreted it as an observational artifact.

A final point about case 4: from one viewpoint, the impact of cancellation on our understanding of the passive mechanism is fairly staggering. Prior to any consideration of internal variance, it is easy and tempting to see the passive mechanism as a kind of null case, one in which no forces act at the large scale. (That's why increases and decreases are equally frequent.) But the internal-variance principle forces us to reconceptualize, to see a passive trend as necessarily the resultant of two equal and opposite vectors, internal variance and selection, both potentially quite strong. Indeed, consider the enormous force that selection must exert in order to cancel the internal-variance vector. If the vast majority of variations arising are increases in complexity, then selection must oppose virtually all of them, while favoring a comparatively large proportion of the small number of decreases arising, in order to leave equal numbers of each. To put it another way, ecological opportunities for more-complex variants must be vanishingly rare, or improbable, compared with those for less-complex variants. Thus, if there was a trend in mean complexity over the history of life, and if we discover that it was passive, we are left with a real puzzle: Why does selection abhor complexity?

Objections

I foresee three objections to the internal-variance principle. First, there is a common notion that losses of parts—and therefore losses of complexity—are more likely than gains because most mutations affecting a given developmental pathway will tend to disrupt it. I think this intuition could be partly right and, insofar, could be fatal to the internal-variance principle. It is right in that truncations of development are probably developmentally easy and can produce reductions in internal variance. Internal variance certainly increases in ontogeny, because organisms go from simple to complex as they develop, and therefore truncations should, on average, leave structure simplified overall. But intuitively anyway (and in the absence of any organized data on the subject), this should be the case only for major truncations that cut short all of development or remove all of some quasi-independent module. Partial truncations that remove arbitrary subsets of features should not affect internal variance, on average, for the simple reason that the expected variance of a set of data (measurements of a set of structures before truncation) is the same as the expected variance of a random subset of those data (those remaining after truncation). Thus, the important empirical question for the internal variance principle is the frequency with which major or modular truncations actually occur, i.e., whether they occur often enough to offset the internal-variance vector.

The idea that loss of structure is developmentally easy may also be partly faulty. The fault lies in a common confusion between loss of function and loss of structure. Random modification must tend toward loss of function, but need not tend toward loss of structure. It is true that a random variation might lead to structural loss via a failure of an inductive tissue contact, elimination of a differentiation event, or a failure of the production of a tissue fold. But given the dynamism of development, would a random variation not be equally likely to lead to structural gain, perhaps by producing a loss of an inductive barrier, an extraneous differentiation, or the failure of a fold-smoothing-mechanism? Are not developmental aberrations, "monstrosities," often more complex structurally? I think these are currently open questions.

Another objection, one might argue, is that selection has favored mechanisms that buffer or canalize development against most of the variations that would increase complexity, because these are mostly deleterious. Thus, arguably, most increases in complexity will never be expressed, perhaps negating the internal-variance vector. However, decreases in complexity are also likely to be deleterious. What buffering mechanisms oppose is departure from a functional norm, not greater complexity per se, and therefore they should oppose the expression of both increases and decreases equally.

Finally, one might object that direct testing of the internal-variance principle is difficult, perhaps impossible. The principle predicts that offspring should be more internally varied than their parents, but only prior to any consideration of selection. And it may be impossible to find circumstances in which selection is truly absent. However, we might look at changes in internal variance in special cases, where the effect of selection should be minimal. For example, we might measure some dimension of each of the teeth in a mammalian adult tooth row and compare the standard deviations of this measure in a large sample of parent-offspring pairs. The teeth do not emerge and function until adulthood, and so arguably, before they do, they ought to be less subject to selection than other structures. Comparable measurements could be obtained from parents and offspring at the same life stage, ideally when the teeth are just erupting. The internal-variance principle predicts that the standard deviation for the offspring will be higher, on average. Of course, the test is imperfect in that teeth are undoubtedly connected to other structures in development and are presumably subject to indirect selection by that route.

Large-Scale Trends of the Fourth Kind

Figure 3 shows a classification of large-scale trends based on two dichotomies, one between alternative mechanisms (rows), and one between alternative underlying causes (columns). A large-scale trend is a long-term directional change in a summary statistic for a clade, such as the mean. Mechanism refers to the pattern of change in the variable in question, that is, whether the trend occurs because of a pervasive tendency in one direction or the other (i.e., a driven trend) or a local inhomogeneity (i.e., a passive trend), such as a boundary (McShea 1994). Causes occur at a lower level, answering the question of whether the drive or boundary is the result of selection or of any of the various types of constraint (e.g.,

Cause

Mechanism	selection	constraint
passive	I	II
driven	III	IV

FIGURE 3. A classification of large-scale trends, according to both mechanism (rows) and underlying cause (columns). See text.

Gould 1989, 2002). (This form of the passive-driven distinction is a crude one; see Alroy 2000; McShea 2000; and Wagner 1996 for more-sophisticated alternatives.)

The four quadrants represent the possible combinations of mechanism and cause, and indeed it seems likely that all are occupied by actual evolutionary trends at some scale. Here I will consider only trends in complexity. Trends in the first quadrant, or "large-scale trends of the first kind," are passive with selection as the underlying cause. These include trends in which increases and decreases occur equally frequently and there is a selective lower boundary. For complexity, such a boundary might arise if organisms less complex than some minimum are not viable. It might be, for example, that life processes cannot be carried out with fewer part types than in a prokaryotic cell, and therefore variants arising at a lower complexity, below the boundary, are selected against. The resulting increase in mean complexity over the history of life would be passive, with selection as the underlying cause.

Large-scale trends of the second kind are passive with constraint as underlying cause. For example, the number of cell types in a eukaryote cannot be less than one, and this mathematical (or logical) constraint imposes a lower bound on number of cell types in metazoans. Valentine et al. (1994) suggested that the increase in number of cell types over the history of animals may have been passive in this way.

Large-scale trends of the third kind are driven with selection as the driving force. The increase in septal-suture complexity in Paleozoic ammonoids may have been a trend of this sort. Saunders et al. (1999) documented that suture complexity increased significantly more often than it decreased among lineages, and they speculated that selection may have been the cause of this upward bias.

Large-scale trends of the fourth kind are driven, so that increases occur more often than decreases among lineages, with constraint producing the upward tendency. Such trends were not well received in evolutionary studies over most of the twentieth century, largely on account of their historical association with orthogenesis. The more extreme of the nineteenth-century orthogeneticists argued that evolution proceeds by acceleration of ontogeny, leaving room for terminal addition of a predictable sequence of increasingly senescent developmental stages. The principal objection to notions of this sort, from the perspective of the Modern Synthesis, is that they seem to justify interpreting certain evolutionary results—such as the massive antlers of the Irish elk—as maladaptive (Gould 2002). But Gould argued that there is nothing in the nature of constraint-driven trends that forces such interpretations, especially if the constraint occurs at the species level. He offered a possible example (citing Strathmann): species with non-planktotrophic larvae can arise from planktotrophic species by the loss of larval swimming ability, but the reverse is much less likely, because regaining a swimming larva is developmentally more difficult, i.e., constrained. The result is not maladaptive design in individuals, just a bias toward the production of non-planktotropic species.

The internal-variance principle is a consequence of mathematical or statistical necessity and therefore constitutes a kind of constraint (what Gould [1989] called a "formal" constraint). And the constraint is expected to operate in all lineages, so any resulting trend must be driven. A difference with the planktotrophy example is that the internal-variance vector acts at the organism level, as in orthogenesis, not the species level. But unlike orthogenesis, the driven variable is a higher-level or abstract property of an organism, its internal variance. So the principle dictates no specific change in structure. And because all variants must pass through the selective filter, the specific structures that survive should all be adaptive.

Conclusion

The internal-variance principle identifies an evolutionary vector, a kind of pervasive force pushing complexity—understood as differentiation among parts—upward. But it is a separate matter whether this vector is actually manifest as a bias in the direction of change in evolutionary lineages, and whether there is an actual trend in, say, the mean complexity of metazoans. If there is a trend, and if it is driven, then the internal-variance vector supplies a likely cause (with or without reinforcement by selection). Or the vector could be opposed by selection to the point where it is overcome, with the result that either no trend or only a passive trend occurs.

Regardless of whether the internal-variance vector is manifest, its existence demands what for some will be a dizzying reversal of intuition. We are used to thinking of complexity as hard to produce. The internal-variance principle shows it to be easy, in particular to be easy to generate in development and therefore in evolution. To see this, we have to purge the conventional notion of complexity of its functional component, and to conceive complexity as purely structural. That is, we have to see complexity—in the relevant sense here—as just structural diversity or differentiation among parts, i.e., internal variance. Having done so, it is easy to see why it arises spontaneously. And it becomes equally easy to see why simplicity, complexity's opposite, is developmentally hard (except perhaps via major truncations of development).

Importantly, the point is not that *adaptive* complexity is easy, not that specific complex structures like eyes and brains spontaneously self-assemble somehow in evolution. Rather, it is that most spontaneous changes produce greater complexity. Of these, the vast majority will be nonfunctional. Eyes and brains represent the tiny subset that are functional. Adaptation stills needs selection. But complexity does not.

Acknowledgments

The idea in this paper owes much to the exhilarating metaphysical thinking of H. Spencer, but also to the more recent inspirational ideas of J. T. Bonner, D. Raup, L. Van Valen, and my colleagues in the biology and philosophy discussion group here at Duke University. Many thanks also to A. Hallam and an anonymous reviewer for useful critiques. I regret that I did not get a chance to run this idea by Stephen Jay Gould. But I want to thank him anyway, for his exuberant iconoclasm, and for writing so well and thinking so clearly about evolutionary trends over the years.

Literature Cited

Alroy, J. 2000. Understanding the dynamics of trends within evolving lineages. Paleobiology 26:319–329.

Bonner, J. T. 1988. The evolution of complexity. Princeton University Press, Princeton, N.J.

Brooks, D. R., and E. O. Wiley. 1988. Evolution as entropy: toward a unified theory of biology. University of Chicago Press, Chicago.

Ciampaglio, C. N., M. Kemp, and D. W. McShea. 2001. Detecting changes in morphospace occupation patterns in the fossil record: characterization and analysis of measures of disparity. Paleobiology 27:695–715.

Fisher, D. C. 1986. Progress in organismal design. Pp. 99–117 *in* D. M. Raup and D. Jablonski, eds. Patterns and processes in the history of life. Springer, Berlin.

Foote, M. 1997. The evolution of morphological diversity. Annual Review of Ecology and Systematics 28:129–152.

Gould, S. J. 1989. A developmental constraint in Cerion, with comments on the definition and interpretation of constraint in evolution. Evolution 43:516–539.

———. 1996. Full house: the spread of excellence from Plato to Darwin. Harmony Books, New York.

———. 2002. The structure of evolutionary theory. Harvard University Press, Cambridge.

Gregory, W. K. 1935. Reduplication in evolution. Quarterly Review of Biology 10:272–290.

McShea, D. W. 1991. Complexity and evolution: what everybody knows. Biology and Philosophy 6:303–324.

———. 1992. A metric for the study of evolutionary trends in the complexity of serial structures. Biological Journal of the Linnean Society 45:39–55.

———. 1993. Evolutionary change in the morphological complexity of the mammalian vertebral column. Evolution 47: 730–740.

———. 1994. Mechanisms of large-scale trends. Evolution 48: 1747–1763.

———. 1996. Metazoan complexity and evolution: is there a trend? Evolution 50:477–492.

———. 2000. Trends, tools, and terminology. Paleobiology 26: 330–333.

———. In press. A universal generative tendency toward increased organismal complexity. *In* B. Hallgrimsson and B. Hall, eds. Variation. Academic Press.

Raup, D. M. 1977. Stochastic models in evolutionary paleobiology. Pp. 59–78 *in* A. Hallam, ed. Patterns of evolution as illustrated by the fossil record. Elsevier, Amsterdam.

Ruse, M. 2003. Darwin and design: does evolution have a purpose? Harvard University Press.
Saunders, W. B., D. M. Work, and S. V. Nikolaeva. 1999. Evolution of complexity in Paleozoic ammonoid sutures. Science 286:760–763.
Schopf, T. J. M., D. M. Raup, S. J. Gould, and D. S. Simberloff. 1975. Genomic versus morphologic rates of evolution: influence of morphologic complexity. Paleobiology 1:63–70.
Sidor, C. A. 2001. Simplification as a trend in synapsid cranial evolution. Evolution 55:1419–1442.
Simpson, G. G. 1967. The meaning of evolution. Yale University Press, New Haven, Conn.
Spencer, H. 1900. The principles of biology. Appleton, New York.
———. 1904. First principles. J. A. Hill, New York.
Stanley, S. M. 1973. An explanation for Cope's rule. Evolution 27: 1–26.
Valentine, J. W., A. G. Collins, and C. P. Meyer. 1994. Morphological complexity increase in metazoans. Paleobiology 20: 131–142.
Wagner, P. J. 1996. Contrasting the underlying pattern of active trends in morphologic evolution. Evolution 50:990–1007.
Weiss, K. M. 1990. Duplication with variation: metameric logic in evolution from genes to morphology. Yearbook of Physical Anthropology 33:1–23.
Wicken, J. S. 1987. Evolution, thermodynamics, and information. Oxford University Press, New York.

Paleobiology, 31(2), 2005, pp. 157–174

Mass turnover and heterochrony events in response to physical change

Elisabeth S. Vrba

Abstract.—The thoughts and writings of Stephen Jay Gould have had an enormous impact on the shaping of macroevolutionary theory. The notion of punctuated equilibria (Eldredge and Gould 1972) remained prominent throughout his work. It also unleashed a storm of debate in paleontology and evolutionary theory. A second theme that recurs throughout Gould's opus is heterochrony (evolution by changes in the rates and timing of ontogenetic events, sensu Gould 1977a). His analyses of these two subjects have inspired many of us to explore further and add to them. My contribution discusses their expansion to encompass large numbers of lineages through long time, and the relationship of punctuated equilibria and heterochrony to physical environmental change, and to each other.

Elisabeth S. Vrba. Department of Geology and Geophysics, Yale University, Post Office Box 208109, New Haven, Connecticut 06520-8109. E-mail: elisabeth.vrba@yale.edu

Accepted: 18 July 2004

Introduction

Themes that recur throughout Stephen Jay Gould's work are punctuated equilibria (Eldredge and Gould 1972) and heterochrony (Gould 1977a, 1979, 1988). His contributions on these subjects have inspired many of us to explore further their theoretical consequences and to add new elements. Two topics in this category will be reviewed and discussed. The first expands punctuated equilibria and its relationship to speciation to encompass large numbers of lineages of diverse taxonomic background. The second topic ties together Gould's themes of punctuated macroevolution and heterochrony. My two topics are linked by a focus on the influence of physical environmental change on the evolution of novelty.

The term "species" will be restricted throughout to a sexually reproducing lineage the members of which share a common fertilization system. "Speciation" refers to the divergence of the fertilization system in a daughter population to reproductive isolation from the parent species (after Paterson 1982). The products of speciation may range from being phenotypically indistinguishable (sibling species) to strongly diverged (Vrba 1980: Fig. 5). "Physical change" refers to the global and local effects from extraterrestrial sources, including the astronomical climatic cycles, and from dynamics in the earth's crust and deeper layers as manifested by topographic changes such as rifting, uplift, sea level change, and volcanism. All such physical change involves climatic change, at least locally; and the term climate will sometimes be used to subsume effects from both sources.

Physical Environmental Change and Speciation

Gulick (1872) argued early on that population separation (or vicariance) brought about by physical change is an important trigger of speciation. He studied the snail fauna on the volcanic slopes of the Hawaiian Archipelago and recorded that lava flows typically separate closely related species of very small total geographic distributions. He noted that these sister taxa, in spite of living in similar environments, are strongly diverged in color, in shape, and in feeding and other behaviors. Gulick (1872) reasoned that because vicariance results in the chance apportionment of different genetic variants among small isolated populations, it can thereby lead to speciation.

Mayr's (1942, 1963) analyses were followed by widespread agreement that allopatric speciation predominates in sexually reproducing lineages. During the 1960s and 1970s some theoretical and empirical claims of sympatric speciation (e.g., Bush 1975) raised the possi-

 0094-8373/05/3102-0012/$1.00

bility that this mode may be quite common. But subsequently many authors argued that the various nonallopatric models are neither supported by empirical evidence nor theoretically likely (Futuyma and Mayer 1980; Templeton 1981; Carson 1982; Paterson 1982, 1985; Brooks and McLennan 1991). More recently there have been additional claims of possible sympatric speciation, mainly in fishes (Schliewen et al. 1994; McKinnon and Rundle 2002) and herbivorous insects (review in Via 2001). These claims are variously based on new models supporting the theoretical plausibility of sympatric speciation and on retrospective phylogenetic and population genetic signatures of such speciation. However, many of these reports acknowledge that the best evidence remains circumstantial and weaker than the abundant evidence supporting allopatric speciation. In Via's (2001: p. 381) words, "it remains difficult to show conclusively that specific pairs of taxa have speciated through sympatric processes alone." The search for well-supported cases of sympatric speciation will continue, and it is possible that particular organismal groups will be found to have a higher incidence of this mode than others. However, it remains fair to say that the notion of *predominant* (not exclusive) allopatric speciation is consistent with the weight of evidence available to date and continues to enjoy widespread consensus.

If allopatric speciation predominates, then so must physical initiation of speciation predominate (Vrba 1980, 1995a). Although the relationship of punctuated equilibria to physical change was not explored by Eldredge and Gould (1972), the pattern they argued for implies independently that the initiation of speciation mostly comes about through physical change (Vrba 1980): if species are in equilibrium for most of their durations, what causal agency of the punctuation can one invoke other than physical change? The consensus on the importance of allopatric speciation, together with the implications of punctuated equilibria and Paterson's (1978) "recognition concept" of species and speciation, led Vrba (1980) to propose that physical change is required for most speciation. Paterson (1978, 1982, 1985) argued that change in the fertilization system, the critical evolutionary change in sexual speciation, is most likely to occur in small, isolated populations that are under selection pressure from new environmental conditions. However, others have proposed that speciation could just as readily occur in large as in small populations (reviewed in Gould 2002).

In the first part of the paper, I discuss two predictions of the notion that physical change initiates most speciation: First, species should generally "start small," namely, in geographic distributions that are more restricted than those they attain later on. Second, speciation events should be concentrated nonrandomly in time in significant association with physical change (a part of the "turnover pulse" hypothesis [Vrba 1985, 1995a]).

Do Species Start Small?

Vrba and DeGusta (2004) studied this question, which had not previously been addressed directly by fossil data, in the record of the African larger mammals of the past 10 Myr divided into 0.5-Myr-long intervals (taxon list, literature sources in Vrba 2000). The number of fossil site records, from which each species is known in an interval, was taken as a proxy for the magnitude of its living geographic range and abundance in that interval. We then tested H_0 (null hypothesis) that the geographic spread of species remained averagely constant across successive survivorship categories, namely from the interval of first appearance to the immediately following one, and so on. The analysis included correction for the "gap bias," the bias introduced by gaps, or at least intervals of very low fossil representation, in the fossil record. Gaps have the effect that each survivorship category as seen in the fossil record is erroneously inflated by species" records that in reality belong to later survivorship categories. That is, the "gap bias" militates against rejection of H_0 (see Vrba and DeGusta for explicit arguments).

The overall result indicated that the mean number of site records increases strongly from the interval of first appearance to the following survivorship interval, followed by a less marked although still significant increase to the next interval, with no significant changes thereafter. Thus Vrba and DeGusta (2004) concluded that the average large African mammal species has indeed started its life in a relatively small population, and thereafter increased in geographic range to reach its long-term equilibrium abundance by approximately 1 Myr after origin. The stress on "*relatively* small" is important. The question of precisely how small the average large mammal species is at origin cannot be resolved by the proxy we used. Our results are probably as compatible with the minuscule founder populations envisioned by Carson (1982) as with extensive populations in large separated geographic areas. The significant increase in fossil site records does indicate that, relative to the species' original size, there is typically a strong size increase toward long-term equilibrium abundance.

This result supports hypotheses of speciation that accord a major role to the formation of isolated populations of reduced size, either by vicariance due to physical change or by dispersal. Vicariance is predominantly produced by tectonic and climatic change. I suggest that incipient speciation that is initiated by dispersal over preexisting barriers also in most cases implies the causal influence of physical change. For instance, chance *Drosophila* fly dispersals over the ocean always occur, whether there are islands within reach or not; but it took the production of the precursor islands of the Hawaiian Archipelago for the founding of those first allopatric populations of Hawaiian drosophilids (Carson et al. 1970). Thus, the chief causes of population size reduction and allopatry in the history of life have probably derived from tectonic and climatic changes; and our finding that a species starts relatively small on its own prompts the hypothesis of speciation pulses, the subject of the next section.

Physical Change and Speciation Pulses

Statement and Predictions.—The hypothesis of speciation pulses forms a part of the turnover pulse hypothesis (Vrba 1985, 1995a). The essential point of this hypothesis is that most speciation minimally requires population isolation, either by vicariance of the species' habitat and thus of its geographic distribution, or following dispersal over a barrier with successful establishment of the dispersal population, both most often initiated by physical change. Extinction is preceded by vicariance and population size reduction initiated by physical change. The larger prediction, stripped to its bare essentials, is that most lineage turnover, speciation, and extinction has occurred in pulses, varying from tiny to massive in scale, across disparate groups of organisms, and in predictable temporal association with changes in the physical environment. Also, suites of species should show parallel persistence, and stasis at least in their fertilization systems, between turnover pulses.

A comparison among turnover pulses is expected to show much heterogeneity—or "mosaic" differences—among pulses. One reason for this is that the environmental changes that trigger turnover are diverse. They vary in nature, intensity, timing (how long they endure, how much fluctuation occurs, and steepness of component changes and net trends), and also in geographic emphasis and extent (from very localized in the ocean or on land to widespread across the earth). Even the turnover response to a major global climatic trend will vary depending on factors such as topography and latitude. (For example, I have proposed that, during intervals when there are ice caps on one or both poles, the low latitudes are most likely to be centers of biotic vicariance and speciation; and that this may have been a contributing factor to the high species richness in the Tropics today [Vrba 1985, 1992].) Also, and perhaps most importantly, the different organismal groups differ sharply in how and when they respond. Organismal groups will vary in whether they respond to a given physical change by speciation, by extinction, by both speciation and extinction, or by no turnover at all. For example, biome generalists are expected to have lower speciation and extinction rates than biome specialists (Vrba 1987, 1992); and many other aspects of biology and life history will affect turnover responses. Lineages that do undergo turnover initiated by a given environmental change may do so at different times during that climatic and/or tectonic trend (see examples in Vrba 1995a,b). Thus, if a turnover pulse is detected in an extensive data set (including numerous lineages in a large area) it is desirable to continue analyses on subdivisions of those data to find out which lineages and areas underwent turnover and to understand the detailed temporal patterns.

The speciation pulse hypothesis has the following additional elements: All species are specific for particular habitats consisting of the physical and biotic resources needed for life and reproduction. Climatic changes result in removal of resources from parts of the species' former geographic distributions and therefore in vicariance. However, we know that vicariance (or dispersal over a barrier) on its own is not sufficient for speciation. Numerous species underwent repeated episodes of geographic shifting, vicariance, and reunion of their distributions (subsumed under the term "distribution drift" in Vrba 1988; also referred to in the literature as habitat tracking), in response to the astronomical climatic oscillations, while maintaining habitat fidelity and without speciation (review in Vrba 1993). For speciation to occur, physical change must be strong enough to produce population isolation, but not so severe as to result in extinction; and the isolated phase must be of sufficiently long duration for the relevant evolutionary changes to occur. Speciation is visualized as involving sustained isolation or near-isolation (without rapid reintegration on the Milankovitch timescale) of shrinking populations in which habitat resources are dwindling, competition increases, and consequently strong natural selection promotes new adaptation to these changes. The change that defines speciation in sexually reproducing lineages is adaptation of the fertilization system to the new environment (Paterson 1982, 1985). If this process does not proceed to reproductive isolation of a viable new reproductive unit, then speciation will not occur.

Thus, in the absence of physical change of appropriate kind and duration, species are buffered against speciation at several levels: The fertilization system involves coadapted signal-response reactions between males and females. It is habitat-adapted and under strong stabilizing selection as long as the habitat persists, and in spite of habitat migration and vicariance (Paterson 1985). Second, distribution drift during climatic oscillations acts as a buffer at the population and species levels because it can allow rapid escape from areas of disappearing habitat, and also repeated reintegration of gene pools over geologically short time intervals as habitats coalesce again. Bennett (1996), Lieberman et al. (1995), and Eldredge et al. (2005 [this volume]) proposed additional ways in which the structure and dynamics of species' geographic distributions promote stasis.

The interactions of one species at a time with the resource variables that constitute its habitat are central to this hypothesis. Of course, the theater of such interactions, such as between a lion and its prey, is the ecosystem; and ecosystems (and communities) differ in physical and biotic variables. Most species live in several different ecosystems in which they experience different interactions (e.g., in Manyara, Kenya, lions mostly prey on buffalo, whereas in the Kruger Park, South Africa, their main prey are impalas). Yet the critical part of the lion's habitat (here the resource class "meaty prey") is represented across the differing ecosystems. Because the focus of the speciation pulse hypothesis is on the species as a whole, and on its habitat, one might call it a "species-based" and "habitat-based" hypothesis, as distinguished from others (discussed below) that are "ecosystem-based" or "community-based" in their theoretical assumptions.

To explore how the predictions of this model differ from those of alternative hypotheses, let us ask, What rhythm of speciation events would we expect, under the different causal hypotheses, if we could see all the events in the real world across the entire area under study, for example all or a part of Africa, and if we plotted their frequencies against time?

1. The null hypothesis, the traditional one that has often been called neoDarwinian, in its most conservative form predicts a random walk in time. In this view turnover is always driven by natural selection but the particular causes of selection are seen as very diverse, the most important being organismal interactions—such as competi-

tion and predation—that can act alone, or in combination with physical change, to drive speciation (and extinction). Under this H_0, selection pressures that cause speciation differ from group to group, and from one local area to the next. Thus the pattern of origination frequencies for large areas, over long time, is predicted to have a constant probability of origination. Examples of such arguments are given by Hoffman (1989), McKee (1993), and Foley (1994).

2. One possible prediction of the turnover pulse hypothesis is that there are many, frequent, small pulses (such as in response to the astronomical climatic cycles, and/or to local tectonic changes). This would be the expected pattern if the data set is too restricted in time and space to encompass the effects of larger physical changes.
3. A second possible turnover pulse prediction is that origination is confined to rare and large pulses in response to large environmental changes. This prediction, which in my view is unlikely, would involve concentration of all turnover during certain intervals, perhaps from less than one to several million years apart. Such large pulses (and to a lesser extent smaller ones as well) may resemble jagged mountain crests, or dissected high plateaus, rather than simple, single peaks, because the timing of turnover responses to climatic or tectonic episodes will differ among organismal groups and local areas.
4. A third possible turnover pulse prediction is of many, frequent, small pulses as under (2) above, interspersed by the less frequent, larger ones described under (3).
5. Additional possible predictions arise from combinations of H_0, the random null model, and the turnover pulse hypothesis. Such predictions could take several forms, including a pattern of a random background of turnover frequency punctuated by rare pulses.

Difficulties of Testing.—Most such difficulties have to do with testing at inappropriate temporal, geographic, and taxonomic scales (Barnosky 2001). Biases and low time resolution can lead to two types of error: inferring pulses that are not there (i.e., erroneous rejection of H_0, Type I error) and failure to detect real pulses (Type II error), as exemplified by the analysis of Signor and Lipps (1982). Take for an example of Type II error an attempt to distinguish between H_0 and real speciation pulses that occurred at the Milankovitch timescale. Such a test using first appearance data with lower time resolution (e.g., the data for 0.5-Myr-long intervals in the African mammal case below) will fail. In general, a fossil record of exceptional time resolution is needed to distinguish between a random turnover pattern, as under H_0 in (1), and the numerous, frequent, small turnover pulses in prediction (2). The same applies to the distinction between a background pattern that is random as in (5) and one that consists of many small turnover pulses as in (4). The best hope, given most available data sets, lies in testing whether or not large turnover pulses as in (3), (4), and (5) can be detected.

Further kinds of Type II errors concern data of inappropriate taxonomic and geographic scales. For instance, the speciation pulse hypothesis cannot be modeled by the interactions in a given ecosystem or community and tested by examining the timing of only the *local* species appearances, many of which represent local immigration of species with longer durations outside the study area. The turnover pulse hypothesis is about what happens to entire species, true speciation and extinction; and to test it in a given area, global first and last appearances of entire species, which happen to be recorded in that area, are the appropriate data. Type II error can also result from data manipulations, such as from the removal of rare species from samples, which is undesirable if most species start and end small.

The main bias that leads to Type I error, seeing pulses that are not there, arises from unequal fossil preservation between time intervals, areas, and groups of organisms, the "gap bias" mentioned earlier. Gaps have the effect that, for instance, a count of first fossil appearances of species in an interval is erroneously inflated by records of species that in reality originated (but were not detected) previously. An early version of a test that corrects for unequal fossil preservation, thus allowing a rigorous test of the pulse hypothesis, was applied to the African larger mammals of the past 20 Myr divided into 1-Myr-long intervals (Vrba 2000). More recently a second, updated form, which additionally corrects for rarity of species in their intervals of origin and extinction (Vrba and DeGusta, 2004), was applied to the past 10 Myr of the same record divided into 0.5-Myr-long intervals (Vrba unpublished). As some results from this study are reported below, a brief review of how the model works is in order.

It consists of two large equations, for origination and extinction respectively, each of which applies to the entire data set (of 20 intervals in my study) and sums the component equations for individual intervals. In the origination equation for a single interval, t_i, $E(\phi_i)$ estimates the expected number of fossil first appearances in t_i as a function of the observed changes in fossil preservation through time (estimated by the numbers of all species recovered across Africa in successive intervals), and under H_0 of random variation about a constant probability of origination, α. The summed equation over all time intervals, which estimates $\Sigma E(\phi_i)$, is solved simultaneously with the constraint that $\Sigma E(\phi_i) = \Sigma O(\phi_i)$, the sum of observed fossil first appearances. The hypergeometric tail probability is calculated for each interval to test whether its $E(\phi_i)$ differs significantly from its $O(\phi_i)$. As intervals that reject H_0 are pointed out, so correspondingly adjusted values are substituted for α in those intervals; and the adjusted model is rerun. This is repeated until no further changes in the significance pattern occur. A similar model for extinction corrects for the bias of last fossil records of species that occur earlier than the later, undetected true extinctions. The two models for origination and extinction are run recursively, at each new round substituting values of origination or extinction rates in intervals that were in the previous round found to differ significantly from H_0, until the entire significance pattern remains stable.

Physical Change and Turnover in Late Neogene African Mammals

The Plio-Pleistocene Climatic Context.—The climatic changes over the past 5 Myr, the period in focus, have been documented in many records, notably in marine oxygen isotope records (see Shackleton et al. 1984 for the North Atlantic; Shackleton 1995 for the South Pacific; deMenocal and Bloemendal 1995 for the Gulf of Aden). Shackleton's (1995) data were used to identify the longer intervals (on the time scale of hundreds of thousands of years) within the past 5 Myr over which the largest net cooling or warming trends occurred (Vrba 2004). Such higher-order climatic translations can drive many species into the long-term isolation needed for divergence to speciation. Cooling trends occurred approximately ~2.9–2.5 Ma (the strongest), ~3.2 Ma, and just after 1 Ma, and brief warming trends about 3.2–3.0 Ma and after 0.5 Ma. Much of the subsequent discussion will be about the evolutionary responses to the cooling toward 2.5 Ma, which was remarkable for several reasons besides its high rate of change and long duration: First, it was severe *relative* to the previous more stable pattern. Second, there was a rare shift about ~2.8 Ma toward climatic dominance of the 41-Kyr astronomical cycle, from previous dominant influence at 23–19-Kyr variance (deMenocal and Bloemendal 1995, who found a second shift to increase in 100-Kyr variance after 0.9 Ma). Third, approximately 2.5 Ma marked the start of the Modern Ice Age with onset of the first extensive Arctic glaciation (Shackleton et al. 1984) and strong effects on African biomes: the Sahara spread about 2.8–2.7 Ma, judging from pollen cores off West Africa (Dupont and Leroy 1995); and open and desert landscapes expanded in the Horn of Africa according to increased dust influx in deep-sea records in the Gulf of Aden (deMenocal and Bloemendal, 1995). In addition to the effects of global climatic changes, there is also evidence of increased tectonism over this period. Partridge and Maud (1987) have argued for a major episode of uplift around 2.5 Ma in Africa.

Results for the Past 5 Myr of African Mammal Evolution.—Time resolution in this record is sufficiently good, with more than 70% of the site records dated by radiometric or paleomagnetic means, that any large speciation (or extinction) pulses spaced at least 1 Myr apart should be detectable. After correcting for the "gap bias" as outlined above, the following results emerged (largely in agreement with those in Vrba 1995b, and 2000, insofar as they can be compared): Over the past 5 Myr the strongest turnover pulse, involving both origination and extinction, occurred in the 3.0–2.5 Ma interval. Between 3.5 and 3.0 Ma there was an origination pulse without an extinction pulse, and 1.0–0.5 Myr ago an extinction pulse without an origination pulse. Significantly low origination and extinction was found in the 5.0–4.5 Ma and 2.5–2.0 Ma intervals, and low origination 1.5–1.0 Myr ago. Where one can compare this set of major turnover events with the strongest climatic trends, the coincidence in time intervals and in intensity is strikingly close: the strongest climatic event, cooling toward about 2.5 Ma, coincides with the strongest turnover pulse, and lesser cooling and turnover events are present in the intervals 3.5–3.0 Ma and 1.0–0.5 Ma. In addition, two periods of high sea level with low polar ice on a warmer Earth overlapped with the 5.0–4.5 Ma and 2.5–2.0 Ma intervals (Haq et al. 1987; Hodell and Warnke 1991), which emerged as significantly low in both origination and extinction rates.

Over the past 5 Myr, global cooling with regression and increased aridity and seasonality in Africa was a more important stimulus of mammalian turnover than was global warming. Of the cooling trends over the past 3 Myr, the one toward 2.5 Ma was the strongest, followed by one lesser trend starting about 1 Ma. Yet individual glacial maxima became colder after 2.5 Ma, especially after 1 Ma (Shackleton 1995). The fact that no further origination pulses occurred after 2.5 Ma suggests that most of the lineages present then were either species that evolved 3.0–2.5 Myr ago with a resistance to cooling, or long-lasting biome generalists that survived right through that cooling event. Stanley's (1985: p. 266) explanation of extinction rates in Atlantic and Mediterranean molluscs during and after the start of the Modern Ice Age is similar: "repeated episodes of cooling caused extinction until the only survivors were species that could tolerate cool intervals." This study indicates that turnover events during the start of the Modern Ice Age affected organismal groups beyond mammals, and in the ocean as well as on land.

The morphologies of species appearing in the African origination pulses (and on other continents [e.g., Archer et al. 1995; Azzaroli 1995; Barry 1995; Bernor and Lipscomb 1995; Webb et al. 1995]) are consistent with increased seasonal cooling and aridification. Examples for the cooling toward 2.5 Ma are micromammals, which turned over in the Shungura Formation, Ethiopia (Wesselman 1995): At 2.9 Ma woodland taxa predominated and even rain forest taxa were present (such as the bushbaby, *Galago demidovii*, which persists as a rain forest form today). These forms were displaced by new grassland-to-semidesert species by 2.4 Ma. The turnover pattern includes terminal extinctions; immigrants from Eurasia such as a hare, *Lepus*; and global first appearances of species, such as a new species of *Heterocephalus*, the genus of desert-adapted naked molerats, and a new species of the ground squirrel genus *Xerus* (Wesselman 1995). This time also marks the first African and global debuts in the record of several species of bipedal, steppe- and desert-adapted rodents, such as the genus *Jaculus* of desert gerboas (Wesselman 1995) and a species of *Pedetes*, the genus of living springhares in South Africa. The first appearances 2.9–2.4 Myr ago also include the earliest African record of *Equus* (Bernor and Lipscomb 1995), a wave of new bovid species (Vrba 1995b), as well as the genus *Paranthropus* (Walker et al. 1986) and the probable direct ancestor of *Homo* (Asfaw et al. 1999).

The limitations of the present African results on turnover pulses should be noted. All the larger mammals across all of Africa were assessed together, and the turnover frequencies were averaged over 0.5-Myr-long intervals. As mentioned before, organismal groups in general differ in whether or not, and how, they respond to any given physical change, and so do geographic areas. A study of the taxonomic subdivisions of the present data set, especially when assessed over shorter intervals, is expected to reveal substantial heterogeneity in turnover patterns among the different mammalian groups. It is possible that some intervals that currently show no significant peaks in speciation or extinction will in reality be found to contain turnover pulses in one or more taxa, pulses that were swamped in the taxonomically diverse samples that were analyzed.

A Disagreement on Turnover in Plio-Pleistocene Mammals, and a Proposed Resolution.—The details associated with biotic turnover responses to physical change in given cases remain subject to disagreement. A close examination of many such disagreements indicates that in most cases the studies are not comparable because differences in geographic, taxonomic, and temporal scale (see Barnosky 2001) separate the assumptions, methods, and predictions of the models.

To exemplify what I mean, consider a comparison of my African studies with those of Behrensmeyer et al. (1997). One aim of these authors was to test Vrba's (1995b) finding of a turnover pulse in African mammals between about 2.8 and 2.5 Ma by examining the past 4.5 Myr in the Turkana Basin (including three formations: the Shungura Formation in the northern part of the basin in Ethiopia, and the southwestern Nachukui and southeastern Koobi Fora Formations in Kenya). They concluded that there was "no major turnover event between 3.0 and 2.5 Myr" (p. 1591) and that this "weakens the case for rapid climatic forcing of continent-scale . . . faunal turnover" (p. 1593). I reexamined Turkana Basin evolution over 4.0–1.0 Ma, using my African mammal data base and the statistical "gap bias" model outlined above. (The species in this analysis were based on references in Appendix A in Vrba 2000, as well as comments from mammal specialists who checked the initial resulting list, as cited in that paper; see also the taxon list for all African larger mammals over the past 20 Myr in Vrba 2000: Table 1.) The results showed a single significant origination pulse in the 3.0–2.5 Ma interval, and no extinction pulses, in the Turkana Basin.

One reason for the differing results is that different questions were asked. I was studying the frequencies of *global* first and last appearances recorded in the Turkana Basin, in an effort to understand the timing of true speciation and extinction. In contrast, Behrensmeyer et al. (1997) were enquiring into the *local* first and last appearances in the area. For instance, their claim that there was no speciation pulse 3.0–2.5 Myr ago was based on a data set in which a substantial proportion of all first appearance data were those of species known from earlier strata outside this basin (roughly 30% of all data in one of their analyses). Thus, this was not a test of a hypothesis of speciation (or extinction) pulses. Second, in their analyses many taxa based on well-preserved, diagnostic fossils were omitted because they were rare (occurring in only one of the three areas) or because the systematists who had described them, while designating them as separate species from anything earlier, described them only as "sp. nov." without naming them. In one of their analyses this procedure eliminated more than one half of all global first appearances (about two-thirds of these omitted species were from the northern Shungura area and mostly 3.0–2.5 Ma in age). As argued earlier, the removal of rare species from a test for turnover pulses is undesirable given that species "start small" (Vrba and DeGusta 2004) and probably "end small" as well. Third, first appearances in members were assigned the dates at the beginnings (or midpoints in one analysis) of those members, discarding the more precise chronological and stratigraphic information available for many of the records. As some of the members are of fairly long duration, this led to distortion of the temporal pattern, including displacement of first appearance records, which the best available dating placed in the 3.0–2.5 Ma interval in my analysis, to earlier intervals. That is, a strong difference in taxonomic scale (parts of species versus entire species), and in geographic and time scaling, affected the results.

A new perspective emerged upon separate examination of the one northern and two southern areas of the Turkana Basin: a strong speciation (and extinction) pulse in the North 3.0–2.5 Myr ago, but none in the combined or separate southern areas. To provide a context for this, consider another aspect of geographic scale: if climatic forcing of continent-scale ecological change and turnover did occur over a given time in Africa, one would still expect refugia, especially in relatively small and ecologically uniform areas such as the Turkana Basin, which had riverine and rift-margin associated deltaic and lake environments throughout the time in question. (The term refugium is used in the broad sense of a biome refugium; e.g., a forest refugium surrounded by more open vegetation might have preserved the characteristic forest vegetation physiognomy, although its detailed taxonomic composition differed from that of the more widespread parent community.) Vrba (1988) had argued that "climatic change in the larger region is recorded in a refugium only close to its ecotonal limits, by the new appearances (or disappearances) of peripheral taxa ... that represent occasional intrusive elements from the alternative biome" (p. 410); and that the evidence "is precisely as predicted if the Omo-Turkana area represented a fluvial-deltaic refugium in the midst of wide-spread climatic change" (p. 413). Thus, finding turnover confined to the northern margin of the basin is consistent with the documented southward spread of the Sahara Desert in the latest Pliocene (Dupont and Leroy 1995), which affected the northern basin more strongly, eliciting significant turnover, whereas the southern deltaic-lacustrine areas behaved more nearly like a biome refugium.

Heterochrony Pulses: Shared Inherited Developmental Responses to Common Environmental Causes

Heterochrony occurs at two levels: in the form of a range of ecophenotypes in response to differing environmental conditions (Hall 2001; see also West-Eberhard 2003), and as novel, genetically based phenotypes. The details of the growth patterns and environmental correlations of these heterochronies tend to be closely comparable between related species, and between ecophenotypes and genetically based phenotypes within monophyletic groups (e.g., Gould 1977a; Wake and Larson 1987; Vrba 1998, 2004). The reason is that heterochronic responses at both levels are influenced by the nature of the ontogenetic trajectories for the characters in question, and aspects of these ontogenies tend to be shared by common inheritance across taxonomic groups (e.g., Gould 1984). A summary and extension of the above is given in the following two statements.

1. Similar environmental changes elicit similar heterochronies in parallel, potentially in numerous lineages across large phylogenetic groups. Such heterochrony often involves change in body size and may be accompanied by large-scale phenotypic reorganization (Alberch 1980; Arnold et al. 1989; Vrba 1998), such that the parallel heterochronies involve concerted evolution of suites of linked characters and "shuffling" among body proportions.
2. At times of widespread climatic change, diverse lineages may show parallel changes in size and in similar kinds of heterochrony associated in time and consistently with the climatic change—a "heterochrony pulse." "Pulse" here does not imply that the lineages responded in unison in a short time, but only that the events are significantly concentrated in time.

The ensuing discussion is about examples that involve climatic cooling, body size increase, and associated heterochrony in Plio-Pleistocene African mammals. The heterochrony pulse hypothesis has not yet been fully addressed for any of these examples. But the patterns to date do suggest that the hypothesis deserves close scrutiny and quantitative testing.

Cooling and Body Size Increase.—Many of the new species with first appearances during cooling 2.9–2.5 Myr ago were larger than their ancestral phenotypes (as cladistically inferred). Vrba (2004) tested H_0 that size changes across lineages are randomly distributed in time in the Alcelaphini (hartebeests, wildebeests) and Reduncini (waterbuck, reedbuck), which together comprise 63 recorded species over the past 5 Myr with a body weight range of approximately 20–250 kg. The result indicated significant peaks in size increase 3.0–2.5 Myr and 1.0–0.5 Myr ago, two of the three periods with strong cooling trends. This result is consistent with Bergmann's (1847) Rule, that larger bodies are associated with colder temperature. Although many exceptions have been noted, in general the predictions are upheld in living mammals (Ashton et al. 2000; Meiri and Dayan 2003) and in the mammalian fossil record (Kurtén 1959; Heintz and Garutt 1965; Davis 1981).

Heterochrony in response to climatic change can involve extensive rearrangement—or "shuffling"—among body proportions, with suites of characters changing in similar ways across related lineages. Take for example change in body size following Bergmann's Rule. Bodies can become enlarged by faster growth relative to the plesiomorphic (or directly ancestral) ontogeny, by prolongation of growth time, or by a combination of both; and the influential factors may include temperature change itself or one of the attendant environmental changes (such as seasonal changes in food and water availability [e.g., Guthrie 1984; Barnosky 1986]). All such changes in growth mode are expected to result in rearrangement of body proportions. I will illustrate this below by using the example of body enlargement by growth prolongation, a mode that I have found to be prevalent among the cases for which growth studies are available. For instance, many African tropical ungulates have shorter growth periods and achieve smaller size in warm lowlands, while their close relatives at higher altitudes and/or latitudes grow for longer and become larger. An example is the variation in the African buffalo (Sinclair 1977): *Syncerus caffer caffer* is much larger (up to 810 kg), grows for longer, and lives at higher latitudes and/or altitudes always near grassland. The smaller and plesiomorphic phenotype *S. c. nanus* (up to 320 kg) with a shorter growth period lives in warmer, more forested regions.

Body Size Increase and "Shuffling" among Body Proportions.—Consider what is expected under the simplest way in which growth prolongation could occur: namely, if all ancestral growth phases for a character become proportionally prolonged (or extended in time by a constant factor) while maintaining both the ancestral number of growth phases and the ancestral growth rates for respective phases (see "sequential hypermorphosis" in McKinney and McNamara 1991). Characters in the same organism have differing growth profiles, in terms of growth timing and rate in relation to age and body weight (e.g., Falkner and Tanner 1986). We can distinguish two major types of allometric growth and associated heterochrony under growth prolongation: in type A heterochrony, characters that grow with negative allometry (with respect to age and body size) will become reduced relative to body size in the prolonged descendant ontogeny (and paedomorphic in that the descendant adult resembles the ancestral juvenile). In type B heterochrony, characters with positively allometric growth may become relatively enlarged. As growth trajectories become prolonged, some characters become relatively reduced and others enlarged, with potentially extensive change among body proportions and substantial evolutionary novelty (Vrba 1998: Fig. 1). Vrba (1998, 2004) predicted that character evolution by Allen's Rule (Allen 1877)—that, in mammals, extremities such as limbs are reduced relative to body size in cooler climates—will often be found to fit type A heterochrony. The very persistence of Allen's Rule in modern biology supports the general hypothesis of similar changes in body proportions across lineages that share inherited developmental responses to common environmental causes. Type B heterochrony, particularly by prolongation of a positively allometric late growth phase, may be how the hypermorphosed antlers of the giant Irish Elk evolved (Gould 1974) and how exaggerated secondary sexual characters in enlarged bodies commonly evolve (Vrba 1998).

Type B heterochrony can also result from prolongation of positively allometric *early* growth, in which case the descendant structure is relatively enlarged and paedomorphic. Vrba (1998) tested, and found supported, the hypothesis that the evolution of the human brain occurred by this mode—in which only one growth parameter has to change, growth length—whereas chimpanzees remained plesiomorphic. In mammals in general, simple growth prolongation is predicted to result in encephalization, as all mammalian brains complete a large proportion of their total growth rapidly early in ontogeny (Count 1947; Holt et al. 1975). This raises the possibility that there were past "encephalization pulses," across many mammalian lineages, in response to cooling over particular intervals (Vrba 1998).

Another example of prolongation of rapid early growth that resulted in enlarged paedomorphic structure may be the hind feet of many of the bipedal rodents that appeared toward 2.5 Ma and at other times of cooling. If the growth of rodents, the juveniles of which in general have relatively large hind feet (Hafner and Hafner 1988), is prolonged, a descendant adult with enlarged hind feet is predicted. Evidence for at least some taxa is consistent with this. For instance, Hafner and Hafner (1988) found that the bipedal kangaroo rats, *Dipodomys*, which inhabit semiarid to arid regions in North America, have longer growth periods and are hypermorphosed in some characters—yet paedomorphosed in others such as the hind limbs—relative to the ancestral ontogeny. They pointed out that the bipedal forms may share suites of characters—including enlarged hind feet, heads, brains, eyes, and auditory bullae—and that this highly specialized body plan is today strongly associated with open, arid habitats and has appeared numerous times independently in rodents (in 24 genera in eight families). The fossil record of bipedal rodent taxa (Lavocat 1978; Savage and Russell 1983; Wesselman 1995) suggests that most may have appeared during times of global cooling and land aridification, during either the late Pliocene or late Miocene (e.g., *Jaculus*, which first appeared in Africa near the start of the Modern Ice Age, and *Gerbillus* and *Alactaga*, which originated in Asia). I do not know how many of the 24 rodent instances of parallel evolution of enlarged hind feet involved growth prolongation. But I suggest that at least some of these appearances of suites of integrated character complexes exemplify coordinated morphological changes, by growth prolongation, within and between lineages in response to a common climatic cause.

In sum, the case illustrates that evolution by growth prolongation, as it acts on characters with different nonlinear growth profiles in the same body plan, can result in a "shuffling" of body proportions. Substantial novelty in form and also in function can result, as in these rodents, which can jump to a height that is 4–25 times their body length, and in the case of the new behaviors that were mediated by human encephalization. It remains to be tested whether these novelties arose as parts of heterochrony pulses.

Discussion

One of my general questions is whether or not physical change has been the predominant pacemaker of speciation. As a context for this modern debate let us revisit the distinctions between the early views of Gulick (1872, also in a lecture, 1853) and Darwin (1859). (Gould was a foremost scholar of the history of evolutionary thought, who in his 1977b paper "Eternal metaphors of paleobiology" discussed the intellectual antecedents of modern environmentalists. It is surprising that he did not include Gulick among them in this or later publications. Carson [1987] recognized Gulick's seminal insight.) For Darwin the initiating causes are located at the level of organisms, competition, and natural selection: "each new species is produced . . . by having some advantage over those with which it comes into competition; and the consequent extinction of the less-favoured forms almost inevitably follows" (Darwin 1859: p. 320). In contrast, Gulick argued that the cause of speciation in the Hawaiian snails is not well explained by selection among competitors (Gulick 1872: p. 224): "The conditions under which they live are so . . . similar, that . . . . it does not follow that the "Survival of the Fittest" . . . . is the determining cause . . . ." Instead, Gulick thought that vicariance was seminal in initiating speciation. Whereas Darwin focused on among-organismal dynamics, Gulick stressed interpopulation dynamics and structure at the species level. A second difference concerns the role of physical change. Although Darwin acknowledged the influence of climate, he stressed its effects on competition rather than on population structure: "in so far as climate chiefly acts in reducing food, it brings on the most severe struggle between the individuals . . . which subsist on the same food" (Darwin 1859: p. 68). In Gulick's view the initiation of vicariance by physical change is crucial. This polarity persists among modern hypotheses of the initiation of speciation, which can roughly be divided into three categories. (My classification overlaps with but differs slightly from that of Barnosky [2001], who gave eight examples of hypotheses that share "the basic tenet that changes in the physical environment rather than biotic interactions . . . are the initiators of major changes in organisms and ecosystems" [p. 172], which he termed "Court Jester hypotheses." Five of his hypotheses are either in my second category [including the turnover pulse hypothesis] or related to it, and the others are the three models cited in the third category.)

1. The first group are "Darwin-descendants" in that the important causal dynamics are ecosystem based: natural selection based on organismal interactions. The various proposals of sympatric speciation that results from resource competition (Rosenzweig 1985; Dieckmann and Doebeli 1999) belong here, as does Van Valen's (1973) Red Queen hypothesis.
2. The second category are "Gulick-descendants" in that physical change and geographic population structures of species are accorded prominence. Examples are hypotheses of Wright (1932, 1967) and Mayr (1942, 1963), and the strong and detailed arguments under the "specific-mate recognition" model of Paterson (1978, 1982, 1985) and the "flush-crash-founder" model of Carson (1982). In 1980, inspired by Mayr (1963), Eldredge and Gould (1972), and Paterson (1978), as well as by the findings that Late Pliocene climatic change coincided with punctuated evolution in southern African hominids and other mammals (Vrba 1974, 1975, reviewed in Vrba 1995b), I proposed that speciation is a function of environmental change (Vrba 1980: p. 72): "the probability of allopatric speciation P(AS) . . . is proportional to the probability of small populations becoming isolated, P(SI), in new environments, P(NE); thus $p(AS) \propto p(SI \bullet .NE)$." The turnover pulse hypothesis went further than the others in this category in predicting significantly time-aggregated punctuated patterns across lineages (Vrba 1985, 1993, 1995a).
3. The third category represents hybrid views: Darwinian in a focus on organismal interactions in a community as a source of stasis, and Gulickian in proposing that physical change is needed to disrupt the coevolutionary stasis and result in turnover events. The hypothesis of "coevolutionary disequilibrium" (Graham and Lundelius 1984) belongs here. Stenseth and Maynard Smith's (1984) "stationary model" is also in this category. Their modeling showed that ecosystems are expected to approach one of two evolutionary modes: (1) evolution according the Red Queen hypothesis; or (2) according to the stationary model: "evolutionary stasis, with zero rate of evolution and no extinction or speciation; evolutionary change occurs only in response to changes in the physical environment" (Stenseth and Maynard Smith 1984: p. 870). The model of "coordinated stasis" (Brett and Baird 1995) belongs here as well.

The proper comparison with the turnover pulse hypothesis is with the last two models, which predict turnover frequencies across lineages through time. The quotation above from Stenseth and Maynard Smith (1984) at first glance suggests a close resemblance between the stationary and turnover pulse hypotheses. But these hypotheses differ in their assumptions and predictions (Vrba 1993). The stationary model is about what happens in one particular ecosystem or community at a time and, because few species are confined to a single ecosystem, much of the turnover in an ecosystem is only local and not equivalent to speciation and extinction. Stasis is a function of the composition of an ecosystem, and of biotic interactions of particular kinds that are likely to be ecosystem-specific and localized within any species. Thus, particular biotic contexts may dictate a steady-state outcome in some parts of a species and a stationary outcome in others. The predictions for phylogenetic patterns of speciations and extinctions are therefore not clear. Only if all ecosystems (over the species in question) follow the stationary mode will the pattern of true speciations and extinctions result that is also predicted by the turnover pulse model.

Some of these arguments also apply to the model of coordinated stasis (Brett and Baird 1995), which similarly invokes within-community interactions to explain stasis. These authors have claimed stasis of species, interrupted by pulses of true speciation and extinction, across all communities in which a set of species occurs. Thus, their predictions are closely comparable with those of the turnover pulse hypothesis, as acknowledged by Brett and Baird (1995: p. 287): "The same term [coordinated stasis] could be used for the blocks of stability in Vrba's (1985) "stability-pulse" hypothesis." Gould (2002) characterized the respective explanations of stasis by Vrba (1985) and Brett and Baird (1995) as "passive consequence or active ecological locking" (p. 921). He wrote, "In Vrba's view . . . rapid physical changes induce the turnover and also engender the subsequent stability as a propagating effect—because interspecies interactions play little role in regulating stability . . . " (p. 919). However, species interactions, at least among subparts of species in the same ecosystem, are important in the maintenance of stasis between pulses. They also play a role in the withdrawal of habitat, for instance as resources are removed by intensifying competition, or as destructive biotic agents—such as disease organisms or predators—increase. But the other species enter into this formulation not as *particular* species, but rather as *classes* of resources and destructive agents. The ecosystem variables that have traditionally featured in ecology are not of primary interest. In contrast, Brett and Baird (1995) did focus on such variables, for example on the total number and identities of species present, and on the overall web of interactions. Thus, although for the most part I agree with Gould's characterization, stasis between turnover pulses is not well described as a "passive consequence" of physical dynamics alone.

Testing of the turnover pulse and related models is still rudimentary in terms of the taxa, areas, time periods, and environmental changes studied. My present African results are consistent with big, irregularly timed turnover pulses in larger mammals in response to major physical changes during the Plio-Pleistocene. However, they do not inform on how individual African taxa and areas responded, on the timing of events on the Milankovitch timescale, and on the generality of the African findings. The hypothesis that physical change has influenced turnover events and concentrated them in time is supported by numerous studies of mammals (e.g., Janis 1993; Archer et al. 1995; Azzaroli 1995; Barry 1995; Bernor and Lipscomb 1995; Webb et al. 1995; Denys 1999; Turner 1999; Barnosky 2001) and some other kinds of organisms (e.g., the marine invertebrate records cited in Brett and Baird 1995). Some of these studies also agree with my finding that some turnover pulses consist mostly of speciation events, others of extinctions, and yet others contain both. But in given cases of generally agreed climatic forcing, there are no tests to date of which theoretical arguments (such as those under the stationary, coordinated stasis, or turnover pulse models) may apply. And in other cases, no agreement has been reached on the detailed patterns associated with biotic turnover through time, and on whether physical change played a significant initiating role at all. As argued above, and exemplified by the disagreement between Behrensmeyer et al.'s (1997) and my African studies, in most cases the studies are not comparable because differences in scale separate their assumptions, methods, and predictions. In sum, to date we have a poor understanding of both the details and the generality of turnover patterns in relation to environmental factors. The summed available evidence does not address the larger question of whether the turnover pulse hypothesis could apply to the biota in general and to diverse marine and terrestrial settings. That question remains open and in need of a more expanded research program.

My second hypothesis of how environmental change affects evolution is the heterochrony pulse hypothesis: the generative properties shared among lineages can result not only in coherence of morphological changes but also in a strongly nonrandom timing of evolutionary events, as diverse lineages respond by similar kinds of heterochrony to the same environmental changes. This has not yet been tested. Of prime interest are studies that involve complex heterochrony, with linked changes in suites of characters in a group of related taxa. Ideally one would like to have a model of how the heterochronic variation is produced during growth, and how it relates to the phylogenetic pattern and to environmental variables, and a good fossil and paleoenvironmental record to test how the heterochronic changes in lineages relate to each other in time and to the environmental changes.

Although no such analyses exist yet, there is already a mosaic of circumstantial evidence that strongly suggests that this hypothesis is worthy of future testing. For instance, as a background for the hypothesis, it has long been recognized that morphogenetic rules are inherited properties of lineages that determine limits on how ontogenies can respond to gene mutations in a given environment. Although the mutations themselves, and the repeated morphogenetic tendencies expressed as parallel phenotypic novelties in a given clade, are random in the sense that they are independent of what natural selection might prefer, they can nevertheless impart a coherent pattern to what can and does evolve in that clade (Alberch 1980). Arnold et al. (1989: p. 408) expressed this in discussing teratology: "The expression of genetic mutations at the phenotypic level is constrained.. . . The fact that the same morphology [e.g., the teratology of two heads known throughout vertebrates] appears recurrently in distantly related lineages is simply a reflection of the generative properties of a developmental process shared by all vertebrates." Gould (1984) provided an invertebrate example of the same principle. He concluded (p. 172) that the parallel occurrences of certain shell morphologies in the snail genus *Cerion* reflect "a structural rule common to the ontogeny of all *Cerion*. It represents a channel preset in the possibilities of *Cerion*'s growth and should not be interpreted simply as a recurrent phenomenon of local adaptation"

There is some support for the heterochrony hypothesis from studies of how body size across lineages has changed through past global temperature changes (e.g., Kurtén 1959; Vrba 2004). There is also a common paleontological pattern, in which diverse lineages show parallel changes in size and in at least aspects of similar kinds of heterochrony over the same time and coincident with global physical change. A famous example of this may be the earliest fossil appearance of the Metazoa during the so-called Cambrian explosion, if Fortey et al.'s (1997) interpretation is correct (reviewed in Vrba 2004). They argued that the Cambrian lineages were there already before 1 Ga, but have not yet been found in the Precambrian because they were minute, soft-bodied, and living between marine sand grains and mud grains. In Fortey et al.'s view, the Cambrian witnessed an explosion in body size increase. That is, larger bodies evolved in parallel in many lineages and areas over the same broad time interval. This example shows evidence of more comprehensive heterochrony than simply change in body size: the general increase in body size in turn resulted in the parallel onset of skeletization as an allometric result of size. Both promoted a higher probability of fossil preservation. Of course, it is possible that influences other than heterochrony were also implicated in this and other cases. For example, Zelditch et al. (2000) concluded from their study of piranhas that heterotopy, evolutionary change in the spatial patterning of development, played a part alongside heterochrony in this group.

The clear implication of such coeval changes in diverse lineages is that global physical change affected all of them similarly. There is widespread consensus that, preceding and across the Precambrian/Cambrian boundary and closely before the Cambrian explosion, there were indeed dramatic fluctuations in global climate and oceanic conditions indicated by excursions in marine isotopes of carbon, strontium, cerium, and other elements (reviewed in Vrba 2004). But the linkages in this case between particular physical changes and body size increase are not yet clear. It also remains to be seen whether groups of characters, besides body size and degree of skeletization, responded in a consistent way across lineages.

Conclusions

Physical change has had a direct influence on the origin of evolutionary novelty at (at least) two levels, and at each of these levels it can produce responses across lineages that are significantly concentrated in time, appearing as "pulses" in the fossil record. At the species level it can do so by its effects on geographic population structures, and on the dynamics among and within populations, with the result of speciation pulses. I would like to persuade colleagues to take seriously the possibility that the pacemaker—the initiator—of speciation is never biotic interaction—but is always physical change. Much of the "fine-tuning" comes from natural selection including that associated with among-organismal interactions. But the *initiation* is reserved for physical change. Recall that the turnover pulse hypothesis predicts a "mosaic" of results in that the kinds of physical changes are heterogeneous in nature and timing, and geographic areas and organismal groups vary strongly in whether and how they respond to a given physical change. That is, under the turnover pulse hypothesis, even if we could record all speciations and extinctions accurately (without taphonomic biases), the resulting turnover pattern would still be one of nonuniform and complexly "mosaic" pulses.

The testing of this hypothesis is still in its infancy. The same is true of my second hypothesis: that at the ontogenetic level, similar environmental changes can elicit broadly coeval, similar heterochronies in parallel in related lineages, namely heterochrony pulses.

The domain of heterochrony pulses is larger than that of speciation pulses because heterochrony can occur without speciation, and much of it is ecophenotypic and reversible (Hall et al. 2004). Yet the likelihood that all metazoan speciation involves heterochrony of one kind or another implies a close relationship, namely an overlapping pattern with the common causal denominator of physical change. Speciation and heterochrony are further linked in that the phylogenetic stability of developmental modes implied by heterochrony pulses is probably implicated at least partly in producing the stasis of species between turnover pulses. In the introduction, I cited two of the basic themes recurring throughout Stephen Jay Gould's work: punctuated macroevolution and heterochrony. Gould (1977b: p. 2) posed the question: "Does the external environment and its alterations set the course of change, or does change arise from some independent and internal dynamic within organisms themselves?" These alternatives are not mutually exclusive. The hypothesis of heterochrony pulses suggests a joint role in determining "the course of change" and ties together Gould's two themes.

In sum, there is some support for the notion that common rules give qualitative and temporal coherence to the evolutionary responses across lineages. These common rules arise from the regularities of physical change and from attributes of ontogenies and species that are widely shared by common inheritance. This approach implies closer linkages between the physical and biotic dynamics on Earth than has traditionally been acknowledged. It contrasts with a neoDarwinian view that selection of small-step random mutations is the vastly predominant evolutionary cause, implying that each evolutionary advance is to a larger extent an independent piece of history. Another implication is that the causal focus of evolutionary theory cannot be restricted to the level of organismal interactions and selection within populations. It needs to include the lower level of morphogenesis, and also the higher one of population structures and dynamics at the species level, as Gould has argued throughout his work up to the end (Gould 2002).

Acknowledgments

I thank Stephen Jay Gould for his inspiration, collaboration, and friendship over many years. I gratefully acknowledge very useful comments from D. Briggs, N. Eldredge, and B. Lieberman.

Literature Cited

Alberch, P. 1980. Ontogenesis and morphological diversification. American Zoologist 20:653–667.

Allen, J. A. 1877. The influence of physical conditions in the genesis of species. Radical Review 1:108–140.

Archer, M., S. J. Head, and H. Godthelp. 1995. Tertiary environmental and biotic change in Australia. Pp. 77–90 *in* Vrba et al. 1995.

Arnold, S. J., P. Alberch, V. Csanyi, R. C. Dawkins, S. Emerson, B. Fritzsche, T. J. Horder, J. Maynard-Smith, M. Starck, and E. S. Vrba, G. P. Wagner, and D. B. Wake. 1989. How do complex organisms evolve? Pp. 403–433 *in* D. B. Wake and G. Roth, eds. Complex organismal functions: integration and evolution in vertebrates. Wiley, Chichester, U.K.

Asfaw, P., T. White, O. Lovejoy, B. Latimer, S. Simpson, and G. Suwa. 1999. *Australopithecus garhi*: a new species of early hominid from Ethiopia. Science 284:629–635.

Ashton, K. G., M. C. Tracy, and A. de Queiroz. 2000. Is Bergmann's rule valid for mammals? American Naturalist 156: 390–415.

Azzaroli, A. 1995. The "Elephant-Equus" and the "End-Villafranchian" Events in Eurasia. Pp. 311–318 *in* Vrba et al. 1995.

Barnosky, A. D. 1986. Big game extinction caused by Late Pleistocene climatic change: Irish Elk (*Megaloceros giganteus*) in Ireland. Quaternary Research 25:128–135.

———. 2001. Distinguishing the effects of the Red Queen and Court Jester on Miocene mammal evolution in the northern Rocky Mountains. Journal of Vertebrate Paleontology 21:162–185.

Barry, J. C. 1995. Faunal turnover and diversity in the terrestrial Neogene of Pakistan. Pp. 115–134 *in* Vrba et al. 1995.

Behrensmeyer, A. K., N. E. Todd, R. Potts, and G. E. McBrinn. 1997. Late Pliocene faunal turnover in the Turkana Basin, Kenya and Ethiopia. Science 278:637–640.

Bennett, K. D. 1996. Ecology and evolution: the pace of life. Cambridge University Press, Cambridge.

Bergmann, C. 1847. Über die Verhältnisse der Wärmeökonomie der Thiere zu ihrer Grösse. Göttinger Studien 31:595–708.

Bernor, R. L., and D. Lipscomb. 1995. A consideration of Old World Hipparionine horse phylogeny and global abiotic processes. Pp. 164–177 *in* Vrba et al. 1995.

Brett, C. E., and G. C. Baird. 1995. Coordinated stasis and evolutionary ecology of Silurian to Middle Devonian faunas in the Appalachian Basin. Pp. 285–315 *in* D. H. Erwin and R. L. Anstey, eds. New approaches to speciation in the fossil record. Columbia University Press, New York.

Brooks, D., and D. McLennan. 1991. Phylogeny, ecology, and behavior. University of Chicago Press, Chicago.

Bush, G. L. 1975. Modes of animal speciation. Annual Review of Ecology and Systematics 6:339–364.

Carson, H. L. 1982. Speciation as a major reorganization of polygenic balances. Pp. 411–433 *in* C. Barigozzi, ed. Mechanisms of speciation. Liss, New York.

———. 1987. The process whereby species originate. BioScience 37:715–720.

Carson, H. L., D. E. Hardy, H. T. Spieth, and W. S. Stone. 1970. The evolutionary biology of the Hawaiian Drosophilidae. Pp. 437–543 *in* M. K. Hecht and W. C. Steere, eds. Essays in evolution and genetics in honor of Theodosius Dobzhansky. Appleton-Century-Crofts, New York.

Count, E. W. 1947. Brain and body weight in man. Annals of the New York Academy of Science 46:993–1122.

Darwin, C. 1859. On the origin of species by means of natural selection, or the preservation of favoured races in the struggle for life. John Murray, London.

Davis, S. J. 1981. The effects of temperature change and domestication on the body size of Late Pleistocene to Holocene mammals of Israel. Paleobiology 7:101–114.

DeMenocal, P., and J. Bloemendal. 1995. Plio-Pleistocene subtropical African climate variability and the paleoenvironment of hominid evolution: a combined data-model approach. Pp. 262–288 *in* Vrba et al. 1995.

Denys, C. 1999. Of mice and men: evolution in East and South Africa during Plio-Pleistocene times. Pp. 226–252 *in* T. G. Bromage and F. Schrenk, eds. African biogeography, climate change, and human evolution. Oxford University Press, New York.

Dieckmann, U., and M. Doebeli. 1999. On the origin of species by sympatric speciation. Journal of Nature 400:354–357.

Dupont, L. M., and S. A. Leroy. 1995. Steps towards drier climatic conditions in North-Western Africa during the Upper Pliocene. Pp. 289–298 *in* Vrba et al. 1995.

Eldredge, N., and S. J. Gould. 1972. Punctuated equilibria: an alternative to phyletic gradualism. Pp. 82–115 *in* T. J. M. Schopf, ed. Models in paleobiology. W. H. Freeman, San Francisco.

Eldredge, N., J. N. Thompson, P. M. Brakefield, S. Gavrilets, D. Jablonski, J. B. C. Jackson, R. E. Lenski, B. S. Lieberman, M. A. McPeek, and W. Miller III. 2005. The dynamics of evolutionary stasis. [This volume.]

Falkner, F., and J. M. Tanner, eds. 1986. Human growth: a comprehensive treatise, Vols. 1, 2. Plenum, New York.

Foley, R. A. 1994. Speciation, extinction and climatic change in hominid evolution. Journal of Human Evolution 26:275–289.

Fortey, R. A., D. E. G. Briggs, and M. A. Wills. 1997. The Cambrian evolutionary explosion recalibrated. BioEssays 19:429–434.

Futuyma, D. J., and G. C. Mayer. 1980. Non-allopatric speciation in animals. Journal of Systematic Zoology 29:254–271.

Gould, S. J. 1974. The evolutionary significance of "bizarre" structures: antler size and skull size in the "Irish Elk," *Megaloceros giganteus*. Journal of Evolution 28:191–220.

———. 1977a. Ontogeny and phylogeny. Harvard University Press, Cambridge.

———. 1977b. Eternal metaphors of Paleontology. Pp. 1–26 *in* A. Hallam, ed. Patterns of evolution. Elsevier, Amsterdam.

———. 1979. On the importance of heterochrony for evolutionary biology. Systematic Zoology 28:224–226.

———. 1984. Morphological channeling by structural constraint: convergence in styles of dwarfing and gigantism in *Cerion*, with a description of two new fossil species and a report on discovery of the largest *Cerion*. Paleobiology 10:172–194.

———. 1988. The uses of heterochrony. Pp. 1–13 *in* M. L. McKinney, ed. Heterochrony in evolution. Plenum, New York.

———. 2002. The structure of evolutionary theory. Harvard University Press, Cambridge.

Graham, R. W., and E. L. Lundelius. 1984. Coevolutionary disequilibrium and Pleistocene extinctions. Pp. 223–249 *in* P. S. Martin and R. G. Klein, eds. Quaternary extinctions: a prehistoric revolution. University of Arizona Press, Tucson.

Gulick, J. T. 1872. On the variation of species as related to their geographical distribution, illustrated by the Achatinellidae. Journal of Nature 6:222–224.

Guthrie, R. D. 1984. Alaskan megabucks, megabulls, and megarams: the issue of Pleistocene gigantism. Bulletin of the Carnegie Museum of Natural History 8:482–510.

Hafner, J. C., and M. S. Hafner. 1988. Heterochrony in rodents. Pp. 217–235 *in* M. L. McKinney, ed. Heterochrony in evolution: a multidisciplinary approach. Plenum, New York.

Hall, B. K. 2001. Organic selection: Proximate environmental effects on the evolution of morphology and behaviour. Journal of Biology and Philosophy 16:215–237.

Hall, B. K., R. D. Pearson, and G. B. Muller, eds. 2004. Perspectives from the fossil record: environment, development, and evolution. MIT Press, Cambridge.

Haq, B. U., J. Hardenbol, and P. R. Vail. 1987. Chronology of fluctuating sea levels since the Triassic. Science 235:1156–1167.

Heintz, A., and V. E. Garutt. 1965. Determination of the absolute age of the fossil remains of mammoth and wooly rhinoceros from the permafrost in Siberia by the help of radiocarbon (C14). Norsk Geologisk Tidsskrift 45:73–79.

Hodell, D. A., and D. A. Warnke. 1991. Climatic evolution of the

Southern-Ocean during the Pliocene Epoch from 4.8 million to 2.6 million years ago. Quaternary Science Reviews 10(2–3): 205–214.
Hoffman, A. 1989. Arguments on evolution. Oxford University Press, Oxford.
Holt, A., D. Cheek, E. Mellits, and D. Hill. 1975. Brain size and the relation of the primate to the nonprimate. Pp. 23–44 *in* D. Cheek, ed. Foetal and postnatal cellular growth: hormones and nutrition. Wiley, New York.
Janis, C. M. 1993. Tertiary mammal evolution in the context of changing climates, vegetation, and tectonic events. Annual Review of Ecology and Systematics 24:467–500.
Kurtén, B. 1959. On the bears of the Holsteinian Interglacial. Stockholm Contributions in Geology 2:73–102.
Lavocat, R. 1978. Rodentia and Lagomorpha. Pp. 69–89 *in* V. J. Maglio and H. B. S. Cooke, eds. Evolution of African mammals. Harvard University Press, Cambridge.
Lieberman, B. S., C. E. Brett, and N. Eldredge. 1995. A study of stasis and change in two species lineages from the Middle Devonian of New York State. Paleobiology 21:15–27.
Mayr, E. 1942. Systematics and the origin of species. Columbia University Press, New York.
———. 1963. Animal species and evolution. Harvard University Press, Cambridge.
McKee, J. K. 1993. Formation and geomorphology of caves in calcareous tufas and implications for the study of the Taung fossil deposits. Transactions of the Royal Society of South Africa 48:307–322.
McKinney, M. L., and K. J. McNamara. 1991. Heterochrony: the evolution of ontogeny. Plenum, New York.
McKinnon, J. S., and H. D. Rundle. 2002. Speciation in nature: the threespine stickleback model systems. Trends in Ecology and Evolution 17:480–488.
Meiri, S., and T. Dayan. 2003. On the validity of Bergmann's rule. Journal of Biogeography 30:331–351.
Partridge, T. C., and R. R. Maud. 1987. South African Journal of Geology 90:179–208.
Paterson, H. E. H. 1978. More evidence against speciation by reinforcement. South African Journal of Science 74:369–371.
———. 1982. Perspective on speciation by reinforcement. South African Journal of Science 78:53–57.
———. 1985. The recognition concept of species. Pp. 21–23 *in* E. S. Vrba, ed. Species and Speciation. Transvaal Museum Monographs No. 4. Pretoria.
Rosenzweig, M. L. 1978. Competitive speciation. Biological Journal of Linnean Society 10:275–289.
Savage, D. E., and D. E. Russell. 1983. Mammalian paleofaunas of the world. Addison-Wesley, London.
Schliewen, U. K., D. Tautz, and S. Paabo. 1994. Sympatric speciation suggested by monophyly of crater lake cichlids. Nature 368:629–632.
Shackleton, N. J. 1995. New data on the evolution of Pliocene climatic variability. Pp. 242–248 *in* Vrba 1995.
Shackleton, N. J., J. Backman, H. Zimmerman, D. V. Kent, M. A. Hall, D. G. Roberts, D. Schnitker, J. G. Baldauf, A. Despairies, R. Homrighausen, P. Huddlestun, J. B. Keene, A. J. Kaltenback, K. Krumsiek, A. C. Morton, J. W. Murray, and J. Westberg-smith. 1984. Oxygen isotope calibration of the onset of ice-rafting and history of glaciation in the North Atlantic region. Nature 307:620–623.
Signor, P. W., and J. H. Lipps. 1982. Sampling bias, gradual extinction patterns and catastrophes in the fossil record. Geological Society of America Special Paper 190:291–296.
Sinclair, A. R. E. 1977. The African buffalo. University of Chicago Press, Chicago.
Stanley, S. M. 1985. Climatic cooling and Plio-Pleistocene mass extinction of molluscs around margins of the Atlantic. South African Journal of Science 81:266.
Stenseth, N. C., and J. Maynard Smith. 1984. Coevolution in ecosystems: Red Queen evolution or stasis? Journal of Evolution 38:870–880.
Templeton, A. R. 1981. Mechanisms of speciation—a population genetic approach. Annual Review of Ecology and Systematics 12:23–48.
Turner, A. 1999. Evolution in the African Plio-Pleistocene mammalian fauna: correlation and causation. Pp. 76–87 *in* T. G. Bromage and F. Schrenk, eds. African biogeography, climate change, and early human evolution. Oxford University Press, Oxford.
Van Valen, L. 1973. A new evolutionary law. Journal of Evolutionary Theory 1:1–30.
Via, S. 2001. Sympatric speciation in animals: the ugly duckling grows up. Trends in Ecology and Evolution 16:381–390.
Vrba, E. S. 1974. CHRON Chronological and ecological implications of the fossil Bovidae at the Sterkfontein Australopithecite Site. Nature 256:16–23.
———. 1975. Some evidence of the chronology and palaeoecology of Sterkfontein, Swartkrans and Kromdraai from the fossil Bovidae. Nature 254:301–304.
———. 1980. Evolution, species and fossils: how does life evolve? South African Journal of Science. 76:61–84.
———. 1985. Environment and evolution: alternative causes of the temporal distribution of evolutionary events. South African Journal of Science 81:229–236.
———. 1987. Ecology in relation to speciation rates: some case histories of Miocene-Recent mammal clades. Evolutionary Ecology 1:283–300.
———. 1988. Late Pliocene climatic events and hominid evolution. Pp. 405–426 *in* F. E. Grine, ed. The evolutionary history of the robust Australopithecines. Aldine, New York.
———. 1992. Mammals as a key to evolutionary theory. Journal of Mammalogy 73:1–28.
———. 1993. Turnover-pulses, the Red Queen, and related topics. American Science 293-A:418–452.
———. 1995a. On the connections between paleoclimate and evolution. Pp. 24–45 *in* Vrba et al. 1995.
———. 1995b. The fossil record of African antelopes (Mammalia, Bovidae) in relation to human evolution and paleoclimate. Pp. 385–424 *in* Vrba et al. 1995.
———. 1998. Multiphasic growth models and the evolution of prolonged growth exemplified by human brain evolution. Journal of Theoretical Biology 190:227–239.
———. 2000. Major features of Neogene mammalian evolution in Africa. Pp. 277–304 *in* T. C. Partridge and R. Maud, eds. Cenozoic geology of southern Africa. Oxford University Press, Oxford.
———. 2004. Ecology, evolution, and development: perspectives from the fossil record. Pp. 85–105 *in* B. K. Hall, R. D. Pearson, and G. B. Muller, eds. Environment, development, and evolution. MIT Press, Cambridge.
Vrba, E. S., and D. DeGusta. 2004. Do species populations really start small? New perspectives from the Late Neogene fossil record of African mammals. Journal of Philosophical Transactions of the Royal Society of London B 359:285–293.
Vrba, E. S., G. H. Denton, T. C. Partridge, and L. H. Burckle, eds. 1995. Paleoclimate and evolution with emphasis on human origins. Yale University Press, New Haven, Conn.
Walker, A., R. E. Leakey, J. N. Harris, and F. H. Brown. 1986. 2.5-Myr Australopithecus boisei from west of Lake Turkana, Kenya. Journal of Nature 322:517–522.
Wake, D. B., and A. Larson. 1987. Multidimensional analysis of an evolutionary lineage. Science 238:42–48.
Webb, S. D., R. C. Hulbert, and W. D. Lambert. 1995. Climatic implications of large-herbivore distributions in the Miocene of North America. Pp. 91–114 *in* Vrba et al. 1995.

Wesselman, H. B. 1995. Of mice and almost-men. Pp. 356–368 *in* Vrba et al. 1995.

West-Eberhard, M. J. 2003. Developmental plasticity and evolution. Oxford University Press, New York.

Wright, S. 1932. The roles of mutation, inbreeding, crossbreeding, and a selection in evolution. Proceedings of the VIth International Congress on Genetics 1:356–66.

———. 1967. Comments on the preliminary working papers of Eden and Waddington. Pp. 117–120 *in* P. S. Moorehead and M. M. Kaplan, eds. Mathematical challenges to the Neo-Darwinian theory of evolution. Wistar Institute Symposium No. 5. Philadelphia.

Zelditch, M. L., H. D. Sheets, and W. L. Fink. 2000. Spatiotemporal reorganization of growth rates in the evolution of ontogeny. Evolution 54:1363–1371.

Paleobiology, 31(2), 2005, pp. 175–191

"Imperfections and oddities" in the origin of the nucleus

Lynn Margulis, Michael F. Dolan, and Jessica H. Whiteside

"Dual terminologies should be reserved for the exclusive use of those who prefer confusion to clarity."
L. R. Cleveland, 1963

Abstract.—We outline a plausible evolutionary sequence that led from prokaryotes to the origin of the first nucleated cell. The nucleus is postulated to evolve after the archaebacterium and eubacterium merged to form the symbiotic ancestor of amitochondriate protists. Descendants of these amitochondriate cells (archaeprotists) today thrive in organic-rich anoxic habitats where they are amenable to study. Eukaryosis, the origin of nucleated cells, occurred by the middle Proterozoic Eon prior to the deposition in sediments of well-preserved microfossils such as *Vandalosphaeridium* and the spiny spheres in the Doushantou cherts of China.

Lynn Margulis and Michael Dolan. Department of Geosciences, University of Massachusetts, Amherst, Massachusetts 01003. E-mail: mdolan@geo.umass.edu
Jessica H. Whiteside. Department of Earth and Environmental Sciences. Lamont-Doherty Earth Observatory of Columbia University. Palisades, New York 10964. E-mail: jhw@ldeo.columbia.edu

Accepted: 26 June 2004

Our evolutionary scenario for the origin of the nucleus in archaeprotist ancestors is based on descriptions by L. R. Cleveland (1892–1971) and Harold Kirby Jr. (1900–1952). Intrinsic to our model is Cleveland's detailed work on cell motility movements of undulipodia, the mitotic spindle and "attraction spheres" (centrosome-like) organelles in mitosis, and Kirby's concept of "mastigont multiplicity."

We agree with Stephen Jay Gould's "panda principle." The "karyomastigont" is an example of a Darwinian "imperfection and oddity" from which a "path of history" may be reconstructed. The significance of the karyomastigont, an organellar system described by Janicki in 1915, in which a nucleus is connected to motility organelles, has not heretofore been evolutionarily explained. Minimally the karyomastigont consists of a nucleus, a nuclear connector (rhizoplast), and an undulipodium (the "9(2)+0" kinetosome/centriole and its shaft, the "9(2)+2"microtubular axoneme). The cell structure and behavior of amitochondriate protists including their karyomastigonts as well as the nature of spirochete motility symbioses are best understood as "imperfections and oddities" in the context of their evolution. The nucleus, in our view, is a product of genetic-level integration of prokaryotic symbionts (archaebacterium+eubacterium) that began as a component of the karyomastigont from which it was released.

No substitute exists for direct study of the microbial communities that serve as modern analogs: bacteria and protoctists in anoxic environments where the processes upon which we base our model still occur. We arrange the observations into a plausible evolutionary sequence. Because the nucleated cell is a coevolved microbial community, the behavior, morphology, metabolism, and ecology of relevant living microbes may provide clues to the ancient origin of the nucleus.

The Model

Our concept of the origin of the nucleus that characterizes modern taxa is summarized in Figure 1A and B. Each step can be seen in live organisms in the video *Eukaryosis: Origin of Eukaryotic Cells* (Margulis and Dolan 2004). We zoom into the hypothesized merger of the archae- and eubacterium to form the archaeprotist with its karyomastigont in Figure 1B. A modern karyomastigont (Fig. 2) is diagrammed for clarity and for comparison with extant microbial associations (Fig. 3). The motile nucleated cell evolved in an anoxic habitat from a chimera that itself originated from archaebacterial-eubacterial fusion. Mitosis and other forms of cell motility evolved in this chi-

 0094-8373/05/3102-0013/$1.00

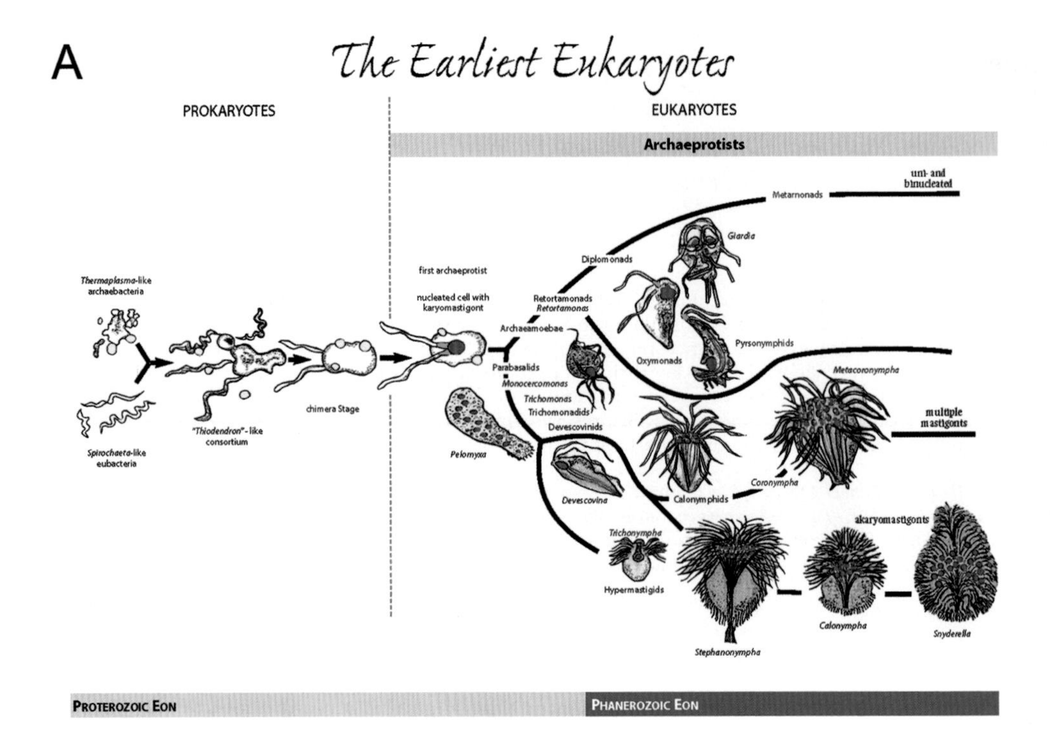
A
The Earliest Eukaryotes
PROKARYOTES
EUKARYOTES
Archaeprotists
uni- and binucleated
Metamonads
Giardia
Diplomonads
first archaeprotist
nucleated cell with karyomastigont
Thermoplasma-like archaebacteria
Retortamonads
Retortamonas
Archaeamoebae
Pyrsonymphids
Oxymonads
Parabasalids
Monocercomonas
Trichomonas
Trichomonadids
Devescovinids
Metacoronympha
multiple mastigonts
chimera Stage
"Thiodendron"-like consortium
Spirochaeta-like eubacteria
Pelomyxa
Devescovina
Calonymphids
Coronympha
akaryomastigonts
Trichonympha
Hypermastigids
Calonympha
Snyderella
Stephanonympha
Proterozoic Eon
Phanerozoic Eon

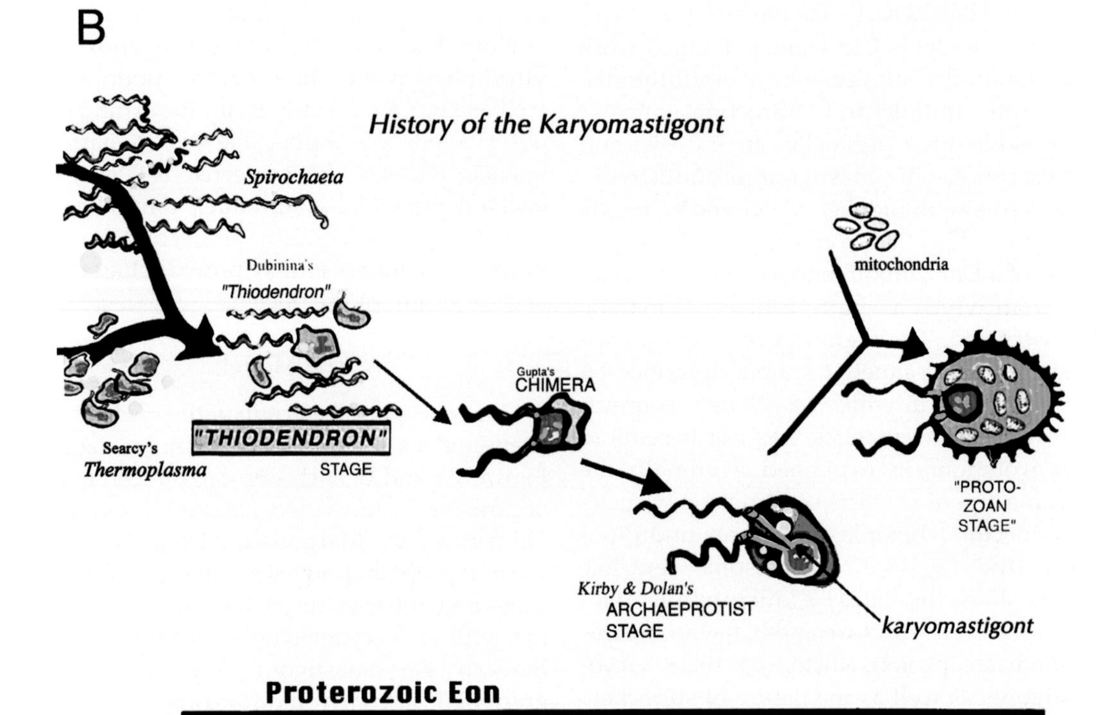
B
History of the Karyomastigont
Spirochaeta
Dubinina's
"Thiodendron"
mitochondria
Gupta's
CHIMERA
Searcy's
Thermoplasma
"THIODENDRON"
STAGE
"PROTO-
ZOAN
STAGE"
Kirby & Dolan's
ARCHAEPROTIST
STAGE
karyomastigont
Proterozoic Eon

mera. Speciation proper began after eukaryosis when the nucleated, mitotic cell left many descendants. The nucleus, mitotic cell divisions, and meiotic sexuality all evolved prior to the symbiotic acquisition of both mitochondria (from α proteobacteria) and chloroplasts (from cyanobacteria).

The nucleus, a membrane-bounded organelle, usually about 6 μm in diameter, is the defining characteristic of all eukaryotic organisms. Nuclei divide. The process, called karyokinesis, is generally followed by cytoplasmic division, called cytokinesis; i.e., the nucleated cell reproduces by mitosis. Mitotic division is entirely absent in all prokaryotes whether eu- or archaebacteria. Although much variation exists in detail, especially in amitochondriate and other protoctists, the mitotic process always requires segregation of chromatin, protein-studded DNA, by microtubules. Chromatin consists of DNA in fibrils that "condense" to form visible, stainable structures recognizable as chromosomes. Microtubules, 24-nm-diameter proteinaceous tubes of variable length, are composed of heterodimeric tubulin protein often accompanied by microtubule-associated proteins (MAPs). MAPs include an ATPase called dynein, another ATPase called kinesin, smaller proteins including γ-tubulin, and a protein of molecular weight of 220 kD called pericentrin (Table 1). Microtubules and MAPs constitute the main structural elements not only of the mitotic spindle but of the interphase cytoskeleton as well. Hence, whether minuscule and invisible in the live cell, as in *Saccharomyces cerevicieae* (yeast), or a dominant and persistent feature in somatic and egg cells and as photographed by Khodjakov in amphibian tissue (Fig. 4), the cytoskeleton is the active organelle system of the cell that segregates chromatin in mitosis in eukaryotes. We conclude that any evolutionary explanation that posits the origin of the nucleus in the ancestor of extant protoctists, animals, fungi, and plants must be simultaneously concerned with the origin of mitosis

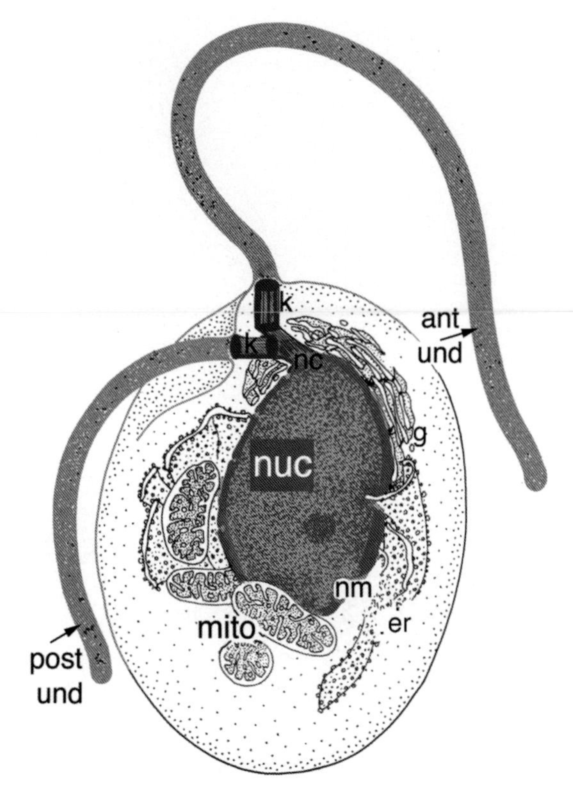

FIGURE 2. Karyomastigont drawing in a generalized protist. (Drawing by Kathryn Delisle.)

with its component cytoskeletal proteins (Dolan 2005). The terms "eukaryosis" or "eukaryogenesis," introduced by John Corliss (1987, 1998), refers to the evolutionary origin of the nucleated cell. Any eukaryosis hypothesis must explain the origin of the double membrane-bounded nucleated cell that divides by the cytoskeletal-mediated mitotic process. Most bacteria reproduce by direct division ("binary fission"). No vestige of chromatin, the mitotic spindle, tubulin-microtubules, or their MAPs has been seen in the cell division process in bacteria (whether eubacteria or archaebacteria). By contrast, fully developed, mitosis usually including its well-known stages (prophase, metaphase, anaphase, telophase, and interphase), characterizes nuclear division in cells of aerobic (mitochondriate) eukaryotes.

←

FIGURE 1. A, Symbiogenetic origin of the earliest eukaryotes: from archaebacterial eubacterial fusion to archaeprotist. B, Postulated history of the karyomastigont.

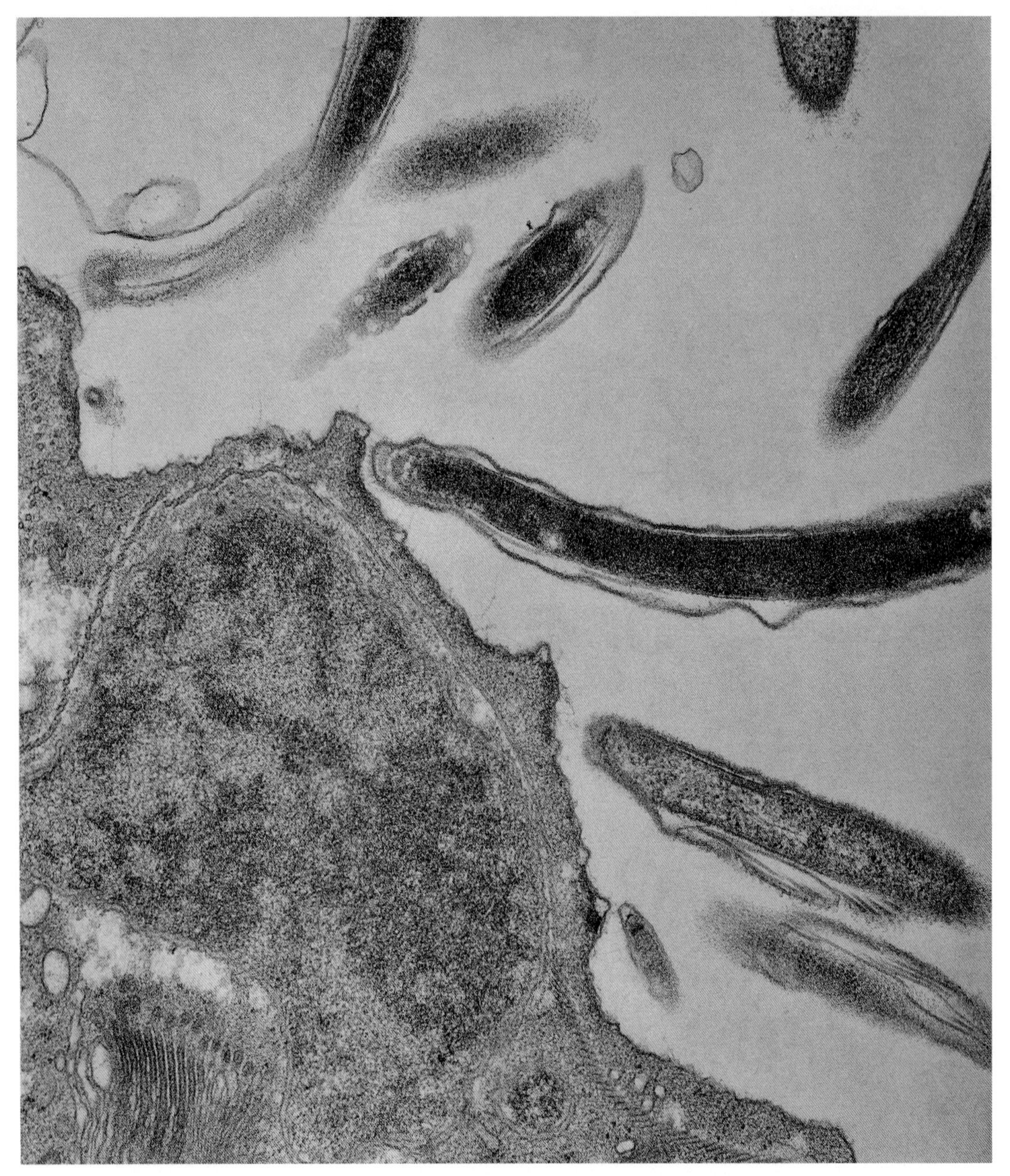

FIGURE 3. Anterior portion of a cell, an unidentified parabasalid, with adhering spirochetes, nucleus, and Golgi apparatus, suggestive of how the early karyomastigont might have evolved (from the dry wood-eating termite *Incisitermes minor*). This is probably a devescovinid. Electron micrograph by David Chase. Scale bar, 1 μm.

In our quest to discern where, when, and in which ancestral organisms the nucleus and its division by the mitotic process could have evolved we have heeded the admonition to seek inherited variations, in Darwin's (1859) words, the "imperfections and oddities."

Darwin answers that we must look for imperfections and oddities, because any perfection in organic design or ecology obliterates the path of history and might have been created as we find it. This principle of imperfections became Darwin's

TABLE 1. Eukaryotic motility proteins: search for spirochete homology.

	Molecular weight, nucleotide	Significance
Protein		
1. astrin	134 kDa	specific association with spindles; may play a role in spindle structure
2. Bub (budding uninhibited by benzimidazole)	~140–150 kDa	activates GTP-ase, mitotic checkpoint component, a kinetochore tension-sensitive protein
3. cenexin	96 kDa	acquired by immature centriole at the transition to mitosis, inner centriole wall
4. CENP-A	17 kDa	centromere protein A, functions in mitotic kinetochore assembly
5. CENP-E	312 kDa, GTP	centromere protein E, assists in binding microtubules to kinetochore, maintains spindle pole structure, kinesin-like
6. centrin	20 kDa	anti-spasmin antibodies recognize it (this cyclin-dependent kinase localizes to centrosome and other microtubular organizing centers
7. dynein (ATPase)	>1000 kDa	motor protein, moves particles toward minus end of microtubular "arms" on axonemal microtubules
8. dynactin	varies	binds membrane, binds NuMa, binds kinetochore
9. kinesin (ATPase) + or − end directed	~120 kDa (varies), ATP, ADP	+ or − end directed movement, ubiquitous motor protein carries materials on microtubules
10. kinectin	160 kDa, GTP	binds membrane, moves vesicles, GTP-activated
11. Mad (mitotic arrest deficiency)	25 kDa ?	mitotic checkpoint component, a kinetochore tension-sensitive protein, tethers microtubules to poles
12. NuMA	240 kDa, GTP	spindle attachment
13. pericentrin	220 kDa, none	lattice in centriole rings of gamma tubulin, it is the major pericentriole-material component, centrosome and mitotic spindle formation and function (main component purified from scleroderma antiserum)
Tubulin proteins		
14. alpha (α) tubulin	50 kDa Ca++ and GTP sensitive	forms walls of microtubules
15. beta (β) tubulin	50 kDa Ca++ and GTP sensitive	forms walls of microtubules
16. gamma (γ) tubulin	50 kDa, none	defines microtubules polarity, acts as nucleating agent for centriolar replication
17. delta (δ) tubulin	51 kDa, ?	forms triplet microtubules of kinetosome-centrioles

Sources: Belgareh et al. 2001; Bermudes et al. 1994; Cole et al. 1998; d'Ambrosio et al. 1999; Daniels and Breyer 1967; Doxsey et al. 1994; Dutcher 2001; Dyson 1985; Echeverri et al. 1996; Erickson 1997; Gaglio et al. 1997; Hall and Luck 1995; Hardwick and Murray 1995; Haren and Merdes 2002; Helenius and Aebi 2001; Jenkins et al. 2002; Khodjakov et al. 2000; Kumar et al. 1995; Lange and Gull 1995; Lindsay et al. 2001; Mack and Compton 2001; Maney et al. 1998; Margulis 2000; Mignot 1996; Moritz et al. 1995; Schopf 1999; Suh et al. 2002; Toyoshima et al. 1992; von Dohlenet et al. 2001; Viscogliosi et al. 1999; Wier et al. 2000; Young et al. 2000; Zimmerman and Doxsey 2000. See Margulis et al. 2000 for summary.

most common guide . . . I like to call it the panda principle . . . (Gould 1986: p. 63)

The membrane-bounded nucleus itself is only one stage, interphase, of the mitotic cycle of eukaryotes. We argue that this conspicuous central organelle evolved simultaneously with the other functional stages, i.e., with the origin of mitosis in single-celled members of the protoctist kingdom, anaerobic protists whose extant descendants still lack mitochondria. Because many of these "oddities and imperfections" in the mitotic division processes have been retained, the details of the evolutionary history of the nucleus can be reconstructed by the study of appropriate protists. Attempts to decipher the evolutionary history of the nucleus in the absence of relevant protistological knowledge are doomed to failure.

Our model for the origin of the nucleus with a list of its salient features is outlined below and documented by reference to an arcane literature that extends from the late nineteenth

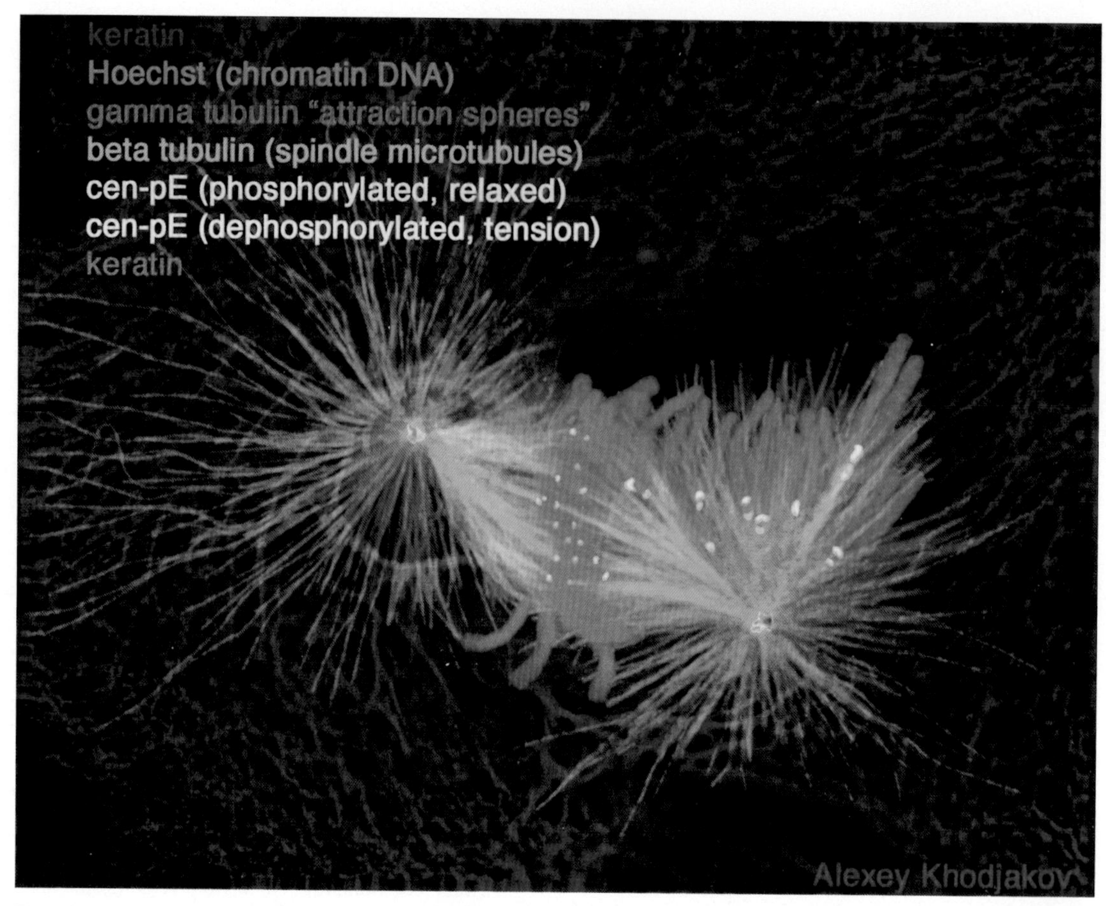

FIGURE 4. Combined fluorescent images of a newt lung cell in mitosis. The light micrographs by Alexey Khodjakov show antigens tagged with the color-coded antibodies that are listed in the upper left. Cen-pE refers to centromeric proteins.

century until the present. Recent laboratory studies are based on the work of illustrious predecessors: imaginative, meticulous, and wide-ranging investigations of live archaeprotists primarily by two outstanding investigators, L. R. Cleveland (1892–1971) and Harold Kirby (1900–1952). Without these industrious scholars and their colleagues we could not have proceeded. We hold with Dobzhansky (1972) and Gould (see Bermudes et al. 1987), that "nothing in biology makes sense except in the light of evolution." Thus any evolutionary scenario must account in detail for the current morphology and behavior of relevant extant organisms. Our view depends heavily on detailed comparative protoctist cell biology and on new findings in microbiology and electron microscopy. Proof requires protein-comparative amino acid sequence studies, and other techniques of chemistry and molecular biology. The scheme is outlined as follows.

Symbiogenetic Ancestry

1. All nucleated cells are heterogenomic; they all evolved from a common ancestor, itself a symbiotic merger of two types of prokaryotes: an archaebacterial *Thermoplasma*-like sulfidogen (Searcy and Lee 1998) and a motile *Spirochaeta*-like eubacterium (Margulis 1996; Margulis et al. 2000).

All Stages Extant

2. No missing links need be postulated: relicts of all stages in eukaryosis exist in living organisms and clues to these events, especially to their timing, can be found in the fossil record (Margulis 1993). The tendency of modern organisms to form associations like those

involved in the origin of eukaryotes can be directly observed. Spirochete attachments with proximity to host cell DNA, we hypothesize, provide analogues of the earliest motility symbioses (Fig. 4).

Sulfur Syntrophy: The Consortium

3. The archaebacterial-eubacterial chimera began in an anoxic, sulfur-rich environment as a syntrophic consortium. At elevated temperatures the consortium stabilized; it was between a sulfide-producing archaebacterium comparable to today's *Thermoplasma* and a highly motile, gram-negative eubacterium, a spirochete that oxidized sulfide to elemental sulfur. Like the modern marine consortium *Thiodendron* (Dubinina et al. 1993a,b), the spirochetal association degraded polymeric carbohydrates (e.g., cellulose) and sugars (e.g., cellobiose). The archaebacterium metabolized other carbon sources such as amino and three-carbon acids. The terminal electron acceptor for the sulfidogenic archaebacterium was elemental sulfur, whereas the eubacterium performed substrate-level phosphorylation by use of organic electron acceptors and elemental sulfur. As a carbohydrate fermenter in the presence of solid sulfur, the archeabacterium produced hydrogen sulfide, and in its absence it gave off hydrogen, CO_2, and small organic fermentation products. The metabolism of both archae- and eubacterial partners was anaerobic and growth was maximal in anoxic environments. The inevitable leaks of cyanobacterial oxygen led the spirochetes to regenerate elemental sulfur that was put to use as terminal electron acceptor by the archaebacterium. Thus the spirochete component of *Thiodendron* is a precise analogue to our model, whereas its partner, *Desulfothiobacter* [Sukov et al. 2001; Searcy and Stein 1980], differs from *Thermoplasma* in its mesothermia, its lack of acidophilia, and its eubacterial status. Just as *Thermoplasma acidophilum* today, in the absence of elemental sulfur, can use ambient oxygen (up to approximately 5%) as a terminal electron acceptor, we envision both anaerobic partners to have been somewhat oxygen tolerant.

Gupta's Chimera: Bacteria that Swim Together Stay Together

4. The stable consortium became the archaebacterial-eubacterial chimera inferred by Radhey Gupta exclusively from protein sequence comparisons of extant prokaryotes (Gupta 1998, 2000). The eubacterial contributor is not identifiable in Gupta's work and he rejects the spirochete hypothesis. However, his analysis requires a eubacterial contribution to eukaryotes prior to the symbiotic acquisition of the α-proteoeubacteria that became the mitochondria. Penetration of the archaebacterium by the spirochete comparable to known current spirochete behavior, including in pathogenesis, gave rise to three evolutionary legacies: (1) internalization of the eubacterium topologically inside the wall-less archaebacterium; (2) permanent attachment of the eubacterium to the archaebacterium via proteinaceous ligation (analogous attachments are observable today as structures on healthy hosts beset with surface spirochetes [Fig. 3]), with resultant motility and mechano- and chemosensitivity of the chimera; and (3) conjugation of the DNA of the two partners. The spirochete attachments evolved into the kinetosome-centrioles of eukaryotes. Bacterial-style recombination resulted in the consolidation of the DNA from both partners in membranous sacs reminiscent of the membrane-bounded nucleoids of the eubacterium *Gemmata obscuriglobus*, its planctomycete relatives, and other free-living prokaryotes (Fuerst and Webb 1991; Wang et al. 2002).

The Karyomastigont Preceded the Untethered Nucleus

5. The karyomastigont evolved as an organellar system that insured motility of the chimera, comparable to the modern analogue in Figure 3. The joint physical continuity of the DNA and protein synthetic system of the merged partners was provided by this conspicuous cell complex. The karyomastigont, an organelle system widespread in many different nucleated cells (e.g., protoctist zoospores, microgametes, plant sperm, mastigote algae) consists of the kinetosome-centriole [eukaryotic "basal bodies" to the 9(2)+2 "fla-

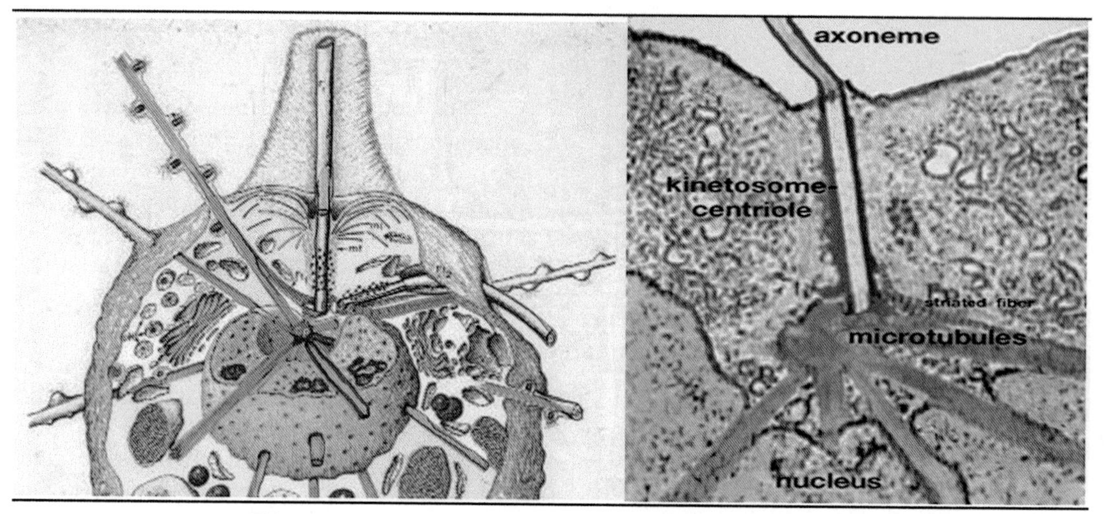

FIGURE 5. *Dimorpha mutans* in division. The distal end of the kinetosome produces the axoneme, whereas the proximal end produces mitotic spindle, paradesmose, or other microtubules that interact with the nuclear membrane. Drawing (left) based on electron micrographs (right) by Guy Brugerolle.

gella''], a fibrous or ribbon-shaped proteinaceous connector, called the rhizoplast in protozoological literature, and the nucleus itself to which the kinetosome-centrioles connect. The interphase karyomastigont of many motile archaeprotists often termed flagellates (mastigotes, in our more precise terminology [Margulis et al. 1990; Dolan et al. 2002]) is separated, during karyokinesis, by a thin mitotic spindle known to protozoologists as the ''paradesmose.'' During nuclear division the developing karyomastigont, through the paradesmose, acts to distribute the newly formed nuclei and associated organelles such as the kinetosome-centrioles and parabasal bodies (Golgi apparatus if present) to two offspring cells. Although first described by Janicki in 1915 and observed by many microscopists, the karyomastigont has not received attention in an evolutionary context. The karyomastigont in our reckoning is the evolutionary precursor to the standard mitotic spindle of animal and plant cells. The kinetosome-centrioles generate the shaft microtubules from their distal parts and the mitotic microtubules from their proximal ends, as can be seen in the karyomastigont of *Dimorpha mutans* (Fig. 5).

Nuclear Liberation

6. The earliest nuclei evolved tethered to the kinetosome-centrioles. The kinetosome-centrioles, relicts of spirochete attachment structures, became both the kinetosomes that organize ciliary and other axonemes and the mitotic centrioles resident at the poles of the cell as the mitotic process evolved and stabilized. The legacy is easily inferred from the many organisms whose kinetosomes behave as centrioles and therefore compromise swimming or feeding during mitotic division. Examples include *Chlamydomonas*, *Dimorpha mutans* (Fig. 5), myriad dinomastigotes (''dinoflagellates''), *Acronema* and other bicosoecids, kinetoplastids, euglenids, and many others. Such one-celled motile organisms tend to have only two undulipodia. Why? Because they divide to make two cells and the kinetosome of the motility organelle is used as the centriole in their mitosis. In all undulipodiated cells from sperm to sensory epithelia to cilia of *Paramecium*, distal growth from each kinetosome produces the shaft (axoneme) with its standard [9(2)+2] structure in transverse electron-microscopical structure. The cilium, sperm tail, or other undulipodium, or even the naked cytoplasmic axoneme, results from distal growth of the kinetosome-centriole. Proximal growth from this structure generates variable and idiosyncratic mitotic structures involved in the segregation of DNA. The use of the same structure (the spirochete attachment

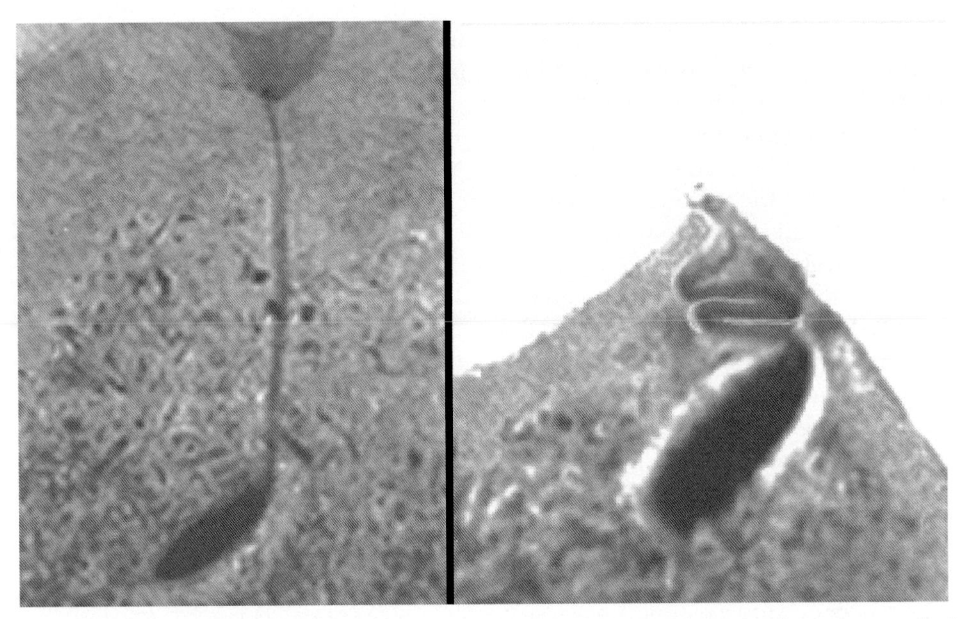

FIGURE 6. Motile karyomastigont of *Mixotricha paradoxa* extended (left), contracted (right); photographed from a permanent preparation by L. R. Cleveland, ca. 1957.

site), now the kinetosome-centriole that segregates joint chimeric DNA, suggests to us that the origin of mitosis was concurrent with the origin of the nucleus itself.

Historical Background and Current Literature

No one who has studied the phenomenon of cell division in live nucleated cells such as whitefish eggs or lily endosperm disagrees with E. B. Wilson's (1925) statement that mitosis consists of two fundamentally separate but correlated series of processes. The behavior of the mitotic spindle is experimentally separable from that of the chromosomes. The "mitotic spindle" is equivalent to the paradesmose of protozoologists and the "achromatic apparatus" of cytologists like L. R. Cleveland. He noted that, unlike the chromatin, the structure took up no Fuelgen or other nuclear stain (Cleveland 1934). Our explanation of the dual nature of mitosis is the minimally dual nature of the nucleated cell: the motile apparatus of the mitotic spindle, the karyomastigont (sometimes motile, as in *Mixotricha*, Fig. 6), the microtubular undulipodia in all their guises, the haptoneme of prymnesiophytes, the blepharoplasts of plant spermiogenic tissue, and a myriad of other homologues are of eubacterial-spirochetal origin. By contrast, the chromatin itself, as repository of the genetic merger of both the thermoacidophil and spirochete, is of joint archaebacterial-eubacterial origin from the onset of eukaryosis. In his laudable compilation of a vast scientific literature on the "centriole as central enigma of the cell" Wheatley (1982) was incapacitated by disciplinary insularity from solving the enigma. All animals, with respect to mitosis and cell motility, especially the role of the kinetosome-centriole, represent a single fundamentally homogenous lineage of organisms. Animals show none of Darwin's oddities and imperfections so important to evolutionary reconstruction. Wheatley, in Simon Robson's (1987) terms, was working with a "data set some billion years or more out of date."

Precisely this awareness of the central importance of motile protists in his search for the origin of mitosis is what drove L. R. Cleveland, Harvard professor of biology (1925–1959), to work for nearly 40 years on the contents of the hindguts of wood-ingesting cockroaches and termites. Although Cleveland's scientific interests and accomplishments were not properly recognized in William Trager's (1980) National Academy of Sciences review of his life, Cleveland's passion is apparent to those who read his work and view his films. He used the roach and termite microorganisms to elucidate general biological principles. The life history of chromosomes, the mitotic process, the origin of the meiotic sexual cycle, and fundamental cell structure were his central concern. Cleveland, it seems to us, solved the problems of the nature and functional importance of the centriole. He distinguished the generative mitotic centriole from its derivative, the kinetosome. He also contributed more than any other investigator to the problem of the origin of eukaryotic sexual systems. He understood how cell and nuclear fission in an abortive cannibalistic fertilization event created the surviving diploid whose progeny were then selected when they succeeded in accomplishing the absolute necessity of the "relief of diploidy" (Cleveland 1947).

Cleveland's scientific journey began when one morning in 1927 he walked out of his cabin at the Mountain Lake Biological Station, Virginia, and tried to discern, in a nearby rotten log, what the colony of wood-dwelling cockroaches were eating and if they could possibly harbor the large motile symbionts known from termites. He knew about the symbionts of wood-ingesting termites from his own previous study. In the United States the microbial inhabitants of termites from decayed railroad ties in New Jersey in the mid–nineteenth century had been first described by Joseph Leidy (1850, 1881). Cleveland was thrilled to see that the *Cryptocercus punctulatus* hindgut inhabitants were every bit as good. The paunch or hypertrophied intestine was replete with hypermastigote parabasalids: *Barbulanympha*, *Trichonympha*, and smaller trichomonads and oxymonads. Because they live in crowded colonies of large-sized individuals, the roaches provide even more abundant, diverse, and generally favorable material than the termites. Cleveland discovered many new protists, he recognized that the life cycle of protists that made resting cysts was correlated with the production of the ecdysone hormone by the insect, and he single-handedly founded the field of wood-feeding roach protozoology. He wrote a statement in his overlooked foreword (1934: p. vii) that qualifies as a miniature masterpiece:

> Shortly after arriving at the laboratory [Mt. Lake, Virginia], I began a search for termites and other wood-feeding insects, wishing to extend my observations on them. I had forgotten that, during discussions in 1923 and 1924 with Dr. T. E. Snyder regarding the possibility of wood-feeding insects other than termites harboring cellulose digesting protozoa, that the roach *Cryptocercus*, and the zorapteran *Zorotypus* were mentioned: and that I was now, for the first time, in a region where *Cryptocercus* had been found. Following our discussions, Snyder (1926), in a general article dealing with the production of castes in termites, mentioned the desirability of examining these and other wood-feeding insects for hypermastigote flagellates; but no one has done so. The readers of Snyder's articles, perhaps like myself, forgot his suggestion. I must confess that I had examined so many genera and species of wood-feeding insects without finding these protozoa, or any forms remotely relating to them, that I, like other workers on this uniquely interesting group of organisms, had become almost reconciled to the generally accepted view that they occurred in termites, where their evolution into many families and genera of highly complex flagellates, had taken place.
>
> The first dead log (which I found within 100 yards of the laboratory) yielded twenty-five or thirty specimens of roach, each of which harbored a fauna of wood-ingesting hypermastigotes far greater than that of any termite. Specimens were sent to Mr. A. N. Caudell of the U.S. National Museum, who identified them as *Cryptocercus punctulatus* Scudder.
>
> Observations were begun immediately on living and fixed preparations of the protozoa and on the roach. Three weeks collecting gave me a supply of four or five thousand specimens which were carried to Boston, where I have studied them and their protozoa for the past three years . . . .

The nuclei with their large chromosomes, even during acts of fertilization and meiosis, could be directly investigated in the teeming crowds of live, translucent mastigotes. The ease of direct observation of cell reproduction motivated Cleveland far more than any claims of medical importance of the studies of so-called parasites. The arcane names and obscure old literature have impeded understanding of the biology of Cleveland's large, amitochondriate motile parabasalids. We suggest that his body of work provides much of the solution to two major evolutionary problems: the significance of the kinetosome-centriole and the origins of eukaryotic sex.

Cleveland described early in his studies the relation between the mitotic process and undulipodia as organelles of motility. He saw that the mitotic movements of cell division required growth of spindle fibers from the "cell centers" and that these same "division centers" (attraction spheres, atractophores, centrosomes, centrioles, blepharoplasts, etc.) were the apparent source of the new undulipodia and other structures for the next generation. He recognized both the similarities in structure and function of cilia and flagella (together undulipodia) yet distinguished the cilia by recognition that they have gained independence from the mitotic process. Although the undulipodia traditionally called "flagella" are involved in mitosis, most undulipodia and all cilia are independent of mitosis. We appreciate the depth of Cleveland's insights and fully agree with his analysis, which can now, nearly 35 years after his death, be put in an evolutionary context. In his review Cleveland (1963: pp. 3–4) wrote:

> The term blepharoplast (Webber 1897), once almost universally employed in the description of the flagellate protozoa as the organelle from which the flagella arise, is an unfortunate one. It was first used by Webber (1897) in a study of pollen cells of *Gingko* [sic] to distinguish it from what Hirasé (1894) had already referred to as the attraction spheres of *Gingko* [sic]. Since organs of locomotion arose from the bodies at the end of the central spindle Webber thought they were not centrioles (or centrosomes) even though their appearance and behavior were typical of these organelles. There are other instances, notably in the pteridophytes *Equisetum* and *Marsilia* (Sharp 1912, 1914), in the collar cells of sponges, in flagellated spermatozoa, and in flagellates, which show clearly that the term blepharoplast as used by Webber, and still sometimes employed in flagellates, as the organelle from which flagella arise, refers to a body which is ontogenetically, phylogenetically, and functionally a centriole. In other words, since centrioles, at some time in their life cycle in so many organisms, are capable of producing flagella, in addition to their more usual function of producing the achromatic figure [mitotic spindle], there is no justification for giving those centrioles a special name such as blepharoplast. [The blepharoplast of plants, by electron microscopy, was later shown to not be a centriole but rather the group of kinetosomes that would generate sperm tails; see Chapter 9 of Margulis 1993 for review.] To do so only creates confusion. In fact, it is quite likely that in the course of time more examples will be found like that of Wolbach (1928) who believes he has seen, in two human tumors of striated muscle, the origin of myofibrils from centrioles. And Costello (1961) believes centrioles play a role in determining cleavage plains [sic] in early embryonic development.
>
> Another reason for not using blepharoplast is that it is also sometimes used synonymously with kinetosome-centriole or basal granule. Dual terminologies should be reserved for the exclusive use of those who prefer confusion to clarity . . . .

The meticulous work of Harold Kirby Jr. on termite microorganisms has guided our own as much as Cleveland's. Kirby, professor of zoology at the University of California, Berkeley, from 1930 until his untimely death in 1952 (when he was chairman), reconstructed the evolutionary history of certain termite protoctist lineages. Especially in studies of the trichomonads Kirby became aware of the evolutionary significance of the karyomastigont. Kirby developed the concept of "mastigont multiplicity," which identified and led us to understand the mastigont in an evolutionary context (Kirby 1994). The mastigont is defined as an organellar system composed of the undulipodia (usually four, each with its underlying kinetosome) and often with other fibers and tubules directly attached to the nucleus. The same mastigont system present in the cell but without the nucleus is what defines the "akaryomastigont," a term Kirby introduced. The concept of mastigont (either with [karyo-] or without [akaryo-] the nucleus) was indispensable to Kirby's analyses of the taxonomic relations between the termite symbionts that he in many cases discovered and named. Kirby, more than any earlier or later investigator, elucidated the family Calonymphidae with more than 20 species in which all genera are characterized by the presence of many more than a single nucleus and by lack of mitochondria. The nuclei tend to be tethered during karyokinesis to centrioles. Kirby, according to his mastigont multiplicity principle, arranged a mass of cell biological data into a series of logical, plausible evolutionary sequences. Whether Kirby, who worked less than Cleveland with live material, and did not work at all with *Mixotricha paradoxa* of Australia, ever knew that the simple four-kinetosome karyomastigont of this hypertrophied spirochete-covered trichomonad itself is motile (Fig. 6) we do not know. But that the mastigont structure is a species-specific organelle system that tends to reproduce as a unit, and in this way to generate new lower taxa, was very clear to him.

Had we not worked directly with Kirby's original permanent preparations of calonymphids and the accompanying manuscript left on his desk at his death (Kirby 1994) one of us (M.D.) would not have realized that the mastigont structure was the earliest form of the nucleus and, indeed, it was from the karyomastigont that the nucleus was liberated. We interpret the first appearance of the karyomastigont as a system selected to maintain connection of the merged genomes (of the archaebacterium and the eubacterium) to motility. From the onset, the karyomastigont ensured that the motility organelles allowed swimming toward food sources as well as proper segregation of the merged genomes at each cell division.

The entire karyomastigont organellar system is singular; i.e., it supports the one nucleus in uninucleate cells such as in *Trichomonas* and in many other smaller trichomonad relatives. In the multinucleate parabasalids, the calonymphids, the system is multiplied. Whereas some species have only eight nuclei (e.g., *Coronympha octonaria*), some genera such as *Snyderella* have dozens to hundreds, and a few (such as *Metacoronympha*, a huge and beautiful calonymphid) can have more than 1000 nuclei (Figs. 7, 8). From Kirby's concept of "mastigote multiplicity" in the ancestral trichomonad from which the large *M. senta* evolved, one infers that some ten karyomastigont reproductions occurred in the absence of cytokinesis. This discussion of the Cleveland-Kirby work follows the Darwin-Gould dictum to look for "imperfections and oddities" in the evolutionary system one seeks to understand. Furthermore we apply a cultural version of Gould's panda principle: we try, as much as we can, to read the original literature of those scientists who were motivated by intellectual curiosity even though they themselves may have been considered imperfect and odd in their day.

Criteria for Proof of the Karyomastigont Model

The criterion for proof of the symbiogenetic karyomastigont model for the origin of the nucleus will be the consistency with it of all scientific observation—bacteriological, protistological, and micropaleontological. Specifically, predictions include the following:

1. Stable *Thermoplasma-Spirochaeta* consortia will be found in nature and/or produced in the laboratory.
2. Certain motility protein active sites, i.e., the relevant functional domains of these proteins, will be more homologous to cytoplasmic (protoplasmic cylinder) proteins in spirochetes than they will to any arbitrarily chosen prokaryotic proteins (for example those from *Escherichia coli* or the cyanobacterium *Synechococcus* [Bermudes et al. 1987]). We briefly summarize below (Table 1) potential candidate eukaryotic cytoskeletal proteins whose domains may have re-

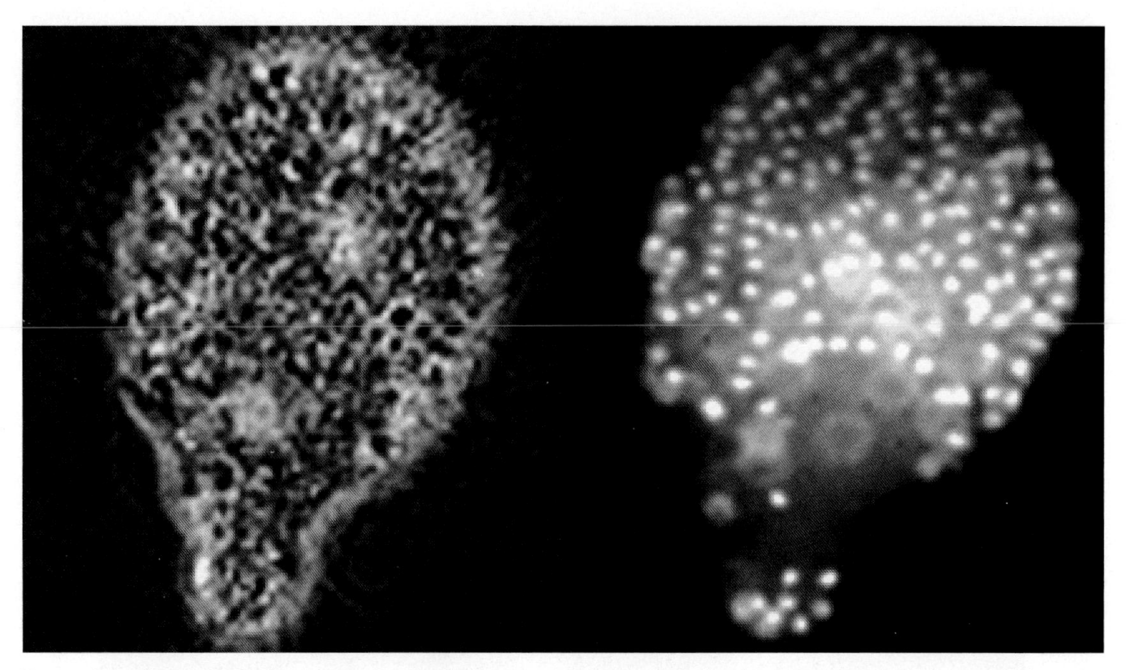

FIGURE 7. The multiple karyomastigonts of *Metacoronympha senta*. Left, phase contrast light micrograph; right, DAPI-stained nuclei fluorescence image. Scale bar, 10 μm. (From Dolan et al. 2000.)

tained motility functions homologous to those in the protoplasmic cylinders of spirochetes.

3. The ancestral spirochete, we postulate, provided motility proteins that eventually became part of the cytoskeletal system and motility apparatus of eukaryotic cells. The modern spirochete most closely related to the merged ancestor, we hypothesize, will possess protein domains with greater sequence homology to ubiquitous eukaryotic motility protein domains than will any other arbitrarily chosen prokaryote. Other details of what would constitute proof of the "spirochete origin of undulipodia" hypothesis were published elsewhere (Margulis 1991, 1993; Margulis and Sagan 1991).
4. The putative mitochondrial genes in the parabasalid nucleus (Martin 2005) will be found to have originated from a separate eubacterial symbiosis.

In homage to the memory of Stephen Jay Gould we invite others to study the oddities and peculiarities of the protoctists in recognition of the fact that their path of history has been obscured by perfection of the nucleus, its mitosis, and derived meiotic sex in both animals and plants. But these processes were not created as we found them. Rather, the secrets of their evolution, results of two billion years of evolution, lie buried in sulfurous organic-rich muds and sloshing around in the cellulolytic hindgut communities of wood-feeding insects, available still for any of us to explore. (Indeed we agree in principle with Knoll's statement (2003: p. 135) that "perhaps the clues [to the origins of eukaryotic cell biology] are hiding in plain sight, awaiting another Lynn Margulis to make sense of them."). A century's worth of clues await discovery in underappreciated literature—generated by Cleveland, Kirby, and many others—in fields such as protistology (protozoology, phycology), mycology, ciliate genetics, and micropaleontology. We suggest that as soon as the great works in these traditions are taught and modern biochemical, cytological, and molecular techniques are applied, the validity of our model, based so heavily on the results of our predecessors, will be obvious.

The wood-ingesting roaches and termites (xylophagous dictyopterans) provide the live material most appropriate to the test of our ideas. Parabasalids, oxymonads, and other

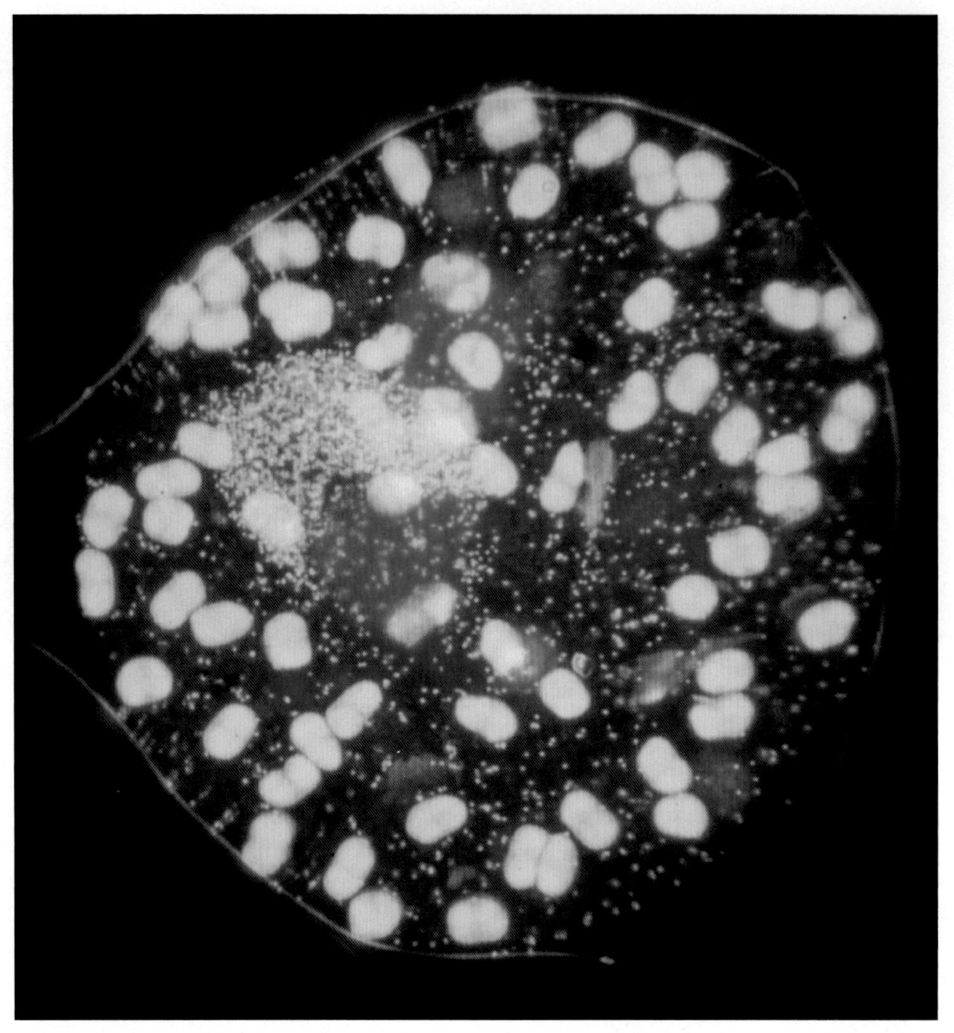

FIGURE 8. Simultaneous karyokinesis of over 50 nuclei in *Snyderella tabogae*. Sytox stain.

amitochondriate eukaryotes, as well as many types of prokaryotes, over 50% of which are spirochetes (Breznak 2000), are invariably present in healthy individuals. Preliminary results indicate that our methods are adequate to test our hypothesis. We apply fluorescent-labeled antibodies to termite hindgut preparations and of course to controls (mammalian cells such as PtK cell lines) where the location and function of the motility proteins such as α-tubulin, β-tubulin, γ-tubulin, kinesin, centrin, pericentrin, CENP-E, and others are known (Chapman et al. 2000; Dolan et al. 2002; Melnitsky and Margulis 2004; Melnitsky et al. 2005). The observation of γ-tubulin fluorescence in the plasma membrane shear zone of *Caduceia versatilis* ("Rubberneckia") where membranous tubules, and probably microtubules too, contact the cell membrane suggests that concerted efforts to find prokaryote homologues of this tubulin should be made (Melnitsky and Margulis 2004). At least 500 specific different proteins are present in axonemes and probably an equal number are characteristic of the mitotic spindle. Accordingly, data from single or a very few proteins or their corresponding DNA sequences will never suffice to distinguish our hypothesis definitively from the many others. Direct DNA sequence comparisons are in principle inadequate as well, because they hardly detect a genetic relation between two very similar spirochetes (e.g., *Borrelia* and *Treponema* [Fraser et al. 1998; Gupta 1998]). Imperative are three-

(or more) way comparisons of those characterized motility proteins of well-established function (i.e., those with tension-sensitivity such as the kinetochore proteins [Kolnicki 2000], motor proteins that transport vesicles directionally over the surface of microtubules, proteins that simultaneously attach membranes to microtubules, vesicle transport along microtubule surfaces, etc.). Of special interest are proteins already shown to be highly conserved in functional domain and structure across a diversity of taxa. Such protein comparisons are advisably made with three kinds of eubacteria, e.g., *E. coli*, cyanobacteria, and spirochetes known to have cytoplasmic tubules and cytoplasmic tubule-associated structures (CTAS; Wier et al. 2000: Fig. 5), as well as with *Thermoplasmas* and other representative archaebacteria. If an organellar system evolved by symbiogenesis and not by random accumulation of mutations or horizontal gene transfer, one gene at a time, then sets of genes with their products will eventually be found that share the same evolutionary history of clear selective advantage. The common evolutionary history and clear selective advantage that have been shown definitively for both chloroplasts and mitochondria, we suggest, will also hold for the suite of proteins and nucleic acids that determine the presence of the microtubule-organizing centers including centriole-kinetosomes of all undulipodia. If we are correct, all nucleated cells evolved by symbiogenesis when an archaebacterium acquired motility by incorporation of a spirochete genome in sulfurous, anoxic environments. This swimming consortium became the ancestor of all mitotic organisms. Eukaryosis, the origin of nucleated cells, probably occurred by the mid-Proterozoic. Certainly eukaryotes thrived prior to the deposition in sediments of well-preserved microfossils such as *Vandalosphaeridium* (Vidal 1998) and the spiny spheres in the Doushantou cherts of China (Knoll 2003).

Acknowledgments

We thank NASA Space Sciences; the College of Natural Sciences and Mathematics and the Graduate School of the University of Massachusetts; Marta Norman of the Lounsbery Foundation; a National Science Foundation Graduate Research Fellowship awarded to J.H.W.; the American Museum of Natural History for support to M.F.D.; and the Howard Hughes undergraduate research award (to our honors student Hannah Melnitsky) for financial support. Aid with research, technical support, and/or manuscript preparation is gratefully acknowledged from C. Asikainen, A. Becerra, M. Chapman, D. Grimaldi, R. Guerrero, B. Hawthorne, W. E. Krumbein, R. Kolnicki, A. Lazcano, J. MacAllister, H. Ritter, D. Sagan, J. Sagan, J. Sapp, and A. Wier. The Hanse Institute for Advanced Study (Delmenhorst, Germany) and the Alexander von Humboldt Foundation (Berlin) provided Lynn Margulis with the opportunity to complete this manuscript. We thank M. Partee (Fig. 1) and K. Delisle (Fig. 2) for help with the drawings, A. Khodjakov (Fig. 4), and G. Brugerolle for permission (Fig. 5). The use of Cleveland's photographs of the *Mixotricha* karyomastigont (Fig. 6) is courtesy of the American Museum of Natural History, New York.

Literature Cited

Belgareh, N., G. Rabut, S. W. Bai, M. van Overbee, J. Beaudouin, N. Daigle, O. V. Zatsepina, F. Pasteau, V. Labas, M. Fromont-Racine, J. Ellenberg, and V. Doyle. 2001. An evolutionarily conserved NPC subcomplex, which redistributes in part to kinetochores in mammalian cells. Journal of Cell Biology 154: 1147–1160.

Bermudes, D., L. Margulis, and G. Tzertzinis. 1987. Prokaryotic origin of undulipodia: application of the panda principle to the centriole enigma. Annals of the New York Academy of Sciences 503:187–197.

Bermudes, D., G. Hinkle, and L. Margulis. 1994. Do prokaryotes contain microtubules? Microbiology Reviews 58:387–400.

Breznak, J. A. 2000. Phylogenetic diversity and physiology of termite gut spirochetes. American Zoology 40:954–955.

Chapman, M. J., M. F. Dolan, and L. Margulis. 2000. Centrioles and kinetosomes: form, function, and evolution. Quarterly Review of Biology 75:409–429.

Cleveland, L. R. 1934. The wood-feeding roach *Cryptocercus*, its protozoa, and the symbiosis between protozoa and roach, in collaboration with S. R. Hall, Elizabeth P. Sanders, and Jane Collier. Memoirs of the American Academy of Arts and Sciences 17:185–342.

———. 1947. The origin and evolution of meiosis. Science 105: 287–288.

———. 1963. Functions of flagellate and other centrioles in cell reproduction. Pp. 3–53 *in* L. Levine, ed. The cell in mitosis. Academic Press, New York.

Cleveland, L. R., and A. V. Grimstone. 1964. The fine structure of the flagellate *Mixotricha paradoxa* and its associated micro-organisms. Proceedings of the Royal Society of London B 159: 668–686.

Cole, D. G., D. R. Diener, A. Himelblau, P. L. Beech, J. C. Fuster, and J. L. Rosenbaum. 1998. *Chlamydomonas* kinesin-II-dependent intraflagellar transport (IFT): IFT particles contain proteins required for ciliary assembly in *Caenorhabditis elegans* sensory neurons. Journal of Cell Biology 141:993–1008.

Corliss, J. O. 1998. Oral communication discussion of his 1987 paper at American Society for Microbiology, Washington, D.C.

———. 1987. Protistan phylogeny and eukaryogenesis. International Review of Cytology 100:319–370.

d'Ambrosio, U., M. F. Dolan, A. M. Wier, and L. Margulis. 1999. Devescovinid trichomonad with axostyle-based rotary motor ("Rubberneckia"): taxonomic assignment as *Caduceia versatilis* sp. nov. European Journal of Protistology 35:327–337.

Daniels, E. W., and E. P. Breyer. 1967. Ultrastructure of the giant amoeba, *Pelomyxa palustris*. Journal of Protozoology 14:167–179.

Darwin, C. 1859. On the origin of species. John Murray, London.

Dobzhansky, T. 1972. Nothing in biology makes sense except in the light of evolution. American Biology Teacher 35:125–129.

Dolan, M. F. 2005. The missing piece: the microtubule cytoskeleton and the origin of eukaryotes. Pp. 281–289 *in* J. Sapp ed. Microbial phylogeny and evolution: concepts and controversies. Oxford University Press, New York.

Dolan, M. F., A. M. Wier, and L. Margulis. 2000. Budding and asymmetric reproduction of a trichomonad with as many as 1000 nuclei in karyomastigonts: *Metacoronympha* from *Incisitermes*. Acta Protozoologica 33:275–280.

Dolan, M. F., H. Melnitsky, L. Margulis, and R. Kolnicki. 2002. Motility proteins and the origin of the nucleus. Anatomical Record 268:290–301.

Doxsey, S. J., P. Stein, L. Evans, P. Calarco, and M. Kirschner. 1994. Pericentrin, a highly conserved protein of centrosomes involved in microtubule organization. Cell 76:639–650.

Dubinina, G. A., M. Y. Grabovich, and N. V. Leshcheva. 1993a. Occurrence, structure, and metabolic-activity of *Thiodendron* sulfur mats in various saltwater environments. Microbiology 62:450–456.

Dubinina, G. A., N. V. Leshcheva, and M. Y. Grabovich. 1993b. The colorless sulfur bacterium *Thiodendron* is actually a symbiotic association of spirochetes and sulfidogens. Microbiology 62:432–444.

Dutcher, S. K. 2001. The tubulin fraternity: alpha to eta. Current Opinion in Cell Biology 13:49–54.

Dyson, F. 1985. Origins of life. Cambridge University Press, Cambridge.

Echeverri, C. J., B. M. Paschal, K. T. Vaughan, and R. B. Vallee. 1996. Molecular characterization of the 50-kD subunit of dynactin reveals function for the complex in chromosome alignment and spindle organization during mitosis. Journal of Cell Biology 132:617–633.

Erickson, H. P. 1997. FtsZ, a tubulin homologue in prokaryote cell division. Trends in Cell Biology 7:362–367.

Fraser, C. M., S. J. Norris, G. M. Weinstock, O. White, G. G. Sutton, R. Dodson, et al. 1998. Complete genome sequence of *Treponema pallidum*, the syphilis spirochete. Science 281:375–388.

Fuerst, J. A., and R. I. Webb. 1991. Membrane-bounded nucleoid in the eubacterium *Gemmata obscuriglobus*. Proceedings of the National Academy of Sciences USA 88:8184–8188.

Gaglio, T., M. A. Dionne, and D. A. Compton. 1997. Mitotic spindle poles are organized by structural and motor proteins in addition to centrosomes. Journal of Cell Biology 138:1055–1066.

Gould, S. J. 1986. Evolution and the triumph of homology, or why history matters. American Scientist 74:60–69

Gupta, R. S. 1998. Protein phylogenies and signature sequences: a reappraisal of evolutionary relationships among archaebacteria, eubacteria, and eukaryotes. Microbiology and Molecular Biology Reviews 62:1435–1491.

———. 2000. The natural evolutionary relationships among prokaryotes. Critical Reviews in Microbiology 26:111–131.

Hall, J. L., and D. J. L. Luck. 1995. Basal body-associated DNA: in situ studies in *Chlamydomonas reinhardtii*. Proceedings of the National Academy of Sciences USA 92:5129–5133.

Hardwick, K. G., and A. W. Murray. 1995. Mad1p, a phosphoprotein component of the spindle assembly in budding yeast. Journal of Cell Biology 131:709–720.

Haren, L., and A. Merdes. 2002. Direct binding of NuMA to tubulin is mediated by a novel sequence motif in the tail domain that bundles and stabilizes microtubules. Journal of Cell Science 115:1815–1824.

Helenius, A., and M. Aebi. 2001. Intracellular functions of N-linked glycans. Science 291:2364–2369.

Janicki, C. 1915. Untersuchungen an parasitischen Flagellaten. Zeitschrift für Wissenschftliche Zoologie 112:573–691.

Jenkins, C., R. Samudrala, I. Anderson, B. P. Hedlund, G. Petroni, N. Michailova, N. Pinel, R. Overbeek, G. Rosati, and J. T. Staley. 2002. Genes for the cytoskeletal protein tubulin in the bacterial genus *Prosthecobacter*. Proceedings of the National Academy of Sciences USA 99:17049–17054.

Khodjakov, A., R. W. Cole, B. R. Oakley, and C. L. Rieder. 2000. Centrosome-independent mitotic spindle formation in vertebrates. Current Biology 10:59–67.

Khodjakov, A., and C. L. Rieder. 1999. The sudden recruitment of gamma-tubulin to the centrosome at the onset of mitosis and its dynamic exchange throughout the cell cycle do not require microtubules. Journal of Cell Biology 146:585–596.

Kirby, H., annotated by L. Margulis. 1994. Harold Kirby's symbionts of termites: Karyomastigont reproduction and calonymphid taxonomy. Symbiosis 16:7–63.

Knoll, A. H. 2003. Life on a young planet: the first three billion years of evolution on Earth. Princeton University Press, Princeton, N.J.

Kolnicki, R. 2000. Kinetochore reproduction in animal evolution: cell biological explanation of karyotypic fission theory. Proceedings of the National Academy of Sciences USA 97: 9493–9497.

Kumar, J., H. Yu, and M. P. Sheetz. 1995. Kinectin, an essential anchor for kinesin-driven vesicle motility. Science 267:1834–1837.

Lange, B. M. H., and K. Gull 1995. A molecular marker for centriole maturation in the mammalian cell cycle. Journal of Cell Biology 130:919–927.

Leadbetter, J. R., T. M. Schmidt, J. R. Graber, and J. A. Breznak. 1999. Acetogenesis from H_2 plus CO_2 by spirochetes from termite guts. Science 283:686–689.

Leidy, J. 1850. On the existence of endophyta in healthy animals as a natural condition. Proceedings of the Academy of Natural Sciences of Philadelphia 4:225–229.

———. 1881. The parasites of termites. Journal of the Academy of Natural Sciences of Philadelphia 8:425–447.

Lindsay, M. R., R. I. Webb, M. Strous, M. S. Jetter, M. K. Butler, R. J. Forde, and J. A. Fuerst. 2001. Cell compartmentalization in planctomycetes: novel types of structural organization for the bacterial cell. Archives of Microbiology 175:413–429.

Mack, G. J., and D. A. Compton. 2001. Analysis of mitotic microtubule-associated proteins using mass spectrometry identifies astrin, a spindle-associated protein. Proceedings of the National Academy of Sciences USA 98:14434–14439.

Maney, T., A. W. Hunter, M. Wagenbach, and L. Wordeman. 1998. Mitotic centromere-associated kinesin is important for anaphase chromosome segregation. Journal of Cell Biology 142:787–801.

Margulis, L. 1991. Symbiosis in evolution: Origins of cell motility. Pp. 305–324 *in* T. Honjo and S. Osawa, eds. Evolution of life: fossils, molecules, and culture. Springer, Tokyo.

———. 1993. Symbiosis in cell evolution, 2d ed. W. H. Freeman, New York.

———. 1996. Archaeal-eubacterial mergers in the origin of Eukarya: phylogenetic classification of life. Proceedings of the National Academy of Sciences USA 93:1071–1076.

———. 2000. Spirochetes. Pp. 353–363 *in* J. Lederberg, ed. Encyclopedia of Microbiology, Vol. 4, 2d ed. Academic Press, New York.

Margulis, L., and M. F. Dolan. 2004. Eukaryosis: origin of eukaryotic cells. Sony U-matic video. 16 minutes. Color. [Unpublished.]

Margulis, L., and D. Sagan. 1991. Origins of sex: three billion years of genetic recombination. Yale University Press, New Haven, Conn.

Margulis, L., J. O. Corliss, M. Melkonian, and D. J. Chapman, eds. 1990. Handbook of Protoctista: the structure, cultivation, habitats and life histories of the eukaryotic microorganisms and their descendants exclusive of animals, plants and fungi. Jones and Bartlett, Boston.

Margulis, L., M. F. Dolan, and R. Guerrero. 2000. The chimeric eukaryote: origin of the nucleus from the karyomastigont in amitochondriate protists. Proceedings of the National Academy of Sciences USA 97:6954–6959.

Martin, W. 2005. Woe is the tree of life. Pp. 134–153 *in* J. Sapp, ed. Microbial phylogeny and evolution: concepts and controversies. Oxford University Press, New York.

Melnitsky, H., and L. Margulis. 2004. Centrosomal proteins in termite symbionts: gamma–tubulin and a scleroderma antigen localize to the bacteria-free cell-rotation zone of the parabasalid *Caduceia versatilis*. Symbiosis 37:323–333.

Melnitsky, H., F. Rainey, and L. Margulis. 2005. The karyomastigont model of eukaryosis. Pp. 261–280 *in* J. Sapp ed. Microbial phylogeny and evolution: concepts and controversies. Oxford University Press, New York (in press).

Mignot, J. P. 1996. The centrosomal big bang: from a unique central organelle towards a constellation of MTOC's. Biology of the Cell 86:81–91.

Moritz, M., M. B. Braunfeld, J. W. Sedat, B. Alberts, and D. A. Agard. 1995. Microtubule nucleation by gamma-tubulin-containing rings in the centrosome. Nature 378:638–640.

Nicklas, R. B., S. C. Ward, and G. J. Gorbsky. 1995. Kinetochore chemistry is sensitive to tension and may link mitotic forces to a cell cycle checkpoint. Journal of Cell Biology 130:929–939.

Robson, S. K. 1987. Review of "Origin of sex: three billion years of genetic recombination" by L. Margulis and D. Sagan. Symbiosis 3:207–212.

Schopf, J. W. 1999. Cradle of life. Princeton University Press, Princeton, N.J.

Searcy, D. G., and S. H. Lee. 1998. Sulfur reduction by human erythrocytes. Journal of Experimental Zoology 282:310–322

Searcy, D. G., and D. B. Stein. 1980. Nucleoprotein subunit structure in an unusual prokaryotic organism: *Thermoplasma acidophilum*. Biochimica et Biophysica Acta 609:180–195.

Suh, M. R., J. W. Han, Y. R. No, and J. Lee. 2002. Transient concentration of a gamma-tubulin-related protein with a pericentrin-related protein in the formation of basal bodies and flagella during the differentiation of *Naegleria gruberi*. Cell Motility and the Cytoskeleton 52:66–81.

Surkov, A. V., G. A. Dubinina, A. M. Lynesko, F. O. Glöckner, and J. Kuever. 2001. *Dethiosulfovibrio russensis* sp. nov., *Dethiosulfovibrio marinus* sp. nov. and *Dethiosulfovibrio acidaminovorans* sp. nov., novel anaerobic, thiosulfate- and sulfur-reducing bacteria isolated from "*Thiodendron*" sulfur mats in different saline environments. International Journal of Systematic and Evolutionary Microbiology 51:327–337.

Toyoshima, I., H. Yu, E. R. Steuer, and M. P. Sheetz. 1992. Kinectin, a major kinesin-binding protein on ER. Journal of Cell Biology 118:1121–1131.

Vidal, G. 1998. Proterozoic and Cambrian bioevents. Revista Española de Paleontología, No. extr. Homenaje al Prof. Gonzalo Vidal, pp. 11–16.

Vidal, G., and M. Moczydłowska-Vidal. 1997. Biodiversity, speciation, and extinction trends of Proterozoic and Cambrian phytoplankton. Paleobiology 23:230–246.

Viscogliosi, E., V. P. Edgcomb, D. Gerbod, et al. 1999. Molecular evolution inferred from small subunit rRNA sequences: what does it tell us about phylogenetic relationships and taxonomy of the parabasalids? Parasite 6:279–291.

von Dohlen, C. D., S. Kohler, S. T. Alsop, and W. R. McManus. 2001. Mealybug beta-proteobacterial endosymbionts contain gamma-proteobacterial symbionts. Nature 412:433–436.

Wang, J., C. Jenkins, R. I. Webb, J. A. Fuerst. 2002. Isolation of *Gemmata*-like and *Isosphaera*-like planctomycete bacteria from soil and freshwater. Applied and Environmental Microbiology 68:417–422.

Wheatley, D. N. 1982. The centriole: a central enigma of cell biology. Elsevier, Amsterdam.

Wier, A., J. Ashen, and L. Margulis. 2000. *Canaleparolina darwiniensis*, gen. nov. sp. nov., and other pillotinaceous spirochetes from insects. International Microbiology 3:213–223.

Wilson, E. B. 1925. The cell in development and heredity. Macmillan, New York.

Young, A., J. B. Dictenberg, A. Purohit, R. Tuft, and S. J. Doxsey. 2000. Cytoplasmic dynein-mediated assembly of pericentrin and gamma tubulin onto centrosomes. American Society for Cell Biology 11:2047–2056.

Zimmerman, W., and S. J. Doxsey. 2000. Construction of centrosomes and spindle poles by molecular motor-driven assembly of protein particles. Traffic 1:927–934.

Paleobiology, 31(2), 2005, pp. 192–210

Mass extinctions and macroevolution

David Jablonski

Abstract.—Mass extinctions are important to macroevolution not only because they involve a sharp increase in extinction intensity over ''background'' levels, but also because they bring a change in extinction selectivity, and these quantitative and qualitative shifts set the stage for evolutionary recoveries. The set of extinction intensities for all stratigraphic stages appears to fall into a single right-skewed distribution, but this apparent continuity may derive from failure to factor out the well-known secular trend in background extinction: high early Paleozoic rates fill in the gap between later background extinction and the major mass extinctions. In any case, the failure of many organism-, species-, and clade-level traits to predict survivorship during mass extinctions is a more important challenge to the extrapolationist premise that all macroevolutionary processes are simply smooth extensions of microevolution. Although a variety of factors have been found to correlate with taxon survivorship for particular extinction events, the most pervasive effect involves geographic range at the clade level, an emergent property independent of the range sizes of constituent species. Such differential extinction would impose ''nonconstructive selectivity,'' in which survivorship is unrelated to many organismic traits but is not strictly random. It also implies that correlations among taxon attributes may obscure causation, and even the focal level of selection, in the survival of a trait or clade, for example when widespread taxa within a major group tend to have particular body sizes, trophic habits, or metabolic rates. Survivorship patterns will also be sensitive to the inexact correlations of taxonomic, morphological, and functional diversity, to phylogenetically nonrandom extinction, and to the topology of evolutionary trees. Evolutionary recoveries may be as important as the extinction events themselves in shaping the long-term trajectories of individual clades and permitting once-marginal groups to diversify, but we know little about sorting processes during recovery intervals. However, both empirical extrapolationism (where outcomes can be predicted from observation of pre- or post-extinction patterns) and theoretical extrapolationism (where mechanisms reside exclusively at the level of organisms within populations) evidently fail during mass extinctions and their evolutionary aftermath. This does not mean that conventional natural selection was inoperative during mass extinctions, but that many features that promoted survivorship during background times were superseded as predictive factors by higher-level attributes. Many intriguing issues remain, including the generality of survivorship rules across extinction events; the potential for gradational changes in selectivity patterns with extinction intensity or the volatility of target clades; the heritability of clade-level traits; the macroevolutionary consequences of the inexact correlations between taxonomic, morphological, and functional diversity; the factors governing the dynamics and outcome of recoveries; and the spatial fabric of extinctions and recoveries. The detection of general survivorship rules—including the disappearance of many patterns evident during background times—demonstrates that studies of mass extinctions and recovery can contribute substantially to evolutionary theory.

David Jablonski. Department of Geophysical Sciences, University of Chicago, 5734 South Ellis Avenue, Chicago, Illinois 60637. E-mail: d-jablonski@uchicago.edu

Accepted: 31 August 2004

Introduction

Mass extinctions and their causes have attracted enormous attention over the past two decades, but relatively few authors have focused on these events from a macroevolutionary standpoint. Steve Gould saw the major mass extinctions as a third evolutionary tier with its own ''predominating causes and patterns.'' The essential property of mass extinctions, he argued, is that these rare, intense events derail the evolutionary patterns shaped by organismic selection, speciation, and species sorting during times of background extinction (operationally defined here simply as the intervals of lower extinction intensity between the ''Big Five'' extinction peaks of the Phanerozoic). Mass extinctions would thus represent another challenge—among several presented by Gould (2002) and others—to *extrapolationist*, non-hierarchical approaches to evolutionary theory, in which large-scale patterns are understood strictly in terms of ''extrapolating from evolutionary change over ecological time frames in local populations'' (Sterelny and Griffiths 1999: p. 305).

0094-8373/05/3102-0014/$1.00

Here I will discuss mass extinctions from an evolutionary perspective, drawing mainly on the heavily studied end-Cretaceous (K/T) event but noting results from other extinctions where comparable analyses are available. I will argue that shifts in extinction selectivity are just as important as variations in intensity, and that the sorting of clades in post-extinction recoveries is also crucial to our understanding the evolutionary role of mass extinctions. I will conclude that simple extrapolationism fails at mass extinction events, whether defined as empirical predictability from background to mass extinction, or in more demanding terms as reducibility to exclusively organism-level processes.

Intensity

Mass extinctions are important episodes in the history of life regardless of their broader evolutionary implications, but they are difficult to study rigorously. The absolute magnitude and temporal fabric of extinctions and subsequent recoveries are almost certainly distorted by the incompleteness and bias of the stratigraphic record (see discussions of Sepkoski and Koch 1996; Kidwell and Holland 2002; Foote 2003). Nevertheless, many lines of evidence refute the extreme Darwin-Lyell claim that the major extinctions are essentially sampling artifacts. This evidence includes, most obviously, the permanent loss of abundant, well-sampled clades that persist until late in the pre-extinction interval (such as ammonoids, rudist bivalves, and globotruncanacean foraminifers at the K/T boundary), qualitative and long-standing shifts in biotic composition, and the temporal sharpening of the events with improved sampling and correlations (e.g., Sepkoski 1984, 1996; Jablonski 1986a, 1995; Hallam and Wignall 1997; Droser et al. 2000; Erwin et al. 2002; Steuber et al. 2002; Bambach et al. 2002; Benton and Twitchett 2003; Erwin 2003; see also the modeling results of Foote 2003, 2005). Taxonomic standardization tends to reduce the intensity of major events relative to estimates from synoptic databases (e.g., Smith and Jeffery 1998; Kiessling and Baron-Szabo 2004; for an exception see Adrain and Westrop 2000), but such revisions rarely encompass the target clade's "background" extinction patterns and so the boundary-focused revisions lack a quantitative context (as also noted by Fara [2000] and Foote [2003]).

Some authors have argued that the apparent continuum of extinction intensities from the most placid stratigraphic intervals to the Big Five events, yielding a right-skewed unimodal distribution when all Phanerozoic stages are plotted as a single population (e.g., Raup 1991a,b, 1996; MacLeod 2003a,b; Wang 2003), undermines any claim for a significant macroevolutionary role for mass extinctions (e.g., Hoffman 1989; Lee and Doughty 2003; MacLeod 2003b). However, this seems an oversimplification, for several reasons.

Most importantly, the seemingly continuous distribution of extinction intensities may arise from grouping heterogeneous data. The high turnover rates of the early Paleozoic, which anchor the secular decline in background extinction intensities documented by Raup and Sepkoski (1982; see also Van Valen 1984; Gilinsky and Bambach 1987; Gilinsky 1994; Sepkoski 1996; Newman and Eble 1999; MacLeod 2003a; Foote 2003), generate a set of high-extinction intervals that bridge the gap between the bulk of background extinction stages and the most extreme of the mass extinctions. If the secular decline is factored out, by omitting the Cambrian–early Ordovician stages and/or by scaling extinction intensities as residuals to a line fit to the Phanerozoic decline, then at least three of the major extinction events form a statistically significant mode in the extinction-frequency distribution (Bambach and Knoll 2001; S. Peters personal communication 2002; Bambach et al. 2004). The end-Permian, end-Cretaceous, and end-Ordovician extinctions constitute that discrete mode in the Bambach et al. analysis, the same three events that emerged in Hubbard and Gilinsky's (1992) bootstrap analysis and in Foote's (2003) models, an outcome that underscores the peculiarity—at least in the synoptic Sepkoski databases—of the Late Devonian and end-Triassic extinctions. These two extinctions seem especially difficult to separate from the effects of facies changes, although the near-extirpation of ammonoids and other taxa and an apparently global upheaval of reef communities lend credibility to some kind of exceptional biotic turnover in both the late Devonian and latest Triassic (McGhee 1996; Page 1996; Hallam and Wignall 1997; Balinski et al. 2002; Copper 2002; Flügel 2002; Racki and House 2002; Stanley 2003). Some of the lesser extinction pulses in the Phanerozoic record may truly reflect variations in preservation rather than true extinction events (e.g., Smith 2001; Peters and Foote 2001; Foote 2003, 2005; Erwin 2004), and removing such artifacts may also tend to set the mass extinctions apart, although this effect needs to be evaluated quantitatively. The apparent continuity of Phanerozoic extinction intensities is therefore debatable at best, and probably obscures more about extinction processes than it reveals.

Selectivity

More important than extinction intensity from a macroevolutionary standpoint is the question of selectivity. A slowly growing set of analyses has found that many traits correlated with extinction risk in today's biota, and with paleontological background extinction, tend to be poor predictors of survivorship during mass extinctions. Factors such as local abundance, reproductive mode, body size and inferred generation time, trophic strategy, life habit, geographic range at the species level, and species richness, which have all been hypothesized or shown to be significant under "normal" extinction intensities, had little effect on genus survivorship during the K/T extinction and were unimportant in one or more of the other mass extinctions as well (Jablonski 1986a,b, 1989, 1995; Jablonski and Raup 1995; Smith and Jeffery 1998, 2000a; Lockwood 2003). (The fact that many of the neontological and microevolutionary patterns are expressed at the species level, whereas most paleontological data are drawn from genus-level data, is a concern only if a hierarchical view is already preferred over an extrapolationist one; under an extrapolationist view patterns should be damped but not qualitatively different across the biological hierarchy.)

One example of this contrast can be seen in marine bivalves, where epifaunal suspension feeders such as scallops have significantly shorter genus durations than infaunal suspension feeders such as cockles in the 140-Myr interval between the end-Triassic extinction and the K/T boundary (Fig. 1). Although this is work in progress, the median durations differ by 50% and the frequency distributions differ significantly ($P < 0.01$), suggesting that the contrast is fairly robust (see also Aberhan and Baumiller 2003, and McRoberts' 2001 report of a smaller but equally significant difference in extinction rates between infauna and epifauna during the Triassic, albeit with a curious reversal in the Norian stage). These results should still be viewed cautiously, because the two functional groups differ mineralogically and thus in preservation potential: infaunal bivalves form less stable, exclusively aragonitic shells whereas the epifauna includes clades with more stable calcitic components. Poor

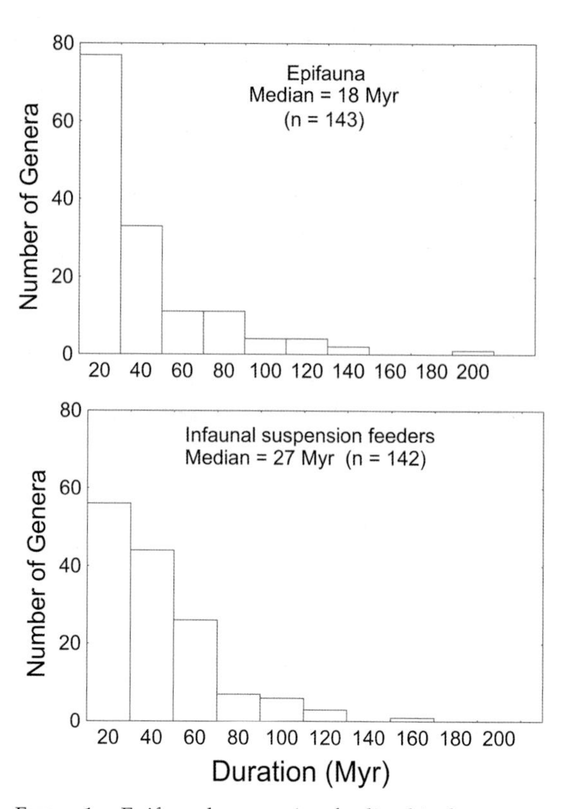

FIGURE 1. Epifaunal suspension-feeding bivalve genera (above) have significantly shorter genus durations than infaunal suspension-feeding bivalves (below) in the 140-Myr interval of background extinction leading up to the end-Cretaceous event (Mann-Whitney *U*-test: $P < 0.01$). As shown in Figure 2, this contrast disappears during each of the last three mass extinctions. (Data from the ongoing revision and ecological characterization of the bivalve portion of Sepkoski 2002 by D. Jablonski, K. Roy, and J. W. Valentine.)

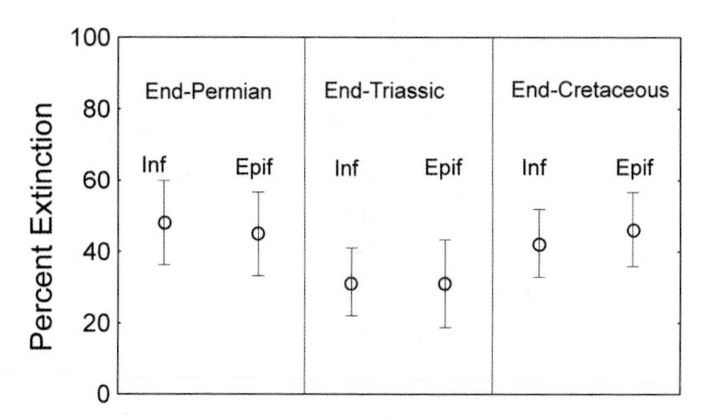

FIGURE 2. Genera of infaunal (Inf) and epifaunal (Epif) suspension-feeding bivalves do not differ significantly in extinction intensity during each of three extinctions: the end-Permian (Jablonski et al. unpublished), end-Triassic (McRoberts 2001), and end-Cretaceous (Jablonski and Raup 1995). (95% confidence intervals following Raup 1991c.)

preservation might impose lower taxonomic resolution on infauna and thus artificially inflate their durations. However, the contrast between mass extinction and background extinction times suggests that preservational effects do not overwhelm other factors (and see Kidwell 2005). The difference in durations disappears in the end-Permian, end-Triassic, and end-Cretaceous extinctions, where both functional groups suffer statistically indistinguishable extinction intensities (Jablonski and Raup 1995; McRoberts 2001; and a very preliminary analysis of end-Permian bivalve extinction) (Fig. 2).

A similar shift in genus survivorship patterns occurs in analyses of species richness, geographic range at the species level, and the interaction of these properties. For example, Late Cretaceous marine bivalve and gastropod genera that contained many, mainly widespread species tend to have significantly greater durations than genera consisting of few, spatially restricted species, and the other combinations tended to show intermediate values (Jablonski 1986a,b) (Fig. 3A). This makes intuitive sense and can readily be modeled in terms of clade demography (e.g., Raup 1985). However, these differences do not predict survivorship patterns during the end-Cretaceous extinction (Figs. 3B, 4) in the North American Coastal Plain (Jablonski 1986a,b) and elsewhere (Jablonski 1989). A partial list of analyses for other groups and events where

A

	Widespread Species	Restricted Species
Species-Rich	130 Myr	48 Myr
Species-Poor	60 Myr	32 Myr

B

	Widespread Species	Restricted Species
Species-Rich	49%	55%
Species-Poor	50%	47%

FIGURE 3. The interaction between species richness and geographic range at the species level in promoting genus survivorship, significant during (A) background times for marine bivalves and gastropods, is not apparent for (B) the end-Cretaceous mass extinction in the Gulf and Atlantic Coastal Plain. In A, values are median durations of genera; in B, values are extinction intensities. Species-rich defined as having three or more species in the study area; widespread as having at least 50% of species with geographic ranges >500 km in the study area; see Jablonski 1986b for details.

species richness was not a buffer against extinction is provided by Jablonski (1995).

Selectivity also shifts at the K/T boundary for developmental modes in marine invertebrates. The Late Cretaceous saw the independent evolution of nonplanktotrophic larvae in numerous gastropod lineages (Jablonski 1986c) and in five of the 14 orders of echinoids across a wide range of habitats and latitudes, presumably in response to pervasive changes in global climate or plankton communities (Jeffery 1997; Smith and Jeffery 2000a). However, the end-Cretaceous mass extinction was

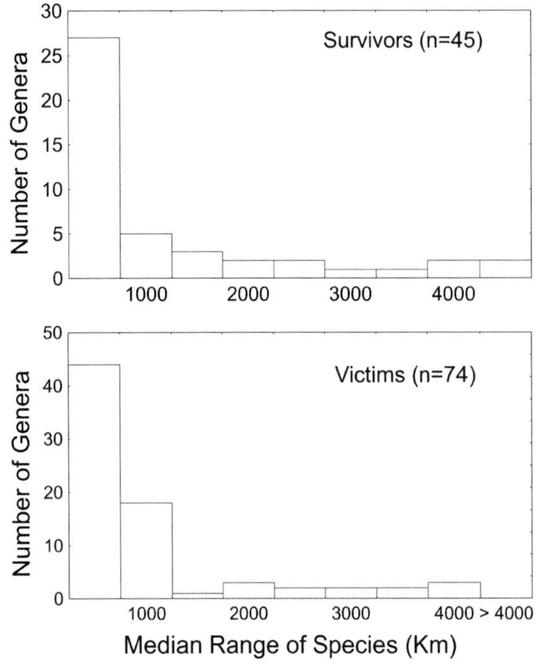

FIGURE 4. A new analysis of gastropods across the K/T boundary in the Gulf and Atlantic Coastal Plain shows no significant difference between the median geographic range of constituent species in victim and surviving genera; Mann-Whitney *U*-test: $P = 0.75$ (Jablonski unpublished).

nonselective with respect to larval types in both mollusks (Jablonski 1986b,c; Valentine and Jablonski 1986) and echinoids (Smith and Jeffery 1998, 2000a; Jeffery 2001). In this instance we can see that the selective regime operating immediately prior to the K/T event was inoperative across the boundary but returned in its aftermath, judging by the early Cenozoic evolution of nonplanktotrophy in a diverse set of gastropod lineages and in at least three additional echinoid orders (Hansen 1982; Jeffery 1997; Smith and Jeffery 1998, 2000a). In contrast, trilobite genera inferred to undergo benthic development fared significantly better than those inferred to undergo planktic development during the end-Ordovician extinction (Chatterton and Speyer 1989), whereas Lerosey-Aubril and Feist (2003) suggested that planktic development favored trilobite survivorship in the late Devonian. This apparent inconsistency for trilobites at different extinction boundaries bears investigation, perhaps including further evaluation of the criteria for developmental modes.

Although these organism-, species-, and clade-level traits lose effectiveness as predictors of survivorship at mass extinction boundaries, survivorship is not completely random. Each event seems to exhibit some form of selectivity, but one factor that promoted survival for most major groups at each of the mass extinctions is broad geographic distribution at the clade level, regardless of species-level ranges (Table 1). Extinction intensities are significantly elevated even for widespread genera during mass extinctions but the differential between widespread and localized taxa remains (e.g., Jablonski and Raup 1995; Erwin 1996; Foote 2003). This provides, among other things, another line of evidence that extinction

TABLE 1. Extinction events and taxa in which broad geographic range at the genus level enhanced survivorship (updated from Jablonski 1995).

End-Ordovician bivalves	Bretsky 1973
End-Ordovician brachiopods	Sheehan and Coorough 1990; Sheehan et al. 1996; Brenchley et al. 2001; Harper and Rong 2001
End-Ordovician bryozoans	Anstey 1986; Anstey et al. 2003
End-Ordovician trilobites	Robertson et al. 1991
End-Ordovician marine invertebrates	Foote 2003
Late Devonian bivalves	Bretsky 1973*
End-Permian bivalves	Bretsky 1973
End-Permian gastropods	Erwin 1989, 1993, 1996†
End-Triassic bivalves	Bretsky 1973; Hallam 1981; Hallam and Wignall 1997: p. 148‡
End-Cretaceous bivalves and gastropods	Jablonski 1986a,b, 1989; Jablonski and Raup 1995
Exception: End-Cretaceous echinoids	Smith and Jeffery 1998, 2000a,b

* Rode and Lieberman (2004) found broad geographic range to promote species survivorship in the Late Devonian but did not provide genus-level analyses.
† Contrary to Smith and Jeffery's (2000b) misreading of these results.
‡ McRoberts and Newton (1995) report no effect of species-level geographic range on species survivorship for European end-Triassic bivalves, consistent with end-Cretaceous results, but they do not provide genus-level statistics.

events are not simply sampling artifacts involving the false disappearance of endemic taxa. It also suggests that McGhee's (1996: p. 125) statement that broad distribution had no effect on survivorship in the Late Devonian extinction(s) should be viewed cautiously: his only evidence is the severe losses suffered by major groups that had widespread members at the time, rather than a quantitative analysis of extinction intensities among geographic-range categories. On the other hand, Smith and Jeffery (1998, 2000a,b) failed to detect differential survivorship of K/T echinoid genera according to geographic range, an anomalous result perhaps deriving from their approach to translating cladistic analyses into a taxonomic classification. A spatially explicit version of Sepkoski and Kendrick's classic (1993) study is sorely needed, to explore how the protocol used to derive taxonomic structure from phylogenetic trees affects not only temporal but also spatial diversity patterns (see also Robeck et al. 2000).

Taken together, most analyses suggest that taxonomic survivorship during the Big Five mass extinctions approaches Raup's (1984) paradigm of "nonconstructive selectivity": not strictly random, but determined in many instances by features that are not tightly linked to traits honed during background times, and thus unlikely to reinforce or promote long-term adaptation of the biota ("wanton extinction" in Raup 1991b; see also Eble 1999; and Gould 2002: pp. 1035–1037, 1323–1324, and elsewhere). This injects what Gould (1985, 1989, 2002) would call a strong element of contingency into macroevolution: even well-established clades and adaptations could be lost during these episodes, simply because they were not associated with the features that enhanced survivorship during these unusual and geologically brief events. This removal of incumbents and the subsequent diversification of formerly marginal taxa is an essential element of the evolutionary role of the major extinctions (Jablonski 1986a,b,d, 2001; Benton 1987, 1991; Jablonski and Sepkoski 1996; Eldredge 1997, 2003; Erwin 1998, 2001, 2004; Gould 2002).

This emerging picture suggests that correlations may often masquerade as direct selectivity. Because biological traits tend to covary, even across hierarchical levels, selection on one feature will tend to drag others along with it. For example, bryozoan taxa with complex colonies are generally more resistant than simple taxa to background levels of extinction but more extinction-prone during Paleozoic mass extinctions, an intriguing shift in apparent selectivity. However, colony complexity is also inversely related to genus-level geographic range, and so the actual basis for differential survival of bryozoan groups in the end-Ordovician extinction is unclear (see Anstey 1978, 1986; Anstey et al. 2003).

Even the often-cited claim for size-selectivity has proven to be questionable in many cases, and the potential examples that remain are complicated by (often nonlinear) covariation of body size with other organism- and species-level traits, from metabolic rate to local abundance to effective population size to genetic population structure to geographic range (Jablonski and Raup 1995; Jablonski 1996; Fara 2000; see also Brown 1995; Gaston and Blackburn 2000; Gaston 2003). For that matter, Fara (2000) argued that body size, diet, habitat, population size, and geographic range *all* covary in tetrapods, undermining attempts to pinpoint the key factor in taxonomic survivorship: was it modest body size, detritus-based food webs, freshwater habit, large population size, or broad geographic range? Large data sets that capture the full range of several variables, and multifactorial approaches that take into account polygonal and other nonlinear relationships, are required here.

Such correlations, whether via chance linkages or from well-tuned adaptive covariation, suggest that many selectivities apparent at the organismal level should be treated as possible indirect effects. For example, what was it about the end-Ordovician extinction that selected against broad apertural sinuses in snails (Wagner 1996) and multiple stipes in graptolites (Mitchell 1990; Melchen and Mitchell 1991); what aspect or driver of the end-Cretaceous extinction selected against schizodont hinges in bivalves, elongate rostra in echinoids, or complex sutures in cephalopods? All of these losses or severe declines more likely represent correlations rather than direct causation, but they had long-term effects on the morphological breadth—and thus the future evolutionary raw material—of their respective clades, and additional examples are plentiful. Perhaps these phenotypes represent energy-intensive metabolisms (Vermeij 1995; Bambach et al. 2002) or taxa with narrow geographic ranges or physiological tolerances, but multifactorial analyses are needed to dissect cause from correlation. Novel methods for testing causation in observational data, developed mostly outside the biological sciences but with considerable potential wherever controlled experiments are impractical, should also be explored (e.g., Shipley 2000).

Part of the difficulty in understanding the role of extinction in macroevolution is that taxonomic extinction intensity need not map closely onto morphological or functional losses. Random species loss can leave considerable phylogenetic or morphological diversity, because evolutionary trees or morphospaces will tend to be thinned rather than truncated (e.g., Nee and May 1997; Foote 1993, 1996, 1997; Roy and Foote 1997; Wills 2001). Thus Smith and Jeffery (2000a: p. 192), finding no significant changes in morphological disparity of echinoids across the K/T boundary, argued that the extinction was neutral with respect to morphology, and Lupia (1999) made the same observation for angiosperm pollen. Of course, individual subclades may suffer selective extinction of certain morphologies even as their large clades show little overall pattern (e.g., Eble 2000 on K/T echinoids; Smith and Roy 1999 on Neogene scallops). Extreme taxonomic bottlenecks will also constrict morphospace occupation and functional variety, as in end-Paleozoic echinoderms (Foote 1999) and ammonoids (McGowan 2002, 2004a,b), if only by sampling error (e.g., Foote 1996, 1997; MacLeod 2002).

On the other hand, several analyses find strong selectivity, in the sense that more morphology, functional diversity, or higher-taxonomic diversity is lost than expected from purely random species removal in the fossil record (e.g., Roy 1996; McGhee 1999; Saunders et al. 1999; McGowan 2004a,b) and among endangered taxa today (Gaston and Blackburn 1995; Bennett and Owens 1997; McKinney 1997; Jernvall and Wright 1998; Russell et al. 1998; Purvis et al. 2000a,b; Cardillo and Bromham 2001; von Euler 2001; Lockwood et al. 2002; Petchey and Gaston 2002; Zavaleta and Hulvey 2004). These nonrandom patterns need not correspond to conventional taxonomic or functional groupings. For example, Triassic ammonoid extinctions are not selective with respect to the basic morphotypes within the clade, but they can leave survivors near the center or around the periphery of a multivariate morphospace, which is then filled in again during the evolutionary recovery phase (McGowan 2004a,b). In attempting to understand losses in morphospace or among higher taxa, indirect correlations with other selective targets such as geographic range again need to be tested (see also Roy et al. 2004). Such indirect selectivity can also arise via strongly unbalanced phylogenetic trees, in which random extinction can remove entire species-poor subclades while only thinning the more profuse ones (Heard and Mooers 2000, 2002; Purvis et al. 2000b). This general role of phylogenetic topology in survivorship patterns at mass extinctions, another form of nonconstructive selectivity, has barely been explored.

"Nonconstructive selectivity" implies survivorship that is indifferent, rather than antithetical, to many of the factors that promote success during background times. This indifference means that some "preadaptation," or more properly exaptation, should occur by chance, when adaptations shaped during background times happen to improve a clade's chances of surviving the particular stresses that triggered a given mass extinction. Statistically, selectivity patterns should not be entirely mutually exclusive during background and mass extinctions. For example, Kitchell et al. (1986) attributed the preferential survival of planktonic centric diatoms at the K/T boundary to the presence of a benthic resting phase selected for during background times by seasonal variations in light, nutrient levels, and other limiting factors (Griffis and Chapman 1988; see also P. Chambers *in* MacLeod et al. 1997, although these new data are averaged over 20 Myr of late Cretaceous–early Tertiary time). Still needed is an analysis of diatom genus or species survivorship during background times relative to other plankton groups, and data on the relation of the resting spore habit to other aspects of diatom biology, including of course geographic range (for example, Barron [1995] noted that shelf species are more likely to have resting spores than are open-ocean species). The fact that no modern resting spores have been shown to remain viable for more than a few years (Hargraves and French 1983; Peters 1996; but see Lewis et al. 1999 for decadal viability) requires an especially sharp and short-lived K/T perturbation for this feature to have played a direct role in taxon survivorship (P. Chambers *in* MacLeod et al. 1997; see also Racki 1999: p. 113). On the other hand, early reports of preferential survivorship of marine detritivores at the K/T boundary, also thought to represent exaptation to an impact-driven productivity crash, have not been corroborated (see Hansen et al. 1993; Jablonski and Raup 1995; Smith and Jeffery 1998; Harries 1999; also Levinton's 1996 review of the coupling of marine planktic and detrital food webs).

Chance exaptation to extinction drivers, retention of morphological and functional breadth when extinction is random at the species level, and the persistence of diffuse ecological interactions such as predation and tiering are just a few of the factors that might explain why the evolutionary clock is not fully reset by mass extinctions. Many large-scale trends and higher taxa persist across the major boundaries, but pinpointing the reasons for that persistence or other aspects of cross-extinction evolutionary trajectories is difficult. For example, the setbacks suffered by many evolutionary trends that cross extinction boundaries (see Jablonski 2001) might be attributable to direct selectivity against the trait that was being maximized under low extinction intensities; to indirect selection owing to correlations with a disfavored trait such as narrow geographic range; or to high extinction intensities alone, with random extinction—relative to the focal trait—clearing recently invaded and thus sparsely occupied morphospace.

Recoveries

The evolutionary role of mass extinctions is not simply to knock the world into a new configuration. The extinction event does have its victims and survivors, and this raw material is crucial in shaping the post-extinction biota. However, the evolutionary novelties and ecological restructuring that typify post-extinction intervals, including the sorting of the survivors into winners and losers, may be as important as the extinction filter in determining the long-term trajectory of individual clades and the nature of the post-extinction world. Recovery intervals generally appear to be significantly longer than the extinction events that precede them (e.g., Erwin 1998, 2001, 2004; and see Sepkoski 1984, 1991a), and if this is true then the greater duration enhances the potential for sorting and biased origination to redirect evolutionary trajectories. But we still know little of the factors controlling recovery dynamics or the evolutionary directions they take.

The comparative study of recoveries is hampered by their notorious association with intervals of poor sampling and preservation. A few evolutionary systems, such as planktic foraminifera after the end-Cretaceous event, lend themselves to detailed analysis, but the fine structure of recoveries and even the timing of evolutionary pulses are generally difficult to retrieve from more than a few localized regions. As Erwin (2001) discussed, geochemical proxies and biomarkers can partially reconstruct short-term ecosystem dynamics (e.g., D'Hondt et al. 1998; Joachimski et al. 2002; and a host of stable-isotope studies), but in our present state of knowledge these tools are vague regarding the behavior of single clades.

Just as survivors of mass extinctions are difficult to predict from pre-extinction successes, long-term post-extinction successes are difficult to predict from the survivors of mass extinctions. Recovery intervals are famously important in opening opportunities for the diversification of once-marginal groups, the mammals and the dinosaurs being the classic example, but not all survivors are winners. Recoveries need not involve an across-the-board expansion of a surviving clade in morphospace. They can instead involve only subregions of the occupied morphospace or be channeled in specific directions (e.g., Dommergues et al. 1996, 2001; Lockwood 2004; McGowan 2004a); in extreme cases, taxa can weather a mass extinction only to disappear a few million years later, or fall into marginal roles (Jablonski 2002). We lack predictive insight into why different clades or subclades follow different trajectories during the recovery phase, however. A simple probabilistic model would treat post-extinction success as a function of the severity of the clade's bottleneck at the boundary: the probability of avoiding stochastic extinction or marginalization in the post-extinction world should be positively related to the number of surviving subtaxa within the clade. However, the post-extinction persistence of the 182 Paleozoic marine invertebrate orders recognized by Benton (1993) is unrelated to the size of the bottleneck suffered by each order, measured in terms of the number of genera known or inferred to have survived a given extinction (Jablonski 2002). Sampling remains a concern for such an analysis, and of course direct or indirect selectivity on the morphological or functional diversity of clades need not be closely tied to raw numbers of survivors. Nonetheless, the data give no support to a model where the success or failure of surviving clades is set by how the clades fared at the extinction event itself. Much more is going on during the recovery phase.

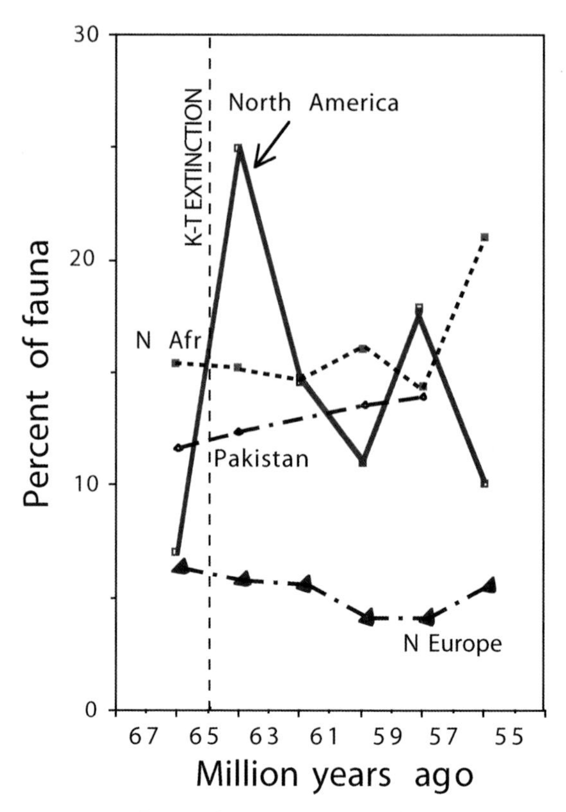

FIGURE 5. Geographic variation in the behavior of "bloom taxa" after the end-Cretaceous mass extinction. The molluscan groups that abruptly diversify and decline immediately after the extinction in North America show significantly less volatile evolutionary behavior in three other regions (North Africa, Pakistan-India, and northern Europe), as a proportion of species in each fauna as shown here, and in raw numbers. (From Jablonski 2003; for details and statistical confidence limits see Jablonski 1998.)

Most analyses have been based on global compendia or local sections and single basins, but interregional comparisons have begun to explore the spatial fabric of recoveries. For example, the molluscan recovery from the end-Cretaceous extinction unfolds differently in North America relative to the other regions studied by Jablonski (1998). The short-lived evolutionary burst of what Hansen (1988) termed "bloom taxa" in North America cannot be seen in northern Europe, north Africa, or India and Pakistan (Fig. 5), and the relative roles of local diversification and invasion also varied among these regions (Jablonski 1998, 2003; see also Heinberg 1999; Stilwell 2003: p. 324; and see Wing 2004 for hints of spatial variation in the plant recovery). These regional differences open a host of new questions on the ecological, biogeographic, and evolutionary assembly of post-extinction biotas. Dissecting the spatial fabric of recoveries, as Miller and colleagues (Miller 1997a, 1998; Miller and Mao 1995, 1998; Novack-Gottshall and Miller 2003) have done so effectively for the Ordovician radiations, is a research direction likely to be full of surprises.

Net diversification rates accelerate for some clades after mass extinctions, providing ample evidence for the macroevolutionary role of incumbency effects and their removal (Sepkoski 1984; Van Valen 1985a; Benton 1987, 1991; Miller and Sepkoski 1988; Patzkowsky 1995; Jablonski and Sepkoski 1996; Erwin 1998,

2001; Jablonski 2001; Foote 1997, 1999, 2003). The extent of the evolutionary opportunities opened by extinction events must depend at least partly on the magnitude of the event (Sepkoski 1984, 1991a; Eldredge 1997, 2002; Solé et al. 2002), but this generalization derives mainly from end-member comparisons and rather simple models. As discussed above, the degree of phylogenetic, morphological, or functional clustering of losses should also be important in giving survivors scope for evolutionary diversification and innovation (e.g., Valentine 1980; Erwin et al. 1987, Valentine and Walker 1987; Brenchley et al. 2001; Sheehan 2001). However, we lack a well-developed calculus for such analyses, and a literal reading of patterns can be misleading or at least open to multiple interpretations. For example, diversification pulses might also be artificially heightened or depressed and smeared by sampling failure in post-extinction intervals (Foote 2003); comparative, spatially explicit analyses of clades having different post-extinction dynamics and different sampling regimens would help to evaluate this effect. The provocative finding that bivalves and mammals produce more new genera relative to estimated speciation rates during post-extinction recoveries than expected from background times (Patzkowsky 1995) should be tested from this sampling standpoint, but it also suggests a much-needed hierarchical, genealogical approach to the dynamics of morphospace occupation. Judging by the few data available, larger divergences or rates of apomorphy acquisition tend to be more clearly concentrated in initial radiations than in recoveries, but other authors have reported post-extinction shifts in the average magnitude of morphological divergence per branchpoint (Anstey and Pachut 1995; Wagner 1995, 1997; Foote 1999, Eble 2000).

Hierarchy

Mass extinctions do appear to disrupt "normal" sorting processes, rendering life's trajectory difficult to predict from the patterns that prevail through the bulk of geologic time. Gould (1985, 2002) codified this view in terms of three evolutionary tiers; Raup (1984, 1991b; 1994) and Jablonski (1986a–d, 1995) were less formal but explored these effects along similar lines, all undermining a purely extrapolationist approach to macroevolutionary patterns.

Some of the debate about the limits of the extrapolationist paradigm probably stems from the application of the term to two different concepts. Mass extinctions demonstrably challenge empirical extrapolationism, i.e., the predictability of long-term outcomes from observations over ecological and short-term geological timescales; mass extinctions so disrupt evolutionary processes that the fates of taxa cannot be predicted from the behavior of populations or species over 95% of geologic time. However, Sterelny and Griffiths (1999) discussed a "fall-back" position that might rescue another version of extrapolationism. If extrapolation simply requires that organism-level selection of any kind—along with the other canonical forces of microevolution—is the prime mover of evolution (a view that might be termed "theoretical extrapolationism," as opposed to a hierarchical view of evolution operating on multiple levels simultaneously), then mass extinctions could still simply represent a relatively brief and intense change in the direction of selection. Under this fall-back position, which goes beyond predictability of outcome to continuity of process in the broadest sense, "Mass extinction by itself, however important, is no threat to the received view of evolution. Mass extinction only undercuts it if it provides crucial evidence for species selection or some other high-level mechanism" (Sterelny and Griffiths 1999: p. 306; see also Grantham 2004).

The most pervasive rule we have for mass extinction selectivity, namely the preferential survival of widespread genera, may be evidence for just such a high-level mechanism and thus support for the hierarchical view (Jablonski 1986a–d, 2000; Gould 2002). An extrapolationist approach to genus-level geographic ranges and survivorship would hold that widespread genera are simply collections of widespread, successful or ecologically generalized species, so that selectivity of genera according to their geographic ranges reflects the capabilities of their component parts.

Several lines of evidence falsify this extrapolation of organismic or species-level properties to the genus (clade) level in this context. As noted above, Jablonski (1986a,b) showed that molluscan genus survivorship at the K/T boundary was unrelated to the geographic ranges of the species within clades. That analysis found no significant difference between the frequency distributions of the geographic ranges, or the median geographic ranges, for all of the species within the surviving genera and those species within the victims (Figs. 3B, 4). The geographic ranges of species are subject to serious sampling problems and the recorded ranges are surely underestimates (Koch 1987; Jablonski 1987), but the nonparametric statistics used here, based on rank order among species ranges rather than absolute values, should be more robust to such sampling biases (Jablonski 1987; Jablonski and Valentine 1990; Gaston et al. 1996; see also Marshall 1991; and recall that Spearman's coefficient gives greater weight to pairs of ranks that are further apart and downplays the effects of ranks that are close together—see Sokal and Rohlf 1995: p. 600). Indirect evidence in the same direction also comes from recent findings that genus survivorship is unrelated to abundance at the K/T boundary for plants and marine bivalves (Hotton 2002; Lockwood 2003; Wing 2004), given the significant relation between species-level geographic range and abundance recorded for many extant groups (Gaston and Blackburn 2000 and references therein). Nor is the basic extrapolationist premise supported for living marine bivalves. For the best-documented marine molluscan biota in the world, on the eastern Pacific shelf from Point Barrow, Alaska, to Cape Horn, Chile, we find no significant relation between the geographic range of genera and the median geographic ranges of their constituent species (Fig. 6) (such analyses of nested taxa are not straightforward, owing to the autocorrelation imposed by the inability of species to be more widespread than their genera).

These paleontological data seem to be telling us, more generally, that the geographic ranges of genera are decoupled from the ranges or tolerances of their constituent species. A reasonable expectation might be that widespread genera consist mainly of environmentally tolerant, high-dispersal, ecologically successful or generalized species that also have broad geographic ranges, so that selectivity of genera according to their geographic ranges simply occurs in the extrapolationist mode. But this does not appear to be the case: the paleontological evidence is not consistent with such an argument and a more direct analysis of present-day molluscan biogeography undermines the basic extrapolationist premise.

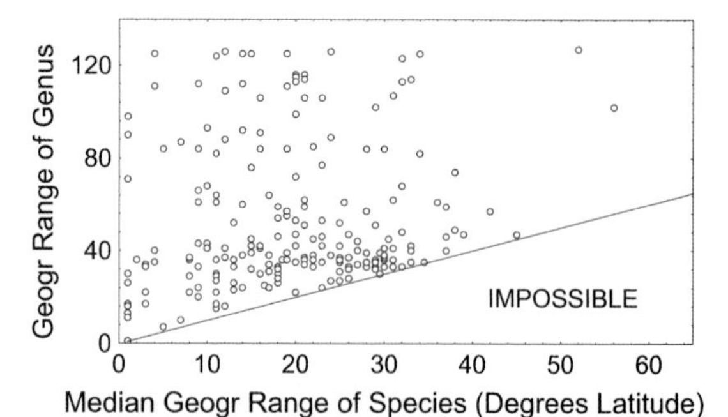

FIGURE 6. In living eastern Pacific marine bivalves, the geographic ranges of genera (in degrees latitude) cannot be predicted by the median geographic ranges of their constituent species ($n = 213$ genera, Spearman's $r = 0.17$; not significant). The lower right corner of the graph represents a field of impossible combinations: species cannot be more widespread than their genera (Jablonski, Roy, and Valentine, unpublished); statistical significance was assessed by resampling the data, using 10,000 repetitions where impossible combinations were discarded with replacement.

In this very concrete sense, geographic range at the genus level is an emergent property. Selectivity during mass extinctions evidently operates directly on the geographic range of the genus and is indifferent to the properties of the component parts at lower hierarchical levels (here, the geographic ranges of species and all of the organismic and species-level features that set those ranges). This difference across the hierarchy implies that geographic range can be an emergent property of clades, at least at the level traditionally ranked as genera, by the same criterion that suggests it can be an emergent property of species (Jablonski 2000, drawing on such related philosophical approaches as multiple-realizability, screening-off, and non-aggregativity; see Brandon 1982, 1988; Brandon et al. 1994; Wimsatt 2000). Such an argument raises the question: what actually determines geographic range at the clade level? If it was simply time since origination (Willis 1922; Gaston and Blackburn 1997; Miller 1997b), then we might expect a simple positive relation between taxon survivorship and taxon age rather than some form of Van Valen's Law (see Van Valen 1973, 1985b, 1987; Gilinsky 1994: p. 453). Dispersal ability and ecological strategies of constituent species must play some role, but interactions with biogeographic context, clade history, and other factors serve to decouple geographic range at the two levels. There are simply too many ways for a genus to become widespread—each apparently equivalent at a mass-extinction event, although this needs further exploration—for smooth extrapolation to apply.

None of these arguments require, or even suggest, that natural selection at the organismal level becomes inoperative during mass extinctions. Instead, they show that selectivity at a higher focal level yielded a predictable, coherent pattern whereas processes at the lower level often did not. Because geographic range at the clade level is probably not heritable as it is for species (Jablonski 1986b, 1987; Hunt et al. 2005—although the results and discussion of Miller and Foote [2003] could be taken to imply otherwise), mass extinctions may best be viewed as selective filters that are nonconstructive in part because they are noncumulative. Without heritability, clades will drift away from the biogeographic and other attributes that got their survivors through the mass extinction and thus become increasingly vulnerable again over time (an argument consistent with Stanley's [1990] "delayed recovery" hypothesis for extinction periodicity).

This is not to say that clade attributes related to extinction vulnerability cannot be stable over long intervals. One attractive explanation for the Phanerozoic decline in background extinction is that class- or order-level turnover rates are set early in the history of each group, so that the overall trend reflects sorting of higher taxa as the most volatile clades drop out of the biota (Sepkoski 1984, 1991a,b, 1999; Valentine 1990, 1992; Gilinsky 1994). The differential sensitivity of high-turnover, Paleozoic-fauna classes and low-turnover, Modern-fauna classes to the major extinction events (Sepkoski 1984, 1991a, 1999; Raup and Boyajian 1988) is another, little-explored selectivity pattern that, as Valentine (1990, 1992) stated, probably bears little relation to ecological adaptedness but nonetheless reflects coherent macroevolutionary processes. This leads to a fascinating set of observations of discordant behavior across hierarchical levels: (1) the relative volatility of class-level taxa during background times correlates with their respective losses during mass extinctions (Sepkoski 1984); (2) subtracting the median extinction rate for each class brings the mass extinction intensities of each to similar magnitudes, so that the most volatile classes do not actually suffer disproportionately, even though each class has a background extinction history more distinct from the others than expected by chance (Raup and Boyajian 1988); and (3) mass extinction selectivity *within* the classes is difficult to predict from background survivorship patterns, so that the evolutionary process at the species and genus level is disrupted and rechanneled by the major extinction events.

Conclusion and Outlook

Mass extinctions have evolutionary effects beyond a simple surge in extinction intensity: survivorship is not strictly random, and selectivity patterns change, so that the rules of extinction and survival are altered for a (geologically) brief time. Because the most pervasive selectivity pattern resides at the clade level, mass extinctions are a challenge to both empirical and theoretical extrapolationism. Evolutionary analyses of mass extinction therefore require a hierarchical framework that incorporates episodic shifts in the intensity and direction of selection at multiple hierarchical levels. Such an approach to the shaping of biological diversity would engender a macroevolutionary theory that is neither rigidly deterministic nor purely stochastic, and would yield a richer understanding of the long-term processes that have shaped clades and biotas.

Plenty of problems remain, of course, and these extend well beyond testing the generality of the patterns documented thus far. To list just a few:

1. The comparison of background and mass

extinction has very logically focused on single events and their immediately adjacent time bins. However, we do not know whether a fuller analysis across extinction events of differing intensities (mindful of the secular trend in background extinction mentioned earlier) would reveal a threshold effect for the shift in selectivity seen at mass extinction boundaries, or a more mixed or gradational change (see McKinney 1995 for a pioneering effort in this direction). Thresholds, if they exist, may be clade specific: volatile groups such as ammonoids might be pushed into a mass extinction selectivity regime while more phlegmatic groups such as bivalves and echinoids are relatively unfazed. This suggests that second-order extinction events, say the Eocene/Oligocene or the mid-Devonian extinctions, may prove to be a predictable, quantitative mix of background and mass extinction. Such an effect would not undermine a hierarchical view, as the differences in species-level and clade-level selectivity would still be a basic component of the evolutionary process, but we would have a deeper insight into how extinction intensity per se scales to the shift in extinction rules seen for the largest perturbations. Differences in survivorship between widespread and endemic genera have been recorded in second-order extinction events and probably hold even during background times (Jablonski 1995; Aberhan and Baumiller 2003). This suggests that the geographic-range factor is always at work at the clade level, but that it becomes most important and clearcut as the array of factors influencing genus duration during normal times dwindles in the face of heightened extinction intensities. More intense extinctions may prove, as a general rule, to be less selective (Jablonski 1986a,b,d, 1995; also Bannerjee and Boyajian 1996; Aguirre et al. 2000; Aberhan and Baumiller 2003), an argument that may also explain the failure of intrinsic factors to predict extinction risk in the most heavily stressed elements of the modern biota, such as freshwater fishes and Australian marsupials (Duncan and Lockwood 2001; Fisher et al. 2003).

2. The imperfect equivalency of taxonomic, morphologic, and functional diversities (see for example Foote 1996; Hulsey and Wainwright 2002; Valentine et al. 2002; Roy et al. 2001, 2004; Neige 2003) suggests that we have much to learn about the scope for indirect effects of extinction—i.e., the termination of genealogical units—on the array of adaptations that are lost or persist to be the springboard for the recovery process, and on the ecological impact of a given extinction intensity (e.g., Ives and Cardinale 2004; Solan et al. 2004; Wootton 2004; Zavaleta and Hulvey 2004). A deeper, predictive understanding of when these three aspects of diversity tend to correspond and when they diverge (see Foote 1996, 1997, 1999; Wills 2001, for a strong start) and the consequences of selectivity in one aspect on the extinction and recovery dynamics of the other two would shed fresh light on the role of contingency in evolution (Gould 2002).
3. We have barely begun to decipher the rules of successful recovery. As with the extinctions themselves, the emergence of winners from among the survivors might be stochastic, or governed by a set of recovery-specific rules, or dominated by processes prevalent during background times. The inordinate production of evolutionary novelties during recovery intervals suggests that post-extinction dynamics do not simply involve an immediate return to business as usual. This probably represents an especially fruitful area at the intersection of macroevolution and ecology (and see Foote 1999 for reasons to keep developmental biology in the mix). As with other diversifications—most famously the Cambrian explosion of metazoans—the cessation of the recovery phase, however defined (Erwin 1998, 2001, 2004), is potentially as interesting an area of inquiry as its onset.
4. The spatial fabric of extinctions and recoveries needs to be explored more rigorously, and in a comparative framework. If global analyses incorporate spatially explicit data, we will have a much better chance of distinguishing facies effects and regional variation in sample quality or density from bi-

otic pattern. Predictive models for the relative roles of local diversification and biotic interchange in the recovery of biodiversity (taxonomic, morphologic, and functional, as discussed above) may have more than academic interest as the present-day biota buckles under human pressure (see Jablonski 2001). The ongoing debate on whether Lazarus taxa represent sampling failure within local assemblages or localized persistence elsewhere (e.g., Jablonski 1986a; Wignall and Benton 1999; Fara 2001; Smith 2001; Twitchett 2001; Rickards and Wright 2002) is just one facet of the spatial dynamics of extinction and recovery.

Despite all of these unknowns and challenges, and I could list many more, the shared survivorship patterns among mass extinctions suggests that we are beginning to pinpoint general biological principles of survivorship and recovery. The K/T and Permo-Triassic events probably had different causes, but the similarities in biodiversity dynamics are all the more intriguing and biologically important for that reason. And as Gould (2002) forcefully argued, rare but intense extinction events under the "different rules" regime, imposed simultaneously on diverse clades, represent distinctive mechanisms operating on a distinctive scale across hierarchical levels. Mass extinctions and recoveries are far from being the entire evolutionary show, but they are vital as windows into the behavior of clades and ecosystems under extreme pressure and its aftermath. A more complete and effective understanding of the evolutionary process must incorporate these major episodes in the history of life.

Acknowledgments

I thank Niles Eldredge and Elizabeth Vrba for inviting me to participate in this tribute to a sorely missed friend and colleague. M. J. Benton, M. Foote, T. A. Grantham, C. H. Jeffery, J. D. Marshall, A. J. McGowan, and especially S. M. Kidwell provided valuable reviews. Thanks also to R. K. Bambach and S. C. Wang, who provided information on their then unpublished work, to C. R. McRoberts for his data on end-Triassic bivalve extinctions, to C. E. Mitchell for advice on graptolites, and to T. Rothfus for performing the resampling analysis of Figure 6. Supported by National Science Foundation grant EAR99-03030 and a Research Associateship at the National History Museum, London.

Literature Cited

Aberhan, M., and T. K. Baumiller. 2003. Selective extinction among Early Jurassic bivalves: a consequence of anoxia. Geology 31:1077–1080.

Adrain, J. M., and S. R. Westrop. 2000. An empirical assessment of taxic paleobiology. Science 289:110–112.

Aguirre, J., R. Riding, and J. C. Braga. 2000. Late Cretaceous incident light reduction: evidence from benthic algae. Lethaia 33:205–213.

Anstey, R. L. 1978. Taxonomic survivorship and morphologic complexity in Paleozoic bryozoan genera. Paleobiology 4: 407–418.

———. 1986. Bryozoan provinces and patterns of generic evolution and extinction in the Late Ordovician of North America. Lethaia 19:33–51.

Anstey, R. L., and J. F. Pachut. 1995. Phylogeny, diversity history, and speciation in Paleozoic bryozoans. Pp. 239–284 *in* D. H. Erwin and R. L. Anstey, eds. New approaches to speciation in the fossil record. Columbia University Press, New York.

Anstey, R. L., J. F. Pachut, and M. E. Tuckey. 2003. Patterns of bryozoan endemism through the Ordovician–Silurian transition. Paleobiology 29:305–328.

Balinski, A., E. Olempska, and G. Racki, eds. 2002. Biotic responses to the Late Devonian global events. Acta Palaeontologica Polonica 47:186–404.

Bambach, R. K., and A. H. Knoll. 2001. Is there a separate class of "mass" extinctions? Geological Society of America Abstracts with Programs 33(6):A-141.

Bambach, R. K., A. H. Knoll, and J. J. Sepkoski Jr. 2002. Anatomical and ecological constraints on Phanerozoic animal diversity in the marine realm. Proceedings of the National Academy of Sciences USA 99:6854–6859.

Bambach, R. K., A. H. Knoll, and S. C. Wang. 2004. Origination, extinction, and mass depletions of marine diversity. Paleobiology 30:522–542.

Bannerjee, A., and G. Boyajian. 1996. Changing biologic selectivity in the Foraminifera over the past 150 m.y. Geology 24: 607–610.

Barron, J. A. 1985. Miocene to Holocene planktic diatoms. Pp. 763–809 *in* H. M. Bolli, J. B. Saunders, and K. Perch-Nielsen, eds. Plankton stratigraphy. Cambridge University Press, Cambridge.

Bennett, P. M., and I. P. F. Owens. 1997. Variation in extinction risk among birds: chance or evolutionary predisposition? Proceedings of the Royal Society of London B 264:401–408.

Benton, M. J. 1987. Progress and competition in macroevolution. Biological Reviews 62:305–338.

———. 1991. Extinction, biotic replacements, and clade interactions. Pp. 89–102 *in* E. C. Dudley, ed. The unity of evolutionary biology. Dioscorides Press, Portland, Oregon.

———., ed. 1993. The fossil record 2. Chapman and Hall, London.

Benton, M. J., and R. J. Twitchett. 2003. How to kill (almost) all life: the end-Permian extinction event. Trends in Ecology and Evolution 18:358–365.

Brandon, R. N. 1982. The levels of selection. Pp. 315–323 *in* P. Asquith and T. Nichols, eds. PSA 1982, Vol. 1. Philosophy of Science Association, East Lansing, Mich.

———. 1988. The levels of selection: a hierarchy of interactors. Pp. 51–71 *in* H. Plotkin, ed. The role of behavior in evolution. MIT Press, Cambridge.

Brandon, R. N., J. Antonovics, R. Burian, S. Carson, G. Cooper, P. S. Davies, C. Horvath, B. D. Mishler, R. C. Richardson, K. Smith, and P. Thrall. 1994. Sober on Brandon on screening-off and the levels of selection. Philosophy of Science 61:475–486.

Brenchley, P. J., J. D. Marshall, and C. J. Underwood. 2001. Do all mass extinctions represent an ecological crisis? Evidence from the Late Ordovician. Geological Journal 36:329–340.

Bretsky, P. W. 1973. Evolutionary patterns in the Paleozoic Bivalvia: documentation and some theoretical considerations. Geological Society of America Bulletin 84:2079–2096.

Brown, J. H. 1995. Macroecology. University of Chicago Press, Chicago.

Cardillo, M., and L. Bromham. 2001. Body size and risk of extinction in Australian mammals. Conservation Biology 15: 1435–1440.

Chatterton, B. D. E., and S. E. Speyer. 1989. Larval ecology, life-history strategies, and patterns of extinction and survivorship among Ordovician trilobites. Paleobiology 15:118–132.

Copper, P. 2002. Reef development at the Frasnian/Famennian mass extinction boundary. Palaeogeography, Palaeoclimatology, Palaeoecology 181:27–65.

D'Hondt, S., P. Donaghay, J. C. Zachos, D. Luttenberg, and M. Lindinger. 1998. Organic carbon fluxes and ecological recovery from the Cretaceous-Tertiary mass extinction. Science 282: 276–279.

Dommergues, J.-L., B. Laurin, and C. Meister. 1996. Evolution of ammonoid morphospace during the Early Jurassic radiation. Paleobiology 22:219–240.

———. 2001. The recovery and radiation of Early Jurassic ammonoids: morphologic versus palaeobiogeographical patterns. Palaeogeography, Palaeoclimatology, Palaeoecology 165:195–213.

Droser, M. L., D. J. Bottjer, P. M. Sheehan, and G. R McGhee Jr. 2000. Decoupling of taxonomic and ecologic severity of Phanerozoic marine mass extinctions. Geology 28:675–678.

Duncan, J. R., and J. L. Lockwood. 2001. Extinction in a field of bullets: a search for causes in the decline of the world's freshwater fishes. Biological Conservation 102:97–105.

Eble, G. J. 1999. On the dual nature of chance in evolutionary biology and paleobiology. Paleobiology 25:75–87.

———. 2000. Constrasting evolutionary flexibility in sister groups: disparity and diversity in Mesozoic atelostomate echinoids. Paleobiology 26:56–79.

Eldredge, N. 1997. Extinction and the evolutionary process. Pp. 60–73 *in* T. Abe, S. A. Levin, and M. Higashi, eds. Biodiversity: an ecological perspective. Springer, Berlin.

———. 2003. The sloshing bucket: how the physical realm controls evolution. Pp. 3–32 *in* J. P. Crutchfield and P. Schuster, eds. Evolutionary dynamics. Oxford University Press, New York.

Erwin, D. H. 1989. Regional paleoecology of Permian gastropod genera, southwestern United States and the end-Permian mass extinction. Palaios 4:424–438.

———. 1993. The great Paleozoic crisis: life and death in the Permian. Columbia University Press, New York.

———. 1996. Understanding biotic recoveries: extinction, survival, and preservation during the end-Permian mass extinction. Pp. 398–418 *in* Jablonski, et al. 1996.

———. 1998. The end and the beginning: recoveries from mass extinctions. Trends in Ecology and Evolution 13:344–349.

———. 2001. Lessons from the past: evolutionary impacts of mass extinctions. Proceedings of the National Academy of Sciences USA 98:5399–5403.

———. 2003. Impact at the Permo-Triassic boundary: a critical evaluation. Astrobiology 3:67–74.

———. 2004. Mass extinctions and evolutionary radiations. Pp. 218–228 *in* A. Moya and E. Font, eds. Evolution from molecules to ecosystems. Oxford University Press, New York.

Erwin, D. H., J. W. Valentine, and J. J. Sepkoski Jr. 1987. A comparative study of diversification events: the early Paleozoic versus the Mesozoic. Evolution 41:1177–1186.

Erwin, D. H., S. A. Bowring, and Y. G. Jin. 2002. End-Permian mass extinctions: a review. Geological Society of America Special Paper 356:363–384.

Fara, E. 2000. Diversity of Callovian-Ypresian (Middle Jurassic-Eocene) tetrapod families and selectivity of extinctions at the K/T boundary. Geobios 33:387–396.

———. 2001. What are Lazarus taxa? Geological Journal 36:291–303.

Fisher, D. O., S. P. Blomberg, and I. P. F. Owens. 2003. Extrinsic versus intrinsic factors in the decline and extinction of Australian marsupials. Proceedings of the Royal Society of London B 270:1801–1808.

Flügel, E. 2002. Triassic reef patterns. *In* W. Kiessling, E. Flügel, and J. Golonka, eds. Phanerozoic reef patterns. SEPM Special Publication 72:391–463.

Foote, M. 1993. Discordance and concordance between morphological and taxonomic diversity. Paleobiology 19:185–204.

———. 1996. Models of morphological diversification. Pp. 62–86 *in* Jablonski et al. 1996.

———. 1997. The evolution of morphological diversity. Annual Review of Ecology and Systematics 28:129–152.

———. 1999. Morphological diversity in the evolutionary radiation of Paleozoic and post-Paleozoic crinoids. Paleobiology Memoirs No. 1. Paleobiology 25(Suppl. to No. 2).

———. 2003. Origination and extinction through the Phanerozoic: a new approach. Journal of Geology 111:125–148.

———. 2005. Pulsed origination and extinction in the marine realm. Paleobiology 31:6–20.

Gaston, K. J. 2003. The structure and dynamics of geographic ranges. Oxford University Press, Oxford.

Gaston, K. J., and T. M. Blackburn. 1995. Birds, body size, and the threat of extinction. Philosophical Transactions of the Royal Society of London B 347:205–212.

———. 1997. Age, area and avian diversification. Biological Journal of the Linnean Society 62:239–253.

———. 2000. Pattern and process in macroecology. Blackwell Science, Oxford.

Gaston, K. J., R. M. Quinn, S. Wood, and H. R. Arnold. 1996. Measures of geographic range size: the effects of sample size. Ecography 19:259–268.

Gilinsky, N. L. 1994. Volatility and the Phanerozoic decline of background extinction intensity. Paleobiology 20:445–458.

Gilinsky, N. L., and R. K. Bambach. 1987. Asymmetrical patterns of origination and extinction in higher taxa. Paleobiology 13: 427–445.

Gould, S. J. 1985. The paradox of the first tier: an agenda for paleobiology. Paleobiology 11:2–12.

———. 1989. Wonderful life. W. W. Norton, New York.

———. 2002. The structure of evolutionary theory. Harvard University Press, Cambridge.

Grantham, T. A. 2004. Constaints and spandrels in Gould's *Structure of Evolutionary Theory.* Biology and Philosophy 19: 29–43.

Griffis, K., and D. J. Chapman. 1988. Survival of phytoplankton under prolonged darkness: implications for the Cretaceous-Tertiary darkness hypothesis. Palaeogeography, Palaeoclimatology, Palaeoecology 67:305–314.

Hallam, A. 1981. The end-Triassic bivalve extinction event. Palaeogeography, Palaeoclimatology, Palaeoecology 35:1–44.

Hallam, A., and P. B. Wignall. 1997. Mass extinctions and their aftermath. Oxford University Press, Oxford.

Hansen, T. A. 1982. Modes of larval development in early Tertiary neogastropods. Paleobiology 8:367–377.

———. 1988. Early Tertiary radiation of marine molluscs and the long-term effects of the Cretaceous-Tertiary extinction. Paleobiology 14:37–51.

Hansen, T. A., B. Upshaw, E. G. Kauffman, and W. Gose. 1993. Patterns of molluscan extinction and recovery across the Cretaceous-Tertiary boundary in east Texas: report on new outcrops. Cretaceous Research 14:685–706.

Hargraves, P. E., and F. W. French. 1983. Diatom resting spores: significance and strategies. Pp. 49–68 *in* G. A. Fryxell, ed. Survival strategies of the algae. Cambridge University Press, Cambridge.

Harper, D. A. T., and J.-Y. Rong. 2001. Palaeozoic brachiopod extinctions, survival and recovery: patterns within the rhynchonelliformeans. Geological Journal 36:317–328.

Harries, P. J. 1999. Repopulations from Cretaceous mass extinctions: environmental and/or evolutionary controls? Geological Society of America Special Paper 332:345–364.

Heard, S. B., and A. Ø. Mooers. 2000. Phylogenetically patterned speciation rates and extinction risks change the loss of evolutionary history during extinctions. Proceedings of the Royal Society of London B 267:613–620.

———. 2002. Signatures of random and selective extinctions in phylogenetic tree balance. Systematic Biology 51:889–898.

Heinberg, C. 1999. Lower Danian bivalves, Stevns Klint, Denmark: continuity across the K/T boundary. Palaeogeography, Palaeoclimatology, Palaeoecology 154:87–106.

Hoffman, A. 1989. Arguments on evolution. Oxford University Press, New York.

Hotton, C. L. 2002. Palynology of the Cretaceous-Tertiary boundary in central Montana: evidence for extraterrestrial impact as a cause of the terminal Cretaceous extinctions. Geological Society of America Special Paper 361:473–502.

Hubbard, A. E., and N. L. Gilinsky. 1992. Mass extinctions as statistical phenomena: an examination of the evidence using χ^2 tests and bootstrapping. Paleobiology 18:148–160.

Hulsey, C. D., and P. C. Wainwright. 2002. Projecting mechanics into morphospace: disparity in the feeding system of labrid fishes. Proceedings of the Royal Society of London B 269:317–326.

Hunt, G., K. Roy, and D. Jablonski. 2005. Heritability of geographic range sizes revisited. American Naturalist (in press).

Ives, I. R., and E. J. Cardinale. 2004. Food-web interactions govern the resistance of communities after non-random extinctions. Nature. 429:174–177.

Jablonski, D. 1986a. Causes and consequences of mass extinctions: a comparative approach. Pp. 183–229 *in* D. K. Elliott, ed. Dynamics of extinction. Wiley, New York.

———. 1986b. Background and mass extinctions: the alternation of macroevolutionary regimes. Science 231:129–133.

———. 1986c. Larval ecology and macroevolution of marine invertebrates. Bulletin of Marine Science 39:565–587.

———. 1986d. Evolutionary consequences of mass extinctions. Pp. 313–329 *in* D. M. Raup and D. Jablonski, eds. Patterns and processes in the history of life. Springer, Berlin.

———. 1987. Heritability at the species level: analysis of geographic ranges of Cretaceous mollusks. Science 238:360–363.

———. 1989. The biology of mass extinction: a paleontological view. Philosophical Transactions of the Royal Society of London B 325:357–368.

———. 1995. Extinction in the fossil record. Pp. 25–44 *in* R. M. May and J. H. Lawton, eds. Extinction rates. Oxford University Press, Oxford.

———. 1996. Body size and macroevolution. Pp. 256–289 *in* Jablonski et al. 1996.

———. 1998. Geographic variation in the molluscan recovery from the end-Cretaceous extinction. Science 279:1327–1330.

———. 2000. Micro- and macroevolution: scale and hierarchy in evolutionary biology and paleobiology. *In* D. H. Erwin, and S. L. Wings, eds. Deep time: *Paleobiology's* perspective. Paleobiology 26(Supplement to No. 4):15–52.

———. 2001. Lessons from the past: evolutionary impacts of mass extinctions. Proceedings of the National Academy of Sciences USA 98:5393–5398.

———. 2002. Survival without recovery after mass extinctions. Proceedings of the National Academy of Sciences USA 99: 8139–8144.

———. 2003. The interplay of physical and biotic factors in macroevolution. Pp. 235–252 *in* L. J. Rothschild and A. M. Lister, eds. Evolution on planet Earth. Elsevier/Academic Press, Amsterdam/London.

Jablonski, D., and D. M. Raup. 1995. Selectivity of end-Cretaceous marine bivalve extinctions. Science 268:389–391.

Jablonski, D., and J. J. Sepkoski Jr. 1996. Paleobiology, community ecology, and scales of ecological pattern. Ecology 77: 1367–1378.

Jablonski, D., and J. W. Valentine. 1990. From regional to total geographic ranges: testing the relationship in Recent bivalves. Paleobiology 16:126–142.

Jablonski, D., D. H. Erwin, and J. H. Lipps. 1996. Evolutionary paleobiology. University of Chicago Press, Chicago.

Jeffery, C. H. 1997. Dawn of echinoid nonplankotrophy: coordinated shifts in development indicate environmental instability prior to the K-T boundary. Geology 25:991–994.

———. 2001. Heart urchins at the Cretaceous/Tertiary boundary: a tale of two clades. Paleobiology 27:140–158.

Jernvall, J., and P. C. Wright. 1998. Diversity components of impending primate extinctions. Proceedings of the National Academy of Sciences USA 95:11279–11283.

Joachimski, M. M., R. D. Pancost, K. H. Freeman, C. Ostertag-Henning, and W. Buggisch. 2002. Carbon isotope geochemistry of the Frasnian-Famennian transition. Palaeogeography, Palaeoclimatology, Palaeoecology 181:91–109.

Kidwell, S. M. 2005. Shell composition has no net impact on large-scale evolutionary patterns in mollusks. Science 307: 914–917.

Kidwell, S. M., and S. M. Holland. 2002. The quality of the fossil record: implications for evolutionary analyses. Annual Review of Ecology and Systematics 33:561–588.

Kiessling, W., and R. C. Baron-Szabo. 2004. Extinction and recovery patterns of scleractinian corals at the Cretaceous-Tertiary boundary. Palaeogeography, Palaeoclimatology, Palaeoecology 214:195–223.

Kitchell, J. A., D. L. Clark, and A. M. Gombos. 1986. Biological selectivity of extinction: a link between background and mass extinction. Palaios 1:504–511.

Koch, C. F. 1987. Prediction of sample-size effects on the measured temporal and geographic distribution patterns of species. Paleobiology 13:100–107.

Lee, M. S. Y., and P. Doughty. 2003. The geometric meaning of macroevolution. Trends in Ecology and Evolution 18:263–266.

Lerosey-Aubril, R., and R. Feist. 2003. Early ontogeny of trilobites: implications for selectivity of survivorship at the end-Devonian crisis. Geological Society of America Abstracts with Programs 34(7):385.

Levinton, J. S. 1996. Trophic group and the end-Cretaceous extinction: did deposit feeders have it made in the shade? Paleobiology 22:104–112.

Lewis, J., A. S. D. Harris, K. J. Jones, and R. L. Edmonds. 1999. Long-term survival of marine planktonic diatoms and dinoflagellates in stored sediment samples. Journal of Plankton Research 21:343–354.

Lockwood, J. L., G. J. Russell, J. L. Gittleman, C. C. Daehler, M. L. McKinney, and A. Purvis. 2002. A metric for analyzing taxonomic patterns of extinction risk. Conservation Biology 16: 1137–1142.

Lockwood, R. 2003. Abundance not linked to survival across the end-Cretaceous mass extinction: patterns in North American bivalves. Proceedings of the National Academy of Sciences USA 100:2478–2482.

———. 2004. The K/T event and infaunality: morphological and ecological patterns of extinction and recovery in veneroid bivalves. Paleobiology 30:507–521.

Lupia, R. 1999. Discordant morphological disparity and taxonomic diversity during the Cretaceous angiosperm radiation: North American pollen record. Paleobiology 25:1–28.

MacLeod, N. 2002. Testing evolutionary hypotheses with adaptive landscapes: use of random phylogenetic-morphological studies. Mathematische Geologie 6:45–55.

———. 2003a. The causes of Phanerozoic extinctions. Pp. 253–277 *in* L. J. Rothschild and A. M. Lister, eds. Evolution on planet Earth. Elsevier/Academic Press, Amsterdam/London.

———. 2003b. [Review of] The structure of evolutionary theory. Palaeontological Association Newsletter 50:40–46.

MacLeod, N., and 21 others. 1997. The Cretaceous-Tertiary biotic transition. Journal of the Geological Society, London 154: 265–292.

Marshall, C. R. 1991. Estimation of taxonomic ranges from the fossil record. *In* N. L. Gilinsky and P. W. Signor, eds. Analytical paleontology. Short Courses in Paleontology 4:19–38. Paleontological Society, Knoxville, Tenn.

McGhee, G. R., Jr. 1996. The Late Devonian mass extinction. Columbia University Press, New York.

———. 1999. Theoretical morphology: the concept and its application. Columbia University Press, New York.

McGowan, A. J. 2002. A morphometric study of the effect of diversity crises on Triassic ammonoid evolution. Geological Society of America Abstracts with Programs 34(6):361.

———. 2004a. Ammonoid taxonomic and morphologic recovery patterns after the Permian-Traissic. Geology 32:665–668.

———. 2004b. The effect of the Permo-Triassic bottleneck on Triassic ammonoid morphological evolution. Paleobiology 30: 369–395.

McKinney, M. L. 1995. Extinction selectivity among lower taxa: gradational patterns and rarefaction error in extinction estimates. Paleobiology 21:300–313.

———. 1997. Extinction vulnerability and selectivity: combining ecological and paleontological views. Annual Review of Ecology and Systematics 28:495–516.

McRoberts, C. A. 2001. Triassic bivalves and the initial marine Mesozoic revolution: a role for predators? Geology 29:359–362.

McRoberts, C. A., and C. R. Newton. 1995. Selective extinction among end-Triassic European bivalves. Geology 23:102–104.

Melchin, M. J., and C. E. Mitchell. 1991. Late Ordovician extinction of the Graptoloidea. *In* C. R. Barnes and S. H. Williams, eds. Advances in Ordovician geology. Geological Survey of Canada Paper 90-9:143–156.

Miller, A. I. 1997a. Dissecting global diversity patterns: examples from the Ordovician Radiation. Annual Review of Ecology and Systematics 28:85–104.

———. 1997b. A new look at age and area: the geographic and environmental expansion of genera during the Ordovician Radiation. Paleobiology 23:410–419.

———. 1998. Biotic transitions in global marine diversity. Science 281:1157–1160.

Miller, A. I., and M. Foote. 2003. Increased longevities of post-Paleozoic marine genera after mass extinctions. Science 302: 1030–1032.

Miller, A. I., and S. Mao. 1995. Association of orogenic activity with the Ordovician radiation of marine life. Geology 23:305–308.

———. 1998. Scales of diversification and the Ordovician Radiation. Pp. 288–310 *in* M. L. McKinney and J. A. Drake, eds. Biodiversity dynamics. Columbia University Press, New York.

Miller, A. I., and J. J. Sepkoski Jr. 1988. Modeling bivalve diversification: the effect of interaction on a macroevolutionary system. Paleobiology 14:364–369.

Mitchell, C. E. 1990. Directional macroevolution of the diplograptacean graptolites: a product of astogenetic heterochrony and directed speciation. *In* P. D. Taylor and G. P. Larwood, eds. Major evolutionary radiations. Systematics Association Special Volume 42:235–264.Clarendon, Oxford.

Nee, S., and R. M. May. 1997. Extinction and the loss of evolutionary history. Science 278:692–694.

Neige, P. 2003. Spatial patterns of disparity and diversity of the Recent cuttlefishes (Cephalopoda) across the Old World. Journal of Biogeography 30:1125–1137.

Newman, M. E. J., and G. J. Eble. 1999. Decline in extinction rates and scale invariance in the fossil record. Paleobiology 25:434–439.

Novack-Gottshall, P. M., and A. I. Miller. 2003. Comparative geographic and environmental diversity dynamics of gastropods and bivalves. Paleobiology 29:576–604.

Page, K. N., 1996. Mesozoic ammonoids in space and time. Pp. 755–794 *in* N. H. Landman, K. Tanabe, and R. A. Davis, eds. Ammonoid paleobiology. Plenum, New York.

Patzkowsky, M. E. 1995. A hierarchical branching model of evolutionary radiations. Paleobiology 21:440–460.

Petchey, O. L., and K. J. Gaston. 2002. Extinction and the loss of functional diversity. Proceedings of the Royal Society of London B 269:1721–1727.

Peters, E. 1996. Prolonged darkness and diatom mortality. 2. Marine temperate species. Journal of Experimental Marine Biology and Ecology 207:43–58.

Peters, S. E., and M. Foote. 2001. Biodiversity in the Phanerozoic: a reinterpretation. Paleobiology 27:583–601.

Purvis, A., K. E. Jones, and G. M. Mace. 2000a. Extinction. BioEssays 22:1123–1133.

Purvis, A., P.-M. Agapow, J. L. Gittleman, and G. M. Mace. 2000b. Nonrandom extinction and the loss of evolutionary history. Science 288:328–330.

Racki, G. 1999. Silica-secreting biota and mass extinctions: survival patterns and processes. Palaeogeography, Palaeoclimatology, Palaeoecology 154:107–132.

Racki, G., and M. R. House, eds. 2002. Late Devonian biotic crisis: ecological, depositional and geochemical records. Palaeogeography, Palaeoclimatology, Palaeoecology 181:1–374.

Raup, D. M. 1984. Evolutionary radiations and extinctions. Pp. 5–14 *in* H. D. Holland and A. F. Trendall, eds. Patterns of change in Earth evolution. Springer, Berlin.

———. 1985. Mathematical models of cladogenesis. Paleobiology 11:42–52.

———. 1991a. A kill curve for Phanerozoic marine species. Paleobiology 17:37–48.

———. 1991b. Extinction: bad genes or bad luck? Norton, New York.

———. 1991c. The future of analytical paleontology. *In* N. L. Gilinsky and P. W. Signor, eds. Analytical paleontology. Short Courses in Paleontology 4:207–216. Paleontological Society, Knoxville, Tenn.

———. 1994. The role of extinction in evolution. Proceedings of the National Academy of Sciences USA 91:6758–6763.

———. 1996. Extinction models. Pp. 419–433 *in* Jablonski et al. 1996.

Raup, D. M., and G. E. Boyajian. 1988. Patterns of generic extinction in the fossil record. Paleobiology 14:109–125.

Raup, D. M., and J. J. Sepkoski Jr. 1982. Mass extinctions in the marine fossil record. Science 215:1501–1503.

Rickards, R. B., and A. J. Wright. 2002. Lazarus taxa, refugia and

relict faunas: evidence from graptolites. Journal of the Geological Society, London 159:1–4.
Robeck, H. E., C. C. Maley, and M. J. Donoghue. 2000. Taxonomy and temporal diversity patterns. Paleobiology 26:171–187.
Robertson, D. B. R., P. J. Brenchley, and A. W. Owen. 1991. Ecological disruption close to the Ordovician-Silurian boundary. Historical Biology 5:131–144.
Rode, A. L., and B. S. Lieberman. 2004. Using GIS to unlock the interactions between biogeography, environment, and evolution in Middle and Late Devonian brachiopods and bivalves. Palaeogeography, Palaeoclimatology, Palaeoecology 211:345–359.
Roy, K. 1996. The roles of mass extinction and biotic interaction in large-scale replacements: a reexamination using the fossil record of stromboidean gastropods. Paleobiology 22:436–452.
Roy, K., and M. Foote. 1997. Morphological approaches to measuring biodiversity. Trends in Ecology and Evolution 12:277–281.
Roy, K., D. P. Balch, and M. E. Hellberg. 2001. Spatial patterns of morphological diversity across the Indo-Pacific: analyses using strombid gastropods. Proceedings of the Royal Society of London B 268:2503–2508.
Roy, K., D. Jablonski, and J. W. Valentine. 2004. Beyond species richness: biogeographic patterns and biodiversity dynamics using other metrics of diversity. Pp. 151–170 *in* M. V. Lomolino and L. R. Heaney, eds. Frontiers of biogeography. Sinauer, Sunderland, Mass.
Russell, G. J., T. M. Brooks, M. L. McKinney, and G. C. Anderson. 1998. Present and future taxonomic selectivity in bird and mammal extinctions. Conservation Biology 12:1365–1376.
Saunders, W. B., D. M. Work, and S. V. Nikolaeva. 1999. Evolution of complexity in Paleozoic ammonoid sutures. Science 286:760–763.
Sepkoski, J. J. Jr. 1984. A kinetic model of Phanerozoic taxonomic diversity. III. Post-Paleozoic families and mass extinctions. Paleobiology 10:246–267.
———. 1991a. Diversity in the Phanerozoic oceans: a partisan review. Pp. 210–236 *in* E. C. Dudley, ed. The unity of evolutionary biology. Dioscorides Press, Portland, Ore.
———. 1991b. A model of onshore-offshore change in faunal diversity. Paleobiology 17:58–77.
———. 1996. Patterns of Phanerozoic extinction: a perspective from global data bases. Pp. 35–51 *in* O. H. Walliser, ed. Global events and event stratigraphy. Springer, Berlin.
———. 1999. Rates of speciation in the fossil record. Pp. 260–282 *in* A. E. Magurran and R. M. May, eds. Evolution of biological diversity. Oxford University Press, Oxford.
———. 2002. A compendium of fossil marine animal genera. Bulletins of American Paleontology 363:1–560.
Sepkoski, J. J., Jr., and D. C. Kendrick. 1993. Numerical experiments with model monophyletic and paraphyletic taxa. Paleobiology 19:168–184.
Sepkoski, J. J., Jr., and C. F. Koch. 1996. Evaluating paleontologic data relating to bio-events. Pp. 21–34 *in* O. H. Walliser, ed. Global events and event stratigraphy. Springer, Berlin.
Sheehan, P. M. 2001. History of marine biodiversity. Geological Journal 36:231–249.
Sheehan, P. M., and P. J. Coorough. 1990. Brachiopod zoogeography across the Ordovician-Silurian boundary. Geological Society of London Memoir 12:181–187.
Sheehan, P. M., P. J. Coorough, and D. E. Fastovsky. 1996. Biotic selectivity during the K/T and Late Ordovician extinction events. Geological Society of America Special Paper 307:477–489.
Shipley, B. 2000. Cause and correlation in biology. Cambridge University Press, Cambridge.
Smith, A. B. 2001. Large-scale heterogeneity of the fossil record: implications for Phanerozoic biodiversity studies. Philosophical Transactions of the Royal Society of London B 356:1–17.
Smith, A. B., and C. H. Jeffery. 1998. Selectivity of extinction among sea urchins at the end of the Cretaceous Period. Nature 392:69–71.
———. 2000a. Changes in the diversity, taxic composition and life-history patterns of echinoids over the past 145 million years. Pp. 181–194 *in* S. J. Culver and P. F. Rawson, eds. Biotic response to global change: the last 145 million years. Cambridge University Press, Cambridge.
———. 2000b. Maastrichtian and Palaeocene echinoids: a key to world faunas. Special Papers in Palaeontology 63.
Smith, J. T., and K. Roy. 1999. Late Neogene extinctions and modern regional species diversity: analyses using the Pectinidae of California. Geological Society of America Abstracts with Programs 31(7):473.
Sokal, R. R., and F. J. Rohlf. 1995. Biometry, 3d ed. W. H. Freeman, San Francisco.
Solan, M., B. J. Cardinale, A. L. Dowling, K. A. M. Engelhardt, J. L. Rusenick, and D. S. Srivastava. 2004. Extinction and ecosystem function in the marine benthos. Science 306:1177–1180.
Solé, R. V., J. M. Montoya, and D. H. Erwin. 2002. Recovery after mass extinction: evolutionary assembly in large-scale biosphere dynamics. Philosophical Transactions of the Royal Society of London B 357:697–707.
Stanley, G. D. 2003. The evolution of modern corals and their early history. Earth-Science Reviews 60:195–225.
Stanley, S. M. 1990. Delayed recovery and the spacing of major extinctions. Paleobiology 16:401–414.
Sterelny, K., and P. E. Griffiths. 1999. Sex and death: an introduction to the philosophy of biology. University of Chicago Press, Chicago.
Steuber, T., S. F. Mitchell, D. Buhl, G. Gunter, and H. U. Kasper. 2002. Catastrophic extinction of Caribbean rudist bivalves at the Cretaceous-Tertiary boundary. Geology 30:999–1002.
Stilwell, J. D. 2003. Patterns of biodiversity and faunal rebound following the K-T boundary extinction event in Austral Palaeocene molluscan faunas. Palaeogeography, Palaeoclimatology, Palaeoecology 195:319–356.
Twitchett, R. J. 2001. Incompleteness of the Permian-Triassic fossil record: a consequence of productivity decline? Geological Journal 36:341–353.
Valentine, J. W. 1980. Determinants of diversity in higher taxonomic categories. Paleobiology 6:397–407.
———. 1990. The macroevolution of clade shape. Pp. 128–150 *in* R. M. Ross and W. D. Allmon, eds. Causes of evolution: a paleontological perspective. University of Chicago Press, Chicago.
———. 1992. Lessons from the history of life. Pp. 17–32 *in* P. R. Grant and H. S. Horn, eds. Molds, molecules and Metazoa: growing points in evolutionary biology. Princeton University Press, Princeton, N.J.
Valentine, J. W., and D. Jablonski. 1986. Mass extinctions: sensitivity of marine larval types. Proceedings of the National Academy of Sciences USA 83:6912–6914.
Valentine, J. W., and T. D. Walker. 1987. Extinctions in a model taxonomic hierarchy. Paleobiology 13:193–207.
Valentine, J. W., K. Roy, and D. Jablonski. 2002. Carnivore/ noncarnivore ratios in northeastern Pacific marine gastropods. Marine Ecology Progress Series 228:153–163.
Van Valen, L. M. 1973. A new evolutionary law. Evolutionary Theory 1:1–30.
———. 1984. A resetting of Phanerozoic community evolution. Nature 307:50–52.
———. 1985a. A theory of origination and extinction. Evolutionary Theory 7:133–142.
———. 1985b. How constant is extinction? Evolutionary Theory 7:93–106.

———. 1987. Comment [on "Phanerozoic trends in background extinction: Consequence of an aging fauna"]. Geology 15:875–876.

Vermeij, G. J. 1995. Economics, volcanoes, and Phanerozoic revolutions. Paleobiology 21:125–152.

von Euler, F. 2001. Selective extinction and rapid loss of evolutionary history in the bird fauna. Proceedings of the Royal Society of London B 268:127–130.

Wagner, P. J. 1995. Testing evolutionary constraint hypotheses: examples with early Paleozoic gastropods. Paleobiology 21: 248–272.

———. 1996. Testing the underlying patterns of active trends. Evolution 50:990–1017.

———. 1997. Patterns of morphological diversification among the Rostroconchia. Paleobiology 23:115–150.

Wang, S. C. 2003. On the continuity of background and mass extinction. Paleobiology 29:455–467.

Wignall, P. B., and M. J. Benton. 1999. Lazarus taxa and fossil abundance at times of biotic crisis. Journal of the Geological Society, London 156:453–456.

Willis, J. C. 1922. Age and area. Cambridge University Press, Cambridge.

Wills, M. A. 2001. Morphological disparity: a primer. Pp. 55–144 *in* J. M. Adrain, G. D. Edegcombe, and B. S. Lieberman, eds. Fossils, phylogeny, and form. Kluwer Academic/Plenum, New York.

Wimsatt, W. C. 2000. Emergence as non-aggregativity and the biases of reductionisms. Foundations of Science 5:269–297.

Wing, S. L. 2004. Mass extinctions in plant evolution. Pp. 61–97 *in* P. D. Taylor, ed. Mass extinctions in the history of life. Cambridge University Press, Cambridge.

Wootton, J. T. 2004. Markov chain models predict the consequences of experimental extinctions. Ecology Letters 7:653–660.

Zavaleta, E. S., and K. B. Hulvey. 2004. Realistic species losses disproportionately reduce grassland resistance to biological invaders. Science 306:1175–1177.